AN ALCHEMICAL QUEST FOR UNIVERSAL KNOWLEDGE

AF737052

History of science credits the Flemish physician, alchemist and philosopher Jan Baptist Van Helmont (1579–1644) for his contributions to the development of chemistry and medicine. Yet, as this book makes clear, focussing on Van Helmont's impact on modern science does not do justice to the complexity of his thought or to his influence on successive generations of intellectuals like Robert Boyle or Gottfried Leibniz.

Revealing Van Helmont as an original thinker who sought to produce a post-Scholastic synthesis of religion and natural philosophy, Georgiana Hedesan reconstructs his ambitious quest for universal knowledge as it emerges from the text of the *Ortus medicinae* (1648). Published after Van Helmont's death by his son, the work can best be understood as a compilation of finished and unfinished treatises, the historical product of a life unsettled by religious persecution and personal misfortune. The present book provides a coherent account of Van Helmont's philosophy by analysing its main tenets.

Divided into two parts, the study opens with a background to Van Helmont's concept of an alchemical Christian philosophy, demonstrating that his outlook was deeply grounded in the tradition of medical alchemy as reformed by Theophrastus von Hohenheim, called Paracelsus (1493–1541). It then reconstitutes Van Helmont's biography, while giving a historical dimension to his intellectual output. The second part reconstructs Van Helmont's Christian philosophy, investigating his views on God, nature and man, as well as his applied philosophy. Hedesan also provides an account of the development of Van Helmont's thought throughout his life. The conclusion sums up Van Helmont's intellectual achievement and highlights avenues of future research.

Georgiana D. Hedesan is a Wellcome Trust Fellow in Medical History and Humanities at University of Oxford, working on a research project on the quest for universal medicine and radical prolongation of life in seventeenth-century alchemy (2013–2016). Prior to this, she held a short-term Cantemir Junior Fellowship at University of Oxford and a Frances A. Yates fellowship at the Warburg Institute, London. A PhD in History from University of Exeter (2012), she is the author of several articles and reviews in scholarly journals dealing with Paracelsus, Van Helmont and alchemy, including articles published in *Medical History* and *Ambix*, as well as chapters in books either published or to be published by University of Chicago Press, Ashgate and Routledge.

Universal Reform: Studies in Intellectual History, 1550–1700

Series Editors
Howard Hotson, St Anne's College, Oxford,
Vladimír Urbánek, Academy of Sciences of the Czech Republic, Prague

The fission of the western Church in the Reformation era released great quantities of energy, not all of which was contained by the confessional churches. Alongside the well-studied process of confessionalisation and the persistence of variety within popular religion, the post-Reformation period witnessed a series of poorly understood attempts by a wide variety of intellectuals to extend the reforming impulse from the spheres of church and theology to many different areas of life and thought. Within these ambitious reforming projects, impulses originating in the Reformation mixed inextricably with projects emerging from the late-Renaissance and with the ongoing transformations of communications, education, art, literature, science, medicine and philosophy. Although specialised literatures exist to study these individual developments, they do not comfortably accommodate studies of how these components were sometimes brought together in the service of wider reforms. By providing a natural home for fresh research uncomfortably accommodated within Renaissance studies, Reformation studies and the histories of science, medicine, philosophy and education, this new series will pursue a more synoptic understanding of individuals, movements and networks pursuing further and more general reform by bringing together studies rooted in all of these subdisciplinary historiographies but constrained by none of them.

An Alchemical Quest for Universal Knowledge

The 'Christian Philosophy' of Jan Baptist Van Helmont (1579–1644)

GEORGIANA D. HEDESAN

Routledge
Taylor & Francis Group
LONDON AND NEW YORK

First published 2016
by Routledge
2 Park Square, Milton Park, Abingdon, Oxon OX14 4RN

and by Routledge
605 Third Avenue, New York, NY 10017

First issued in paperback 2022

Routledge is an imprint of the Taylor & Francis Group, an informa business

© 2016 Georgiana D. Hedesan

The right of Georgiana D. Hedesan to be identified as author of this work has been asserted by her in accordance with sections 77 and 78 of the Copyright, Designs and Patents Act 1988.

All rights reserved. No part of this book may be reprinted or reproduced or utilised in any form or by any electronic, mechanical, or other means, now known or hereafter invented, including photocopying and recording, or in any information storage or retrieval system, without permission in writing from the publishers.

Trademark notice: Product or corporate names may be trademarks or registered trademarks, and are used only for identification and explanation without intent to infringe.

Publisher's Note
The publisher has gone to great lengths to ensure the quality of this reprint but points out that some imperfections in the original copies may be apparent.

British Library Cataloguing in Publication Data
A catalogue record for this book is available from the British Library

Library of Congress Cataloging-in-Publication Data
The LOC data has been applied for.

ISBN 13: 978-1-4724-6916-8 (hbk)
ISBN 13: 978-1-03-240254-3 (pbk)
ISBN 13: 978-1-315-56673-3 (ebk)

DOI: 10.4324/9781315566733

Typeset in Garamond Premier Pro
by Apex CoVantage, LLC

Contents

List of Figures

Acknowledgements

I am thankful to a number of wonderful scholars that have supported me in the arduous process of writing this book. The book grew out of my PhD thesis at University of Exeter, supervised by the regretted Professor Nicholas Goodrick-Clarke and by Professor Mark Wynn. I am grateful to both for their tremendous support and guidance throughout the PhD years. I am also thankful to Dr Stephen Clucas and Professor Jonathan Barry for their comments and advice as examiners of my thesis. I would not have been able to finish my PhD without the partial financial support of the University of Exeter and my Teaching Assistant position at its EXESESO Centre.

In the process of devising this manuscript, I have also benefitted from inestimable advice from a number of scholars. I would first single out Dr Margaret Pelling, the unsung editor of Walter Pagel's book on Van Helmont, who provided invaluable comments on chapters of my book manuscript. I am also thankful to Prof Howard Hotson, Prof Robert Halleux, Dr Jennifer Rampling, Dr Anna Marie Roos, Dr Erica Charters, Dr Guido Giglioni and the series' anonymous reviewers for their guiding advice at various stages of book development.

This book has greatly benefitted from a research trip at the Mechelen Archives in Belgium. In this sense, I am grateful for the financial support of the Society for the History of Alchemy and Chemistry's 'New Scholar Award' (2011). I wish to also thank the Diocesan Pastoral Centre in Mechelen for hosting me, and archivist Dr Gerrit Vanden Bosch for his support in the Mechelen archives.

Finally, I would like to express my deepest gratitude to my family: my sister Andreea, who has unconditionally supported me throughout the PhD and early career, and my parents, Petruta and Gheorghe. I would not be here without their love.

Preface

On 4 March 1634, the Flemish philosopher, physician and alchemist Jan Baptist Van Helmont (1579–1644)[1] was arrested by the Mechelen ecclesiastical court on suspicion of heresy. His imprisonment, occurring barely a year after the sentencing of Galileo Galilei (1564–1642) to indefinite house arrest, was also incited by the Inquisition. It occurred in a period when Counter-Reformation efforts were being directed toward challenges of orthodoxy originating within Catholic lands themselves. Like Galileo, Van Helmont was a born Catholic espousing a 'new philosophy' that significantly deviated from traditional Aristotelian Scholasticism. Like Galileo's, Van Helmont's natural philosophy also appeared to stand in contradiction with Catholic theology. Like Galileo, Van Helmont was also put under house arrest; however, he was eventually set free in 1637 without being charged.

Despite such similarities, far fewer people remember Van Helmont today than Galileo. Standard accounts of the 'birth of modern science', the 'Scientific Revolution' or 'the birth of modern philosophy' still pay limited attention to Van Helmont, if at all.[2] The Flemish philosopher fares better in histories of chemistry, where he is mainly remembered for his quantitative experimentation and for defining the notion of 'gas'. Even in this specific field he has long been upstaged by Robert Boyle (1627–1691), whom posterity has found more congruent to modern science.

Yet Van Helmont was not always so little known. His masterpiece, *Ortus medicinae*, went through numerous editions and translations in the latter half of the seventeenth century and early eighteenth century. Many luminaries of the period read, owned and commented on Helmontian treatises, including Boyle, John Locke

---

1 There is no generally accepted spelling of Van Helmont's name. In respect to his two first names, some scholars have used the French 'Jean-Baptiste', others the Latin 'Johannes' or 'Joannes Baptista'; Walter Pagel and J.R. Partington preferred 'Joan Baptista'. I have employed 'Jan Baptist', which is in line with the Flemish origins of the philosopher. In regard to the last name, both 'van Helmont' and 'Van Helmont' have been used in scholarship. I have adopted the latter spelling which seems relatively more common in scholarship, for instance in the writings of Pagel, Robert Halleux and Steffen Ducheyne. It is also a simpler notation for English-language readers. It must be pointed out, for the sake of historical accuracy, that Van Helmont did not call himself 'van' at all, but 'de' in the French style; his letters are signed 'J.B. de Helmont' or 'J.B.D.H'.

2 For instance, there is no mention of Van Helmont in Steven Shapin's popular *The Scientific Revolution* (Chicago, IL: University of Chicago Press, 1996); there is less than one page treatment of Van Helmont in Paolo Rossi's *The Birth of Modern Science* (Oxford: Blackwell, 2000), and limited discussion of him in the otherwise outstanding *Cambridge History of Science: Vol. 3 Early Modern Science*, ed. Katharine Park and Lorraine Daston (Cambridge: Cambridge University Press, 2006). John Henry's *The Scientific Revolution and the Origins of Modern Science*, 2nd edition (Basingstoke: Palgrave Macmillan, 2001) makes some mention of Van Helmont and Helmontianism, but describes the Flemish philosopher as a 'lesser figure' in the Scientific Revolution (86). *The Cambridge Companion to Early Modern Philosophy*, ed. Donald R. Rutherford (Cambridge: Cambridge University Press, 2006) does not mention Van Helmont at all.

(1632–1704), Gottfried Leibniz (1646–1716), and Isaac Newton (1642–1727). The 'father of modern chemistry', Antoine Lavoisier (1743–1794) praised Van Helmont as the most important chemist of the seventeenth century.[3]

In the past two decades, Van Helmont has thankfully begun to emerge from a long period of obscurity. Scholarship is now better aware of the significant influence of Helmontian thought on the next generations of natural philosophers and scientists. William R. Newman and Lawrence M. Principe have argued that 'Van Helmont's writings constituted what was probably the most wide-ranging and influential chymical theory of the second half of the seventeenth century'.[4] This is very likely true, yet Van Helmont's influence cannot be reduced only to 'chymistry'.[5] The works of the Flemish philosopher led to a new questioning of medical and academic learning, as well as renewed enthusiasm for empirical investigation and alternative philosophical speculations.[6] Although most recent scholarship on Van Helmont's influence has been centred on England, we are now better aware that he had an impact in France, Germany, Holland, Austria, Italy and even as far as Greece, Moldavia and the American colonies.[7] The nature and extent of this influence is yet to be charted.

---

3 On Lavoisier's praise for Van Helmont and his indirect Helmontian inheritance, see William R. Newman and Lawrence M. Principe, *Alchemy Tried in the Fire: The Fate of Helmontian Chymistry* (Chicago, IL: Chicago University Press, 2002), 298–309.

4 Newman and Principe, *Alchemy Tried in the Fire*, 296.

5 Newman and Principe describe 'chymistry' as an early modern field that encompassed 'alchemy' and 'chemistry', William R. Newman and Lawrence M. Principe, 'Alchemy vs Chemistry: The Etymological Origins of a Historiographic Mistake', *Early Science and Medicine* 3 (1998): 41.

6 There is significantly more scholarship on Helmontianism than on Van Helmont. Besides Newman and Principe's work, it is also worth highlighting Antonio Clericuzio, 'From Van Helmont to Boyle. A Study of the Transmission of Helmontian Chemical and Medical Theories in Seventeenth-Century England', *The British Journal for the History of Science* 26 (1993): 303–34; Steffen Ducheyne, 'A Preliminary Study of the Appropriation of Van Helmont's Oeuvre in Britain in Chymistry, Medicine and Natural Philosophy', *Ambix* 55 (2008): 122–35; Peter R. Anstey, 'John Locke and Helmontian Medicine', in *The Body as Object and Instrument of Knowledge: Embodied Empiricism in the Early Modern Science*, ed. Charles T. Wolfe and Ofer Gal (Dordrecht: Springer, 2010), 93–121; Anna-Marie Roos, *The Salt of the Earth: Natural Philosophy, Medicine, and Chymistry in England, 1650–1750* (Leiden: Brill, 2007), particularly 47–107; Ole Peter Grell, 'Plague, Prayer and Physic: Helmontian Medicine in Restoration England', in *Religio Medici: Medicine and Religion in Seventeenth-Century England*, ed. Ole Peter Grell and Andrew Cunningham (Aldershot: Ashgate, 1996), 204–27.

7 See for instance Bernard Joly, 'La réception de la pensée de Van Helmont dans l'oeuvre de Pierre Jean Fabre', in *Alchemy Revisited: Proceedings of the International Conference on the History of Alchemy at the University of Groningen 17–19 April 1989*, ed. Z.R.W.M. von Martels (Leiden: Brill, 1990), 206–14; Victor D. Boantza, 'Alkahest and Fire: Debating Matter, Chymistry and Natural History at the Early Parisian Academy of Sciences', in *The Body as Object and Instrument of Knowledge*, 75–92; Georgiana D. Hedesan, 'Reproducing the Tree of Life: Radical Prolongation of Life and Biblical Interpretation in Seventeenth-Century Medical Alchemy', *Ambix* 60 (2013): 354–56 on Franz Oswald Grembs; Vlad Alexandrescu, introduction to *L'immagine irraffigurabile della Scienza Sacro-Santa* (Milan: Mondadori, 2013), 1–61 on Prince Cantemir; Walter W. Woodward, *Prospero's America: John Winthrop, Jr, Alchemy, and the Creation of New England Culture, 1606–1676* (Chapel Hill: University of North Carolina Press, 2010), particularly 203–06. More examples can be found in the classical account of Lynn Thorndike, *History of Magic and Experimental Science* (New York: Columbia University Press, 1958), VII, 230–40.

Despite this important scholarly work on Helmontian influence, historians do not yet have a deep insight into the nature of Van Helmont's own worldview. There is only one major monograph on the Flemish physician: Walter Pagel's *Joan Baptista Van Helmont: Reformer of Science and Medicine* (Cambridge: Cambridge University Press, 1982). Pagel's work is now fairly obsolete, as it follows the presentist assumption of classical history of science: its overall intention is to emphasise Van Helmont's contributions to modern medicine and science.

This work seeks to move away from a positivistic history of science by focusing on Van Helmont's intentions in framing his work. It will argue that Van Helmont had highly ambitious aims for his philosophy. In his posthumous masterpiece, *Ortus medicinae* (1648),[8] he laid out a project for a profound reformation of thought. I will show that Van Helmont sought nothing less than to lay the path to universal knowledge, which in his mind depended on two chief pillars: Christian thought and medical alchemy.

Indeed, despite accusations of heresy, Van Helmont was a profoundly devout man. His religiosity went beyond private belief and worship: it was imbued in the very substance of his worldview. His Christian thought included theology, mystical practices, specific attitudes toward the natural world, as well as theories about medicine and the divine status of the physician.

Van Helmont's views were also deeply permeated with a philosophical outlook inspired by medical alchemy. By 'medical alchemy', I am referring to a complex of ideas and practices related to the employment of alchemical processes and products for medical purposes.[9] It is important to realise that the term 'alchemy' is more suitable to Van Helmont's thought and practice than 'chemistry', since his views can easily be connected to those of medieval and early modern alchemy. By comparison, the advent of modern chemistry occurred in the late eighteenth century. The term proposed by Newman and Principe, 'chymistry', would also be valid for this analysis. Nevertheless, I use the term 'alchemy' in this book to emphasise the direct link between Van Helmont and the alchemical tradition.[10]

---

[8] Jan Baptist Van Helmont, *Ortus medicinae, id est, Initia physicae inaudita. Progressus medicinae novus in morborum ultionem ad vitam longam* (Amsterdam: Ludovic Elzevir, 1648). For page references I have used the Latin version of 1652 (Amsterdam: Ludovic Elzevir); the main reason for this lies in the fact that the editors' integration of the *Opuscula medica inaudita* in the *Ortus* was incomplete and muddled in the 1648 edition: while the *Opuscula* appears in the Table of Contents, the pagination is garbled and the *Opuscula* is ultimately appended with separate page numbers. Moreover, the 1648 *Ortus* lacks an index. The editors correct these matters and give coherence to the *Ortus* beginning with the second edition. Still, to assist the reader, I have made references to chapter titles, and where available, section numbers. Chapter numbers are given in order and in accordance with the list in the *Appendix*. There are several seventeenth-century translations of Van Helmont's work in English, French and German, but no authoritative recent edition. The only complete English translation is the 1662 *Oriatrike or Physick Refined: the Common Errors Therein Refuted and the Whole are Reformed and Rectified*, trans. John Chandler (London: Lodowick Lloyd, 1662), but this should be consulted with caution. The translations are hence my own. My approach to translating has been to emphasise intelligibility over word accuracy. In the case of quotes and specific references, I have included the Latin original in the footnote.

[9] I am subsuming within this framework such terms as spagyrics, chemiatria, medical chymistry, iatrochemistry, etc. Van Helmont generally referred to this field as *spagyria*, *chymia* or *pyrotechnia*.

[10] 'Alchemy' should be understood in a wide sense to mean a variety of theories and practices that focussed on the artificial perfecting of such things as metals, medicine, gems or even plants.

Moreover, the term 'medical alchemy' must be understood to imply a strong philosophical component. For many medical alchemists of the era, alchemy was equally a practical art, or *technē*, as well as a fully developed philosophy: it was a *scientia* that included both theory and practice. This view was fully embraced by Van Helmont, who called himself a 'philosopher by fire' (*philosophus per ignem*) as well as an adept of the 'art of the fire' (*pyrotechnia*).[11] He believed that alchemy was the only science that could offer true knowledge of causes and of the structure of all beings.[12] In this sense, he was eager to contrast the 'certain' understanding of alchemy with the unreal and 'sophistical' knowledge offered by the Scholastic tradition still prevalent in his age.

Van Helmont has long been associated with the medical-alchemical current inspired by the Swiss physician Theophrastus Bombastus von Hohenheim, or Paracelsus (1493–1541), despite his criticism of the latter's theories. I will indeed show that many of Van Helmont's ideas originate either with Paracelsus or the Paracelsian current. Nevertheless, the complexity of Van Helmont's thought transcended Paracelsianism, becoming an original and controversial 'new philosophy' of the second half of the seventeenth century.

'Christian Philosophy' and the Alchemists

Van Helmont perceived his union of natural philosophy and religion as a 'Christian philosophy'.[13] At first glance, the term 'Christian philosophy' does not suggest a reforming spirit. The notion was current in the period and referred to the reconciliation of Christian thought and pagan philosophy (usually of ancient Greek origin).[14]

---

[11] In the period, the term *pyrotechnia* often had alchemical connotations; for instance, Paracelsian physician William Davisson published a book called *Philosophia Pyrotechnica* in 1633. The term undoubtedly reveals the popularity of Vanuccio Biringuccio's work on metallurgy and metalworking, *Pirotechnia* (1540).

[12] See the discussion below, Chapter 6, under heading 'The Possibility of *Scientia* of Nature'.

[13] Thus, in the chapter-treatise 'Ortus formarum', §20, 108, he affirms that his theory according to which forms are infused directly by God is dictated by 'Christian philosophy'. In 'Natura contrariorum nescia', §17, 134, he similarly claims that, according to 'Christian philosophy', the seeds always obey goals known to and directed by God. In the next chapter, 'Blas humanum', §4, 145, Van Helmont condemns the Aristotelian theory of the unmovable mover as incongruent with 'Christian philosophy'. In a similar vein, in ch. 91, 'Mortus introitus in naturam humanam decus Virginum', 513, he argues that 'Christian philosophy' rejects the Aristotelian concept of final cause.

[14] I am not referring here to the modern scholarly definition of 'Christian philosophy'; instead, I am investigating what an early modern philosopher like Van Helmont would have thought a 'Christian philosophy' was. On the topic of 'Christian philosophy' as a modern scholarly construct, see the useful review of Jorge J.E. Garcia, 'Does Philosophy Tolerate Christening? Thomas Aquinas and the Notion of Christian Philosophy', in *Philosophy of Religion for a New Century: Essays in Honor of Eugene Thomas Long*, ed. Jeremiah Hackett and Jerald Vallulis (Dordrecht: Kluwer, 2004), 37–63 (especially 37–41). As Garcia points out, the modern concept was introduced by the historian Etienne Gilson and his mostly Catholic Thomist supporters.

Some of the early Fathers of the Church including Paul, Augustine (354–430) and Pseudo-Dionysius the Areopagite (fl. 400–500) were the first to attempt it.[15]

In the medieval and Renaissance periods, the model of synthesis between Christianity and philosophy was represented by the work of Thomas Aquinas (1225–1274) and the Scholastics.[16] Scholasticism focussed on harmonising the thought of Aristotle (received mainly via Arabic philosophy), with Christian religion. Yet this process could only be achieved by eliminating such Aristotelian tenets as the eternity of the world or the naturalistic view of the soul. This compromise highlighted the limits of the Christian Aristotelian synthesis.

The Renaissance and the Reformation brought two important developments: rediscovered ancient philosophies as well as new forms of religious fervour. The humanist and religious movements often led not only to a vigorous criticism of Scholasticism, but also to attempts at novel syntheses between Christianity and philosophy. It is during this period that the term 'Christian philosophy' and associated phrases such as 'pious philosophy', 'Christian physics' and 'Mosaic philosophy' came into wider usage.[17] They denoted new harmonising attempts that were often explicitly anti-Scholastic and anti-Aristotelian.[18]

Early modern alchemical philosophers were particularly supportive of the idea of 'Christian philosophy'. During the Middle Ages, alchemy emerged as a body of knowledge and practices that had an ambiguous relationship with standard Scholasticism.[19] Alchemy was not accepted as an object of study in universities, but some Scholastics recognised it as both an 'art' and a 'science' (*scientia*).[20] Alchemists clearly perceived themselves as 'philosophers', possessing a superior and abstruse philosophy that drew on Hermetic, Aristotelian and Neoplatonic ideas.

The late Middle Ages marked a period where alchemy became increasingly intertwined with Christian belief. Authors like Petrus Bonus of Ferrara (fl. 1330), George

15 In the Late Antiquity and the Middle Ages, the term 'theology' was preferred to the phrase 'Christian philosophy', which was only rarely used by thinkers such as Justin (c. 100–165), Augustine, and Petrarch (1304–1374). See Etienne Gilson, 'Notes bibliographiques pour servir à l'histoire de la notion de philosophie chrétienne', in *L'esprit de la philosophie medievale*, 2nd ed. (Paris: Vrin, 1944), 413, and Paul Oskar Kristeller, 'Thomism and the Italian Thought of the Renaissance', in *Medieval Aspects of Renaissance Learning: Three Essays by Paul Oskar Kristeller*, ed. and trans. Edward P. Mahoney (New York, NY: Columbia University Press, 1992), 34, n. 7.

16 Hence Etienne Gilson's branding of Thomism as 'Christian Philosophy', despite the fact that Thomas never called it by that name; see Garcia, 'Does Philosophy Tolerate Christening?', 38.

17 On examples of these, see Ann Blair, 'Mosaic Physics and the Search for a Pious Natural Philosophy in the Late Renaissance', *Isis* 91 (2000): 32–58. Her discussion of 'Mosaic' and 'Christian philosophy' is commendable, but overreliant on the biased testimony of early modern Protestant historian Johann Jakob Brucker, who distinguished 'Mosaic and Christian philosophers' like Jan Amos Comenius and Johann Heinrich Alsted from 'theosophers' like Paracelsus, Robert Fludd and Jacob Boehme.

18 Blair, 'Mosaic Physics', 34.

19 A general introduction on the history of alchemy is given in my article 'Alchemy' in *The Occult World*, ed. Christopher Partridge (Oxford: Routledge, 2015), 552–63.

20 Jean Marc Mandosio, 'L'alchimie dans les classifications des sciences et des arts à la Renaissance', in *Alchimie et Philosophie à la Renaissance*, ed. Jean-Claude Margolin and Sylvain Matton (Paris: Vrin, 1993), 11–41.

Ripley (d. c. 1490), Lodovico Lazzarelli (1447–1500), and Giovanni Mercurio da Correggio (fl. 1451) emphasised the connections between Christianity and alchemy. Following in medieval footsteps, Paracelsus and his supporters developed their own versions of a religiously grounded philosophy based on medical alchemy.[21]

Van Helmont followed the path laid out by other alchemists to affirm the profound complementarity of Christianity and alchemy. Yet his ambitions went much further. He sought to elevate alchemical thought to the status of a universal science that would provide the philosophical basis for a true 'Christian philosophy'. Indeed, he believed that true harmonisation could be achieved by unifying Christian thought with alchemical philosophy.

The union would be achieved by operating from both directions. On one hand, Christianity had to return to its core roots, the faith of the Apostles and the first Fathers. In this sense, Van Helmont had three guiding lights: Paul, Augustine, and Pseudo-Dionysius the Areopagite. To this he added the medieval mystical tradition, which to him reflected pure Christian faith. In practice, the 'pure' Christianity Van Helmont sought to restore was imbued by Platonic and Neoplatonic thought. On the other hand, alchemical philosophy had to be 'fleshed out' in such a way as to offer an all-encompassing and coherent vision of the natural world that would rival that of the 'Philosopher' himself, Aristotle. This was, of course, no easy task, but Van Helmont believed that in his endeavour he could rely on alchemical and Paracelsian thought, as well as on his fervent belief in empirical study and divine insight.

The main purpose of this book, then, is to understand how Van Helmont attempted to construct his 'Christian philosophy' as a synthesis that could challenge and even replace Scholasticism as a form of universal knowledge. I will also try to disentangle the ways alchemy and Christian thought interacted in his mind. It is one of my contentions that religion and alchemy were entwined in his thought to such extent that their discourses were not clearly separate. This juxtaposition originated from a belief that the source of all knowledge (*scientia*) was God, and hence both the Book of Grace and the Book of Nature were intimately related.

Past Scholarly Treatments of Van Helmont

Scholarship on Van Helmont has mostly originated from the field of history of science and medicine. Classical historians of science were concerned with demonstrating Van Helmont's relevancy for modern science. As such they generally focussed on Van Helmont's 'discoveries', such as the acidity of the gastric fluid and his invention and description of 'gas'.[22] Van Helmont's analysis of Spa waters also drew praise, as has his research on the urinary calculi in 'Tractatus de lithiasi' ('Treatise on the Disease of the

---

21 On current scholarship in regards to the relationship between alchemy and religion, see Tara Nummedal, 'Introduction', Special Issue on Alchemy and Religion in Christian Europe, *Ambix* 60 (2013), 311–22.

22 James R. Partington, *A History of Chemistry*, 4 vols (London: Macmillan, 1961), II, 227–32, 237.

Stone').[23] His ideas on ferments were often noted for their impact on later chemists such as Georg Ernst Stahl (1659–1734) or Hermann Boerhaave (1668–1738).[24] Classical historians of science also paid particular attention to Van Helmont's experiments and practical laboratory techniques.[25] Amongst the experiments most looked into was the so-called 'willow tree experiment', which is still being praised as either the beginning of experimental plant physiology or the first quantitative experiment in plant nutrition.[26]

Despite acknowledging these positive scientific contributions, classical scholarship was generally ambivalent about Van Helmont, as historians variously saw him as the last of the scorned alchemists or the first of the modern chemists. As James R. Partington observed in his painstaking survey of the history of chemistry, references to Van Helmont between 1710 and 1950 tended to be positive, although rarely enthusiastic.[27] There can be no greater contrast than the claim of James Campbell Brown that Van Helmont was 'a great chemist, undoubtedly the greatest prior to Lavoisier',[28] and Herbert Butterfield's verdict that Van Helmont

> made one or two significant chemical discoveries, but these are buried in so much fancifulness . . . that even twentieth-century commentators on Van Helmont are fabulous creatures themselves, and the strangest things in [Francis] Bacon seem rationalistic and modern in comparison.[29]

One such 'fabulous creature' probably targeted by Butterfield was Walter Pagel, who has written the best-known monograph on Van Helmont. As already noted, Pagel deserves praise for his valiant effort to bring Van Helmont back into academic scholarship, but the modern reader must be wary of the book's claims. Perhaps the strangest of Pagel's conclusions is that Van Helmont was a closet Aristotelian vitalist.[30] The German-English scholar believed that the Flemish philosopher's rejection of Aris-

---

23 Partington, *A History of Chemistry*, II, 228, 233.

24 Partington, *A History of Chemistry*, II, 236.

25 T.S. Patterson, 'Van Helmont's Fire and Water Experiments', *Annals of Science* 1 (1936): 463–4.

26 The quantity of scholarship on this Helmontian topic is large: see Hebbel E. Hoff, 'Nicolaus of Cusa, van Helmont and Boyle: The First Experiment of the Renaissance in Quantitative Biology and Medicine', *Journal of the History of Medicine and Allied Sciences* 19 (1964): 115–16; H.M. Howe, 'A Root of van Helmont's Tree', *Isis* 56 (1965): 409–19; David R. Hershey, 'Digging Deeper into Helmont's Famous Willow Tree Experiment', *American Biology Teacher* 53 (1991): 458; Douglas Allchin, 'Reassessing van Helmont, Reassessing History', *BioScene: Journal of College Biology Teaching* 19 (1993): 3; David R. Hershey, 'Pseudohistory and Pseudoscience: Corrections to Allchin's Historical, Conceptual and Educational Claims', *Science & Education* 15 (2006): 123.

27 Partington, *A History of Chemistry*, II, 209.

28 James Campbell Brown, *History of Chemistry* (London: J. & A. Churchill, 1913), 202.

29 Herbert Butterfield, *The Origins of Modern Science, 1500–1800* (New York, NY: Macmillan, 1951), 98.

30 It is possible that Pagel was influenced by the revival of vitalism in the early twentieth century, and particularly by the influential *History and Theory of Vitalism* of Hans Driesch (London: Macmillan, 1914). According to Driesch, Van Helmont's theories were an inferior version of those of Aristotle: 'This [Van Helmont's idea of the *Archeus*] is really and unmistakably the Aristotelian teaching—only less profound', 25.

totle was only apparent, and associated Van Helmont's ideas with those of William Harvey (1578–1657) and Francis Glisson (1599–1677).[31] Yet, as I will show, Pagel's attempt to paint Van Helmont as an Aristotelian vitalist is historically inaccurate. Van Helmont would not have been pleased to be described as an Aristotelian, when in fact his writings were dedicated to eliminating Aristotelian speculation from natural philosophy. Moreover, the term 'vitalist' would have made little sense to him, being an anachronistic effort to mould an early modern thinker in the guise of a twentieth-century scientist.[32]

Besides Pagel, who manifested a life-long interest in Van Helmont, other scholars began to pay attention to Van Helmont in the 1960s. This 'new wave' of interest was arguably prompted by greater scholarly awareness of Boyle's indebtedness to his work. Primarily, Charles Webster and Allen Debus analysed some of the Helmontian background of Boyle's *Sceptical Chymist*.[33] Webster followed this with his epoch-making *Great Instauration*, which brought Van Helmont's influence in mid-seventeenth-century England to greater attention.[34] In turn, Allen Debus's *The Chemical Philosophy: Paracelsian Science and Medicine in the 16th and 17th Centuries* (1977) dedicated a chapter to Van Helmont's ideas, thus spurring further interest in his work.[35] While Debus made an important contribution in the integration of Van Helmont's thought within the larger perspective of Paracelsianism, his approach was unfortunately marred by his positivistic concern with the impact of Van Helmont's notions on future scientific chemistry.

Since Pagel and Debus, more scholars have been drawn to the study of Van Helmont's ideas, either directly or by means of ascertaining the root of Helmontianism. A series of articles by Robert Halleux have illuminated some of the historical circumstances of Van Helmont's writing.[36] Halleux's most important contribution,

---

31 Walter Pagel, *Joan Baptista Van Helmont: Reformer of Science and Medicine* (Cambridge: Cambridge University Press, 1982), 36–8, 42, 120–24. The forced association between Van Helmont and Glisson has also been criticised by Guido Giglioni, *Immaginazione e malattia: Saggio su Jan Baptista Van Helmont* (Milan: Francoangeli, 2000), 43.

32 As André Pichot points out, one can only refer to vitalism from the eighteenth century onwards, *Histoire de la notion de vie* (Paris: Gallimard, 1993), 453.

33 Charles Webster, 'Water as the Ultimate Principle of Nature: The Background to Boyle's *Sceptical Chymist*', *Ambix* 13(1966): 96–107; and Allen Debus, particularly in 'Fire Analysis and the Elements in the 16th and 17th Centuries', *Annals of Science* 23 (1967): 128–9.

34 Charles Webster, *The Great Instauration* (London: Duckworth, 1975).

35 Allen Debus, *The Chemical Philosophy: Paracelsian Science and Medicine in the 16th and 17th Centuries*, 2nd edition (Mineola, NY: Dover Publications, 2002), 295–381.

36 Robert Halleux, 'Helmontiana I', *Academiae analecta* (Brussels: Koninklijke Academie voor Wetenschappen, Letteren en Schone Kunsten van België, 1983), 33–63, 'Helmontiana II: Le prologue de l'*Eisagoge*, la conversion de Van Helmont au Paracelsisme et les songes de Descartes', *Academiae analecta* (Brussels: Koninklijke Academie voor Wetenschappen, Letteren en Schone Kunsten van België, 1987), 17–36; 'Theory and Experiment in the Early Writings of Johan Baptist Van Helmont', in *Theory and Experiment*, ed. Diderick Batens (Dordrecht: Kluwer, 1988), 93–101; 'Le procès d'inquisition du chimiste Jean-Baptiste Van Helmont (1578–1644): les enjeux et les arguments', *Académie des inscriptions et belles-lettres* 148 (2004): 1062–78.

'Helmontiana II' (1987), concentrates on the little-known manuscript of the *Eisagoge* found in the Archives of the Archbishopric of Mechelen. Halleux emphasises the originality of the Prologue, proceeding to contextualise its dream-vision not only within classical alchemical literature (such as the *Physika kai Mystika* of Pseudo-Democritus) but also contemporaneous accounts of initiatory dreams, such as those of René Descartes (1596–1650).[37]

In an indication of the maturation of Helmontian scholarship, scholars have begun to analyse specific topics present in Van Helmont's writings. A short German-language monograph by Berthold Heinecke, based on his doctoral dissertation at the Technical University of Dresden in 1990, analysed Van Helmont's 'mysticism' and his scientific experimentalism.[38] Heinecke defined Van Helmont as a 'philosophical mystic', drawing on Alois Maria Haas's definition of mysticism as a 'strict unity between subject and object'.[39] Heinecke argued that the Flemish philosopher clearly differentiated between mysticism and experiment.[40]

William R. Newman's article, 'The Corpuscular Theory in J.B. Van Helmont and its Medieval Sources' (1993), clarified the corpuscularianism present in Van Helmont's work. Newman called Van Helmont's ideas on the topic 'vital corpuscularianism', describing them as a synthesis of the ideas of Geber (fl. thirteenth century) and Paracelsus.[41] Further, Newman pointed out Van Helmont's indebtedness to Scholastic idea of the *minima naturalia*, which was linked with his ideas on the action of ferments and *semina*.[42] Newman also argued that Van Helmont's concept of the liquor *Alkahest* was inspired by Geber.[43]

Paulo Porto's work has also focussed on certain notions of Van Helmont's philosophy. In his monograph *Van Helmont e o conceito de gás: química e medicina no século XVII* (1995), Porto shed further light on the Helmontian idea of gas, while his praised article on the universal solvent *Alkahest* (2002) emphasised its importance for medical purposes.[44]

---

37 Halleux, 'Helmontiana II', 25–8, 30–33.

38 Berthold Heinecke, *Mystik und Wissenschaft bei Johan Baptista Van Helmont (1579–1655)* (Bern: Peter Lang, 1995), and 'The Mysticism and Science of Johann Baptista Van Helmont (1579–1644)', *Ambix* 42 (1995): 65–78.

39 Heinecke, *Mystik und Wissenschaft*, 73; 'Mysticism and Science', 66.

40 Heinecke, 'Mysticism and Science', 69.

41 William R. Newman, 'The Corpuscular Theory in J.B. Van Helmont and its Medieval Sources', *Vivarium* 31 (1993): 174–5.

42 Newman, 'The Corpuscular Theory', 176–81.

43 Newman, 'The Corpuscular Theory', 185. On the liquor *Alkahest* in Van Helmont, see also Paulo Porto, *Summus atque felicissimus salium*": The Medical Relevance of the Liquor Alkahest', *Bulletin of the History of Medicine* 76 (2002): 1–29; Bernard Joly, 'L'alkahest, dissolvent universel, ou quand la theorie rend pensable une pratique impossible', *Revue d'Histoire des Sciences* 49 (1996): 305–44; Ladislao Reti, 'Van Helmont, Boyle and the Alkahest', in Ladislao Reti and William C. Gibson, *Some Aspects of Seventeenth-Century Medicine and Science* (Los Angeles, LA: William Andrews Clark Memorial Library, 1969), 3–19. The *Alkahest* is further discussed in Chapter 7 of this book.

44 Paulo Porto, *Van Helmont e o conceito de gás: química e medicina no século XVII* (São Paolo: EDUC-EDUSP, 1995).

Guido Giglioni's monograph *Immaginazione e malattia: Saggio su Jan Baptiste Van Helmont* (2000) is the most significant recent contribution to Van Helmont scholarship. The book focussed on the role of imagination in Van Helmont's medical concepts of disease and healing. According to Giglioni, *Ortus medicinae* is the last great medical, theological and philosophical synthesis on the subject of imagination before 'the Cartesian fracture'.[45] Giglioni's perspective is drawn from the history of medicine and imagination, presenting Van Helmont as a Renaissance physician, albeit one for whom the role of God dwarfs that of Nature.[46]

Newman and Principe's *Alchemy Tried in the Fire: The Fate of Helmontian Chymistry* (2004) contains a chapter that focuses on Van Helmont's experimental techniques. According to Newman and Principe, Van Helmont integrated Paracelsian approaches with medieval metallurgical techniques to design a new concept of mass balance as a tool for chemical demonstration.[47] The book shows rather persuasively that Van Helmont was a gifted practitioner of 'chymistry' and that his ideas had a strong impact on George Starkey, Robert Boyle, and other chemists.

Structure of the Book

I have written this book with the conviction that a new synoptic analysis of Van Helmont's work is necessary in order to grasp the complexity of his thought. Despite the articles and books mentioned above, Pagel's 1982 monograph is still the only comprehensive analysis of Van Helmont's philosophy, and continues to be the main reference point for scholars seeking an understanding of Van Helmont's ideas. Yet, as I have pointed out, Pagel's book provides a shaky foundation for Helmontian scholarship because of its Whiggish approach and interpretations.

My approach is fundamentally historical, focusing on Van Helmont's ideas and aims in the context of his time. Of course, no overview can claim to exhaust Van Helmont's thought in itself, and, indeed, future avenues of research are identified and highlighted throughout the book. Within these limitations, the study will attempt to be as comprehensive and non-judgmental as possible.

An Alchemical Quest for Universal Knowledge: The 'Christian Philosophy' of Jan Baptist Van Helmont (1579–1644) is divided into two main parts. The first part places Van Helmont in historical context. Chapter 1 investigates the influence of early modern medical alchemy and Paracelsianism on Van Helmont. Chapter 2 analyses Van Helmont's life in correlation with his intellectual output, while Chapter 3 focuses on the structure and historical circumstances of his masterpiece *Ortus medicinae*.

---

45 Giglioni, *Immaginazione e malattia*, 10.

46 Giglioni, *Immaginazione e malattia*, 60–67, 94.

47 Newman and Principe, *Alchemy Tried in the Fire*, 90–91.

The second part of the book contains the investigation of Van Helmont's 'Christian philosophy' based on the *Ortus medicinae*. The chapters follow a philosophical structure, addressing the main topics of his thought (God, Nature and Man). The chapter on God shows that Van Helmont envisaged God as simultaneously transcendent and immanent in nature; in particular, the Flemish philosopher argued that God was the direct bestower of form and life on all creatures. Van Helmont also envisaged God's relationship with human beings in an idiosyncratic, even heterodox fashion.

The chapter on Nature discusses Van Helmont's dualist matter-spirit theory, his two-element view, and the vast array of spiritual entities he considered to exist in nature, including the *Archeus*, *Gas* and the formal lights. The chapter on Man focusses on the notion of the human being as 'image of God', his ambiguous status after the Fall of Adam and Eve, and the restoration of lost knowledge via mystical techniques and empirical inquiry. A separate chapter is dedicated to Van Helmont's applied philosophy, with a focus on his medical-alchemical pursuits. The book is brought to a close by a general conclusion that reviews the achievements of Van Helmont's thought and suggests further directions of research on the phenomenon of Helmontianism.

Part I
Van Helmont in Context

Chapter 1
The Medical Alchemical Background of Van Helmont's Christian Philosophy

The early modern period saw a resurgence of interest in alchemy, mainly prompted by the reinterpretation provided by Paracelsus. Nowadays, alchemy is often perceived as *chrysopoeia*, or the art of transmuting base metal into gold. In fact, making gold was only one of the possible goals of alchemy. As Chiara Crisciani and Michela Pereira have pointed out, medical applications of alchemy had become much more popular than *chrysopoeia* by 1400.[1] The late medieval emphasis on medical alchemy was eagerly seized upon by Paracelsus. In parallel, he also drew upon a rich tradition that viewed alchemy as a religious or sacred philosophy.[2]

The present review of the medical alchemical background to Van Helmont's thought is not meant to be comprehensive, due to the overwhelming complexity of the topic. Instead, it attempts to highlight those aspects of Van Helmont's medical-alchemical heritage that would have been conducive to his Christian philosophical worldview. Moreover, my analysis is chiefly focussed on the Paracelsian phenomenon; however, as I will further point out, the medieval alchemical influence on Van Helmont's Christian philosophy should not be discounted.

Paracelsus and His *Astronomia Magna*

Paracelsus was a physician who rejected the traditional medical framework based on the work of Galen and Aristotle. Instead of Scholastic books, he preferred to draw knowledge from 'experience' and folk practices.[3] He advocated the providential figure of the physician, who had to base his knowledge on four pillars: alchemy, philosophy, astronomy and the virtues (or ethics).[4]

---

1 Chiara Crisciani and Michela Pereira, *L'Arte del Sole e della Luna: Alchimia e filosofia nel Medioevo* (Spoleto: Centro Italiano di Studi sull'Alto Medioevo, 1996), 72.

2 On the medieval Christian alchemical tradition, see also Chiara Crisciani, 'Note Sull'Alchemia "Francescana" nel sec XIII', in *Atti del XXV Congresso Nazionale di Filosofia* (Roma: Societa filosofica italiana, 1980), 214–20; Antoine Calvet, 'Alchimie et Joachimisme dans les *Alchimica* Pseudo-Arnaldiens', in *Alchimie et Philosophie à la Renaissance*, 93–109; Chiara Crisciani, *Il Papa e l'Alchimia* (Rome: Viella, 2002).

3 Andrew Weeks, *Paracelsus: Speculative Theory and the Crisis of the Early Reformation* (Albany, NY: SUNY Press, 1997), 7.

4 Walter Pagel, *Paracelsus: An Introduction to Philosophical Medicine in the Era of the Renaissance* (Basel: Karger, 1958), 222–3; Theophrastus Bombastus von Hohenheim, called Paracelsus, 'Paragranum', in *Paracelsus: Essential Theoretical Writings*, ed. and trans. Andrew Weeks (Leiden: Brill, 2008), 61–296.

Yet Paracelsus's speculations were not limited to medical ideas, as he wrote on natural philosophy and on theology as well. He sought a comprehensive view of the world that was rooted in Christian truth and included natural knowledge based on keen observation.[5] Paracelsus's later writings in particular demonstrate his intention to achieve a unification of religious and secular knowledge.

As this book will show, Van Helmont was a thorough reader of Paracelsus. Amongst Paracelsus's writings, Van Helmont's concept of a Christian philosophy draws most heavily on his *Astronomia magna* (1537/8), a treatise which is deeply imbued with Christian themes.[6] Here Paracelsus proclaimed himself a Christian and rejected any accusations of heathen teaching.[7] Paracelsus taught that philosophy should originate from the Bible, not from pagan philosophers:

> Holy Scripture represents the beginning of all philosophy and natural science . . . Consequently, if a philosopher is not born out of theology, he has no cornerstone upon which to base his philosophy. For truth springs from theology, and cannot be discovered without its help.[8]

These ideas were accompanied by Paracelsus's theory of the two Lights, the Light of Grace and the Light of Nature. He posited that both lights came from the same source, God, who is essential Light.[9] The correspondence of the two lights provided the theoretical basis for Paracelsus's attempt to reconcile the study of nature with religion. In his vision, religion and science pursued the same divine truth. Such a perspective justified the concept of a Christian philosophy, and provided a background for the quest for sacred wisdom. The doctrine of the two lights was followed by many enthusiastic Paracelsians, to the point where some of them, like Adam Haslmayr and Benedict Figulus, embraced it as a type of new religion, the *Theophrastia sancta*.[10]

Astronomia magna further affirms that God created all things *ex nihilo* and argues that man has an immortal soul, which is *imago Dei*. The treatise rejects

---

5 Debus, *The Chemical Philosophy*, 61.

6 See Hartmut Rudolph's remark that this work offers a natural philosophical viewpoint 'that is theological in a Christian sense', 'Hohenheim's Anthropology in the Light of his Writings on the Eucharist', in *Paracelsus: The Man and his Reputation, His Ideas and their Transformation*, ed. Ole Peter Grell (Leiden: Brill, 1998), 199; see also Rudolph's 'Kosmosspekulation und Trinitätslehre: Ein Beitrag zur Beziehung zwischen Weltbild und Theologie bei Paracelsus', *Salzburger Beiträge zur Paracelsusforschung* 21 (1980): 33. For more on *Astronomia magna* see also Dane T. Daniel, 'Paracelsus' Astronomia Magna (1537/38): Bible-Based Science and the Religious Roots of the Scientific Revolution' (PhD diss., Indiana University, 2003). I am grateful to Daniel for sharing this work with me.

7 Theophrastus Bombastus von Hohenheim, called Paracelsus, 'Astronomia Magna', in *Paracelsus: Essential Readings*, ed. Nicholas Goodrick-Clarke (Wellingborough: Aquarian, 1990), 111–12.

8 Paracelsus, 'Astronomia Magna', 116.

9 Weeks, *Paracelsus: Speculative Theory*, 183. See Paracelsus, 'Paragranum', 269: 'the Holy Spirit is the igniter of the light of nature'.

10 Carlos Gilly, '*Theophrastia sancta*: Paracelsianism as a Religion in Conflict with the Established Churches', in *Paracelsus: The Man and His Reputation*, 161–5, 166, calls it the 'religion of two lights'.

Adamic flesh, which it deems mortal and perishable, and argues that the eternal body will be created through fiery purification. It also states that Christ was born without an Adamic seed, and that through baptism, human beings can also be regenerated in the shape of Christ.[11] As we shall see in Chapter 4, on God, all these ideas are very similar to Helmontian arguments.

To these religious ideas is appended an entire philosophy whereby Paracelsus attempts to place magic within a Christian framework.[12] Like many Renaissance thinkers, he supports the idea of natural magic; however, he affirms that celestial magic is superior to that, being a type of divine power. The celestial Christian magus must be reborn through the Holy Spirit, and can effect wondrous things on earth.[13]

A history of the influence of the *Astronomia magna* has not been undertaken. However, the research of Dane T. Daniel, Carlos Gilly and Owen Hannaway has indicated the importance of this treatise in the context of later Paracelsian thought.[14] Even so, the impact of the *Astronomia magna* on Van Helmont has not been discussed until now.

Early Paracelsianism: Petrus Severinus (1542–1602)

Petrus Severinus, a Danish physician educated at the University of Paris, was the first commentator of Paracelsus to bring an aura of relative respectability to Paracelsian philosophy.[15] He emphasised the medical works of Paracelsus rather than his theological ones. Despite Thomas Erastus's (1524–1583) attack on Severinus, the Danish physician was highly influential, and many post-1570 medical alchemists were influenced by his interpretation of Paracelsus at least to some degree.[16]

---

11 Dane T. Daniel, 'Medieval Alchemy and Paracelsus' Theology: Pseudo-Lull's *Testamentum* and Paracelsus' *Astronomia Magna*', *Nova Acta Paracelsica* 22/23 (2008/2009): 134; Ute Gause, 'On Paracelsus's Epistemology in his Early Theological Writings and in his *Astronomia Magna*', in *Paracelsus: The Man and his Reputation*, 219; Paracelsus, 'Astronomia Magna', 111; see also Theophrastus Paracelsus, 'Astronomia magna sive Tota philosophia sagax', in *Operum medico-chimicorum sive paradoxorum*, 12 vols (Frankfurt: Collegio Musarum Palthenianarum, 1605), X, 266–70, 279–81.

12 For Paracelsus's views on magic, see also Charles Webster, *Paracelsus: Medicine, Magic and Mission at the End of Time* (London: Yale University Press, 2008), 153–68.

13 Arlene Miller Guinsburg, 'Paracelsian Magic and Theology: A Case Study of the Matthew Commentaries', *Medizinisches Journal* 16 (1981): 135; see also Rudoph, 'Hohenheim's Anthropology', 204. Paracelsus, 'Astronomia magna sive Tota philosophia sagax', X, 99, 296–7, 320–21.

14 See above on Daniel and Gilly; for the *Astronomia Magna's* influence on Oswald Croll please refer to Owen Hannaway, *The Chemists and the Word* (Baltimore, MA: Johns Hopkins University Press, 1975), 7–8.

15 Jole Shackelford, *A Philosophical Path for Paracelsian Medicine: The Ideas, Intellectual Context, and Influence of Petrus Severinus, 1540–1602* (Copenhagen: Museum Tusculanum Press, 2004), 209–49; Ole Peter Grell, 'The Acceptable Face of Paracelsianism: The Legacy of *Idea medicinae* and the Introduction of Paracelsianism in Early Modern Denmark', in *Paracelsus: The Man and his Reputation*, 245–69.

16 Shackelford, *A Philosophical Path*, 209–49.

Severinus's philosophy is a potent mixture of Paracelsian ideas and Renaissance medical concepts. The Danish philosopher attempted to synthesise and rationalise the vast Paracelsian corpus, but in doing so he coloured it in a distinctive Platonic tint. He harmonised it with the contemporary trends in Renaissance medicine, particularly the work of the French physician Jean Fernel (1497–1558). Furthermore, he associated Paracelsus with the Hippocratic tradition of medicine as well as the Renaissance concept of *prisca sapientia* or *prisca theologia*.[17] Thus, he argued that Paracelsus was the restorer of ancient medicine and philosophy, which had been tainted by Galen and his followers.

It is known that the young Van Helmont was a Severinian admirer. His unpublished *Eisagoge in artem medicam à Paracelso restitutam* (1607), purported to be a defence of Paracelsus, is largely a reiteration of Severinus's ideas. The *Supplementum de Spadanis fontibus* (1624) further betrays the ongoing influence of the Danish physician.[18] Jole Shackelford has also argued that Van Helmont's later thought in the *Ortus medicinae* bears the mark of the continuing influence of Severinus.[19] However, I will show in Chapter 5 that the key Severinian concepts are altered in the *Ortus*.

The speculation of the Danish physician was less religious in tone than Paracelsus's. Hiro Hirai pointed out that Biblical themes did not entice Severinus as they did Paracelsus and that he was not interested in arguing on behalf of such Christian ideas as *creatio ex nihilo* or the Word of God.[20] Still, one can find traces of Christian thought filtered through Renaissance Neoplatonic ideas. Severinus supported the notions of the harmony of nature, which he called 'the divine analogy', and the Augustinian concept of the seminal reasons planted by God through the virtue of the Word and Spirit of God hovering over the waters.[21]

Middle Paracelsianism: Joseph Du Chesne (1544–1604)

The French physician Joseph Du Chesne (Lat. Quercetanus) belonged to the generation of Paracelsian philosophers that followed in Petrus Severinus's footsteps. Together with Oswald Croll and Michael Sendivogius, Du Chesne represented the

---

17 On Severinus's interpretation of Hippocrates, see Jole Shackelford, 'The Chemical Hippocrates: Paracelsian and Hippocratic Theory in Petrus Severinus' Medical Philosophy', in *Reinventing Hippocrates*, ed. David Cantor (Aldershot: Ashgate, 2002), 59–88. For classic treatments of the topic of *prisca theologia*, see D.P. Walker, *The Ancient Theology* (London: Duckworth, 1972) and Charles Schmitt, '*Prisca Theologia* e *Philosophia Perennis*: due temi del Rinascimento italiano e la loro fortuna', in *Il pensiero italiano del Rinascimento e il tempo nostro: Atti del V Convegno Internazionale di Centro di studi umanistici, Montelpuciano, 1968* (Florence: Leo S. Olschki, 1970), 211–36.

18 Hiro Hirai, *Le concept de semence dans les théories de la matière à la Renaissance, de Marsile Ficin à Pierre Gassendi* (Turnhout: Brepols, 2005), 445–50.

19 Shackelford, *A Philosophical Path*, 247.

20 Hirai, *Le concept de semence*, 242.

21 For the Renaissance influence, see Hirai, *Le concept de semence*, 222, 220, 224–5; Petrus Severinus, *Idea medicinae philosophicae* (Basel: Henricpetri, 1571), 41–42. On this, see further Chapter 5, Nature.

high point of what may be termed 'middle Paracelsianism' at the turn of the seventeenth century. He was the leader of the Paracelsian faction in France, which became embroiled in a lengthy dispute with the medical faculty at the University of Paris.[22]

Even more than the ideas of Severinus, Du Chesne's views were eclectic. He sought the conciliation of Paracelsianism with the Greek medical-philosophical tradition, including Aristotle, Plato and Galen. Du Chesne also believed that there was an alchemical *prisca sapientia* originating from Hermes Trismegistus and adopted by Hippocrates.[23] This was a lineage of 'Hermetic philosophers' that stretched to the present day.

Du Chesne extended this wisdom tradition to Moses and Genesis. Consequently, he wrote an alchemical interpretation of Creation that attempted to show the concordance between alchemical philosophy and the Bible.[24] Du Chesne's God was immanent in the universe as *natura naturans* and Universal Nature. This Universal Nature, which he equated with the Platonic Soul of the World and the Spirit of God hovering over the waters, was a divine virtue implanted in all things.[25]

Van Helmont referred to Joseph Du Chesne in several of his writings. He seems to have been well informed of the debates that were going on at the time at the University of Paris.[26] In a letter to Marin Mersenne, he mentions the 1604 University of Paris injunction against Paracelsus caused by the writings of Du Chesne.[27] Nevertheless, his references to Du Chesne tend to be negative. Van Helmont particularly mistrusts Du Chesne's claim that the figure of herbs can be found in salt, as he believes that the herbal spirit is hidden in the invisible seeds (*seminaria*).[28] Instead, the presence of the spirit can be detected in the gas that is emanated by fermenting cheese.

At the same time, Van Helmont's criticism of Du Chesne is subdued. Thus, he describes the French physician as a 'good man' (*bonus vir*) who was simply deceived by his own experiment.[29] Indeed, there was much in Du Chesne that Van Helmont

22 On Du Chesne's career, see Didier Kahn, *Alchimie et Paracelsisme in France* (Paris: Droz, 2007), 233–78.

23 Joseph Du Chesne, *The Practice of Chymicall and Hermeticall Physicke for the Preservation of Health*, trans. Thomas Tymme (London: Thomas Creede, 1605), chap 1–2 (there are no page numbers); see also Hirai, *Le concept de semence*, 70.

24 For an exhaustive analysis of Du Chesne's alchemical interpretation of Genesis, see Didier Kahn, 'L'intérpretation alchimique de la Genèse chez Joseph Du Chesne dans le context de ses doctrines alchimiques et cosmologiques', in *Scientiae et artes: Die Vermittlung alten und neuen Wissens in Literatur, Kunst und Musik*, ed. Barbara Mahlmann-Bauer (Wiesbaden: Harrassowitz, 2004), 682–9.

25 Du Chesne, *The Practice of Chymicall and Hermeticall Physicke*, ch 2.

26 Allen Debus, *The French Paracelsians* (Cambridge: Cambridge University Press, 1991), 46–65.

27 'Jean-Baptiste van Helmont, à Bruxelles, à Mersenne, à Paris, 19 Décembre 1630', in *Correspondance du Marin Mersenne*, ed. Cornelis de Waard, 17 vols (Paris: Gabriel Beauchesne, 1936), II, 585. This censorship took place on 12 April 1604; see also Didier Kahn, 'L'intérpretation alchimique de la Genèse', 688.

28 'Jean-Baptiste van Helmont, à Bruxelles, à Mersenne, à Paris, 26 Septembre 1630', II, 532. The point of Van Helmont here seems to be that the spirit cannot be seen in any condition.

29 Ch. 60, 'Pharmacopolion ac Dispensatorium modernorum', §13, 368.

would have found congruent, including his religious views and his search for the *prisca sapientia* roots of alchemical philosophy.

Middle Paracelsianism: Oswald Croll (1563–1609)

A physician at the court of the Holy Roman Emperor Rudolf II of Habsburg (1552–1612) in Prague, Croll became famous for his posthumously published work *Basilica chymica* (1609). This contains one of the best-known defences of Paracelsian ideas, and as such was made the object of a classical 1975 study by Owen Hannaway.[30]

Croll's 'Admonitory Preface' to the *Basilica chymica* advocates a Paracelsianism with strong Christian overtones.[31] In a manner reminiscent of Paracelsus's *Astronomia magna*, Croll affirms that 'true Phylosophy should be grounded on the Scriptures' and that 'Christians should not be ruled by Heathen Phylosophy'.[32] Croll firmly believes that the theological and philosophical truth correspond, as does the Light of Grace with that of Nature.[33]

Oswald Croll dealt at length with the theory of the two lights, which is an important key to understanding his ideas of the unification of knowledge. As he argued,

> The Light of Grace begetteth a true Theologer; yet not without Phylosophy: the Light of Nature, which is the Treasury of God confirmed in the Scriptures, maketh a true Phylosopher, yet not without Theologie, which is the Foundation of true Wisdom . . . Hence it is plaine that every true Theologer is a Phylosopher, and every true Phylosopher is a Theologer.[34]

Hence Croll advocated that the theological and philosophical truths are the same, and that the pursuit of both lights was ultimately sanctioned by their origin in the Holy Spirit.

With Croll we see a coherent attempt at grounding his ideas in a Christian thought of mystical persuasion. For instance, he insists on the fundamental incomprehensibility of God. According to Croll, God is the 'fountaine of the Abysse' and the 'Sea of Nihilitude or Nothingnesse'.[35]

Croll's anthropology focuses on the dichotomy of the inner (immortal) and outer (mortal) man and the tripartite anthropology of soul, spirit and body.[36] He puts a special emphasis on Man being the image of God, an idea directly linked with the

---

30 Hannaway, *The Chemists and the Word*, 1–74.

31 Oswald Croll, 'The Admonitory Preface of Oswald Crollie, Physitian: to the Most Illustrious Prince Christian Anhaltin', in *Philosophy Reformed & Improved in Four Profound Tractates*, trans. Henry Pinnell (London: Lodowick Lloyd, 1657), 1–226.

32 Croll, 'Admonitory Preface', 51; the second quotation is on the margin.

33 Croll, 'Admonitory Preface', 135; for this see also Hannaway, *The Chemists and the Word*, 7–10.

34 Croll, 'Admonitory Preface', 135.

35 Croll, 'Admonitory Preface', 215.

36 Croll, 'Admonitory Preface', 45 on margin; also 60–61.

philosophical theme of *gnothi seauton*, 'know thyself', interpreted in a mystical sense as knowing the God within.[37] The highest mystical state is that whereby one would

> know God himself the maker of all things, and passe into him with a full image of his likenesse, as with a kind of essentiall touch without a bond, whereby thou mayest be transform'd, and made (as it were) a God.[38]

All these are themes that can also be found in Van Helmont's thought as well.

For Croll, philosophy found its pious consummation in medicine. Indeed, the German physician advocated a union of theory and practice, which involved religion, philosophy and laboratory work. He did not shy away from sharing numerous alchemical recipes as part of his *Basilica chymica*. However, the highest secret that alchemists could attain, Croll maintains, is the universal medicine of long life, which is the true 'adept philosophy'.[39] This medicine was known to Adam, but the knowledge of it had become hidden and could be unveiled only by diligent searchers into Nature.[40] As we will see, the acquisition of a long-life medicine would form the highest ideal of Van Helmont's own medical practice.

Alchemical Trends in the Early Seventeenth Century

At the turn of the seventeenth century, alchemy occupied a prominent but controversial position in the European intellectual landscape. Powerful patrons like Emperor Rudolf II, Landgrave Moritz of Hesse-Kassel (1572–1632) and Archbishop Ernest of Bavaria (1554–1612) created large hubs of alchemical practitioners. Alchemical sympathisers began to penetrate the jealously guarded university environment, particularly in the German lands: Joachim Tanckius (1557–1609) in Leipzig, Johannes Hartmann (1568–1631) in Marburg and Daniel Sennert (1572–1637) in Wittenberg were all proponents of alchemical philosophy. Yet, as alchemy attracted more and more sponsorship and more and more acceptance in learned circles, it also fragmented and diversified in ways that have only been partially explored.

This period saw the powerful growth of Christian alchemy. Prominent figures such as Heinrich Khunrath (c.1560–1605), Clovis Hesteau de Nuysement (c.1660–c.1623), Pierre-Jean Fabre (c.1588–1658), Jean d'Espagnet (1564–c.1637), Robert Fludd (1574–1637) and others affirmed the fundamental compatibility of Christianity and alchemical philosophy. Van Helmont himself was one of the chief exponents of this belief: he was convinced that Christian thought and alchemy were naturally aligned, and sought to express this in the *Ortus medicinae*.

---

37 Croll, 'Admonitory Preface', 47–8, 57.

38 Croll, 'Admonitory Preface', 213.

39 Croll, 'Admonitory Preface', 97.

40 Croll, 'Admonitory Preface', 218–19.

Seventeenth-century Christian alchemists were influenced not only by Paracelsian thought but also by medieval alchemy. The early modern period was marked by the publication of many medieval alchemical treatises, hitherto only available in manuscript form. The landmark publication of the two-volume compendium *Artis auriferae quae alchemia vocant* (1593) was quickly followed by the first five volumes of the *Theatrum chemicum* of Lazarus Zetzner of Strasbourg (1604–1622).

Medieval alchemy contained many references to the sacred nature of the alchemical 'science' (*scientia*). Treatises with real or supposed authors such as Hermes Trismegistus, Zosimos, Morienus, Geber, Arnau of Villanova, Aristotle or Thomas Aquinas accredited the idea that alchemy was a form of divine knowledge bestowed by God. Works such as the *Turba philosophorum* and the *Rosarium philosophorum* postulated the existence of an alchemical lineage of ancient philosophers, a *prisca sapientia* that included Hermes Trismegistus, Pythagoras, Plato, and others. The powerful topos of Christ as the philosophers' stone, introduced by pseudo-Arnaldian texts, conferred a particularly Christian tone to alchemical knowledge. The later Middle Ages, in particular, were marked by influential attempts to ally natural philosophy with Christian faith. Such was Pseudo-Lull's *Testamentum*, which posited a harmonious unity between Christianity and a natural philosophy conceived in Aristotelian terms, but also imbued with Platonic views.[41] Petrus Bonus of Ferrara's *Pretiosa margarita novella* (*New Pearl of Great Price*, early 1300s) insisted on divine revelation and higher knowledge as the only paths to the attainment of the *lapis philosophorum*.[42]

The 'rediscovery' of medieval alchemy in the early modern period led to attempts at reaffirming or reinterpreting it. Michael Sendivogius (1566–1636) was particularly successful at using medieval ideas mixed with Paracelsianism to establish a new and highly influential philosophy. Sendivogius's utmost goal was the philosophers' stone, or the elixir. Paracelsus had of course condemned gold-making in his authentic treatises, but pseudo-Paracelsian treatises justified the search for the philosophers' stone in terms of its medical uses. In turn, alchemical philosophers such as Joseph Du Chesne, Pierre-Jean Fabre and Heinrich Nollius (1590–1630) advocated the pursuit of the *lapis* as universal medicine.

Seventeenth-century alchemical philosophy was seldom a purely theoretical undertaking; most alchemical philosophers during this period combined theory with practice. Alchemy had always defined itself as a form of applied philosophy that included a theoretical (*theorica*) and a practical (*practica*) part. This tradition was also followed by Paracelsus, whose entire philosophy was directed toward medical applications. In the seventeenth century, medicine continued to be the most accepted purpose of alchemy, but the application of alchemical principles to a host of disciplines including agriculture, husbandry, mining, painting, metallurgy or gunnery

---

41 Michela Pereira, *Oro dei Filosofi: Saggio sulle idee di un alchimista del Trecento* (Spoleto: Centro Italiano di Studi sull'Alto Medioevo, 1992), 174, 178.

42 Petrus Bonus of Ferrara, *Pretiosa margarita novella*, ed. Janus Lacinius (Venice: Aldus, 1526). See Chiara Crisciani, 'The Conception of Alchemy as Expressed in the "Pretiosa Margarita Novella" of Petrus Bonus of Ferrara', *Ambix* 20 (1973): 165–81.

became more prominent. Beyond the arts, alchemy sometimes became mixed with religious and political goals, as evinced by certain alchemical sympathisers of the Rose Cross manifestos (1614–1617).[43] In turn, the perceived link between alchemy and unorthodox religion alarmed many theologians and philosophers and led to the condemnation of several Christian alchemical thinkers such as Heinrich Khunrath and Robert Fludd.

The backlash against alchemy was not only led by fears of religious nonconformism, but also by the perceived menace to the medical establishment. Threatened by the ascendancy of Paracelsianism, the University of Paris condemned several Paracelsian physicians such as Roch le Baillif (1540–1598), Du Chesne and Théodore Turquet de Mayerne (1573–1655). In the German lands, Andreas Libavius (c.1550–1616) sought to dissociate Paracelsianism from alchemy in order to salvage the latter.[44]

Fears of alchemical ideas and practices also extended to alchemy's perceived association with increasingly reviled magic. In particular, the Jesuits began a hard-fought battle against the use of the Paracelsian weapon-salve cure and other 'sympathetic' remedies. Since these claimed to cure at a distance, the Jesuits maintained that they employed demonic magic.

It was in this environment of opportunity and threat that Van Helmont wrote his work. To young Jan Baptist, medical alchemy must have seemed a relatively risky, yet highly rewarding field to specialise in. He hence approached not only with enthusiasm but also some naiveté on how acceptable alchemy might become to established religion and philosophy. By the 1620s and 1630s, however, the dangers of pursuing an uncompromising Paracelsian path became all too real to him. As the mood of the Counter-Reformation became increasingly somber, Van Helmont was ultimately forced to reconsider his early enthusiasm and reformulate his doctrines in light of the new reality. Yet, as we shall see, his faith in medical alchemy never wavered.

---

[43] On this topic, see Gilly, *Theophrastia sancta*, and Carlos Gilly, *Adam Haslmayr. Der erste Verkünder der Manifeste der Rosenkreuzer* (Amsterdam: Bibliotheca Philosophica Hermetica, 1994).

[44] On these disputes, see Kahn, *Alchimie et Paracelsisme*, 353–411, and Bruce T. Moran, *Andreas Libavius and the Transformation of Alchemy: Separating Chemical Cultures with Polemical Fire* (Sagamore Beach: Science History Publications, 2007), 83–188.

Chapter 2
Van Helmont's Life

The details of Jan Baptist Van Helmont's life are available from several primary sources. For the first part of his life (1579–1616), one has to mainly rely on Van Helmont's own autobiographical accounts, particularly the treatises 'Studia authoris' ('The Author's Studies') and 'Tumulus pestis' ('The Plague's Grave'). Between 1616 and 1637, the period of his involvement in the weapon-salve controversy, the information is drawn from the preserved correspondence with Marin Mersenne (1588–1648) and Pierre Gassendi (1592–1655), as well as from the archived trial records, the three-volume 'Causa J. B. Helmontii medici', stored in the Archives of the Archbishopric of Mechelen.[1] For the latter years, from 1638 to 1644, some autobiographical information is again available, as well as Van Helmont's son Franciscus Mercurius's testimony in the introduction to *Ortus medicinae* (1648). Further information was unearthed through the archival work of Belgian scholars Henri De Waele, Robert Halleux, G. Desmarez, Louis Stroobant and J. Nauwelaers.

Relying so much on autobiographical writings is not ideal. On one hand, Van Helmont wrote his recollection when he was in his 60s, so time may have blurred some of his memories. On the other hand, an autobiography can always be suspected of 'forgetting' some less pleasant incidents or putting a spin on others in the hope posterity would take them at face value. Indeed, we can legitimately wonder why Van Helmont wrote an autobiography at all. His unfortunate encounter with the Inquisition may provide part of the answer: surely Van Helmont did not want to be remembered as a would-be heretic. In this sense, he was clearly following the example of Augustine, famous for his autobiographical *Confessions*, to prove that he was a faithful Christian. Another reason is linked with his alchemical interests. This was a period when alchemical autobiographies were becoming popular. The template was the 'autobiography' of Bernard Trevisan, an obscure late-medieval alchemist.[2] Like Trevisan, Van Helmont was trying to establish his credentials as an alchemical Adept to the reader. There is a strong element of self-fashioning to Van Helmont's work, particularly to the *Ortus medicinae*, where his main claim is that of being an exceptional and unduly persecuted thinker.

---

1 Mechelen (Malines), Archives of the Archbishopric of Mechelen, MS 'Causa J.B. Helmontii medici', 3 vols.

2 Bernard Trevisan, 'Livre de venerable docteur allemant Messire Bernard Conte de la Marche Trevisane', in *Opuscule tres-excellent de la vraye philosophie naturelle des Metaulx* (Lyon: Benoist Rigaud, 1574), 183–208.

Certainly, Van Helmont's self-presentation is an uncanny mix of Christian humility and ambitious boasting. On one hand, he often talks about his hesitation at publishing his works, claiming to have prayed to God to give the gift of knowledge to someone else: 'I seriously begged the Lord to choose someone more worthy than myself' to unveil the secrets of medicine.[3] On the other hand, he proclaims that only he has found the keys to natural philosophy: 'I come in the deserted house to sweep the dirt. Very many things will be searched into, and unknown things will be taught; those badly treated will be eliminated, those unclean will be wiped away, while false ones will be cast aside.[4] The elaborate tomb allegory in 'Tumulus Pestis' is intended to show that he was the only physician to have understood the plague and hence its cure.[5] Throughout *Ortus medicinae*, Van Helmont claims that he is bringing to light ideas that were 'unheard of', paradoxical, yet true.[6]

The extremes of boastfulness and humility are mediated by the Christian argument of charity. Van Helmont argues that the desire to help others determined him to make his knowledge manifest: 'thus are the gifts of knowledge given: without envy, deceit, expectation of profit, or the small glory of vain ambition'.[7] He claims to have received an inspired dream whereby he became aware that his hesitation to publish was due to arrogance and to the selfish fear of acquiring a bad reputation.[8] Consequently, he realised that the duty of charity required that he would not bury his talent in the earth. His spin on the notion of 'charity' hence sought to defuse possible accusations of self-aggrandisement.

This chapter focuses on Van Helmont's life but also attempts to correlate it with his intellectual development. Van Helmont's thought evolved from his first work, the *Eisagoge* (1607), to the *Ortus medicinae*, published posthumously in 1648. Scholars have only recently become more aware of this development, particularly due to the work of Robert Halleux. For the purposes of this book, I propose a periodisation of the writings of Van Helmont as follows: the *early period*, roughly corresponding with his youth and comprising the *Eisagoge*; the *middle period*, covering his mature writings (1616–1634), and the *late period*, starting with his house arrest and finishing with his death in 1644. The middle period includes two published treatises, *De magnetica vulnerum curatione* (1621) and *Supplementum de Spadanis fontibus* (1624), the extant correspondence with Mersenne and Gassendi, as well as several treatises remaining in manuscript: the three commentaries on Hippocrates, the apologetical

[3] *Opuscula medica inaudita*, 'Philiatro lectori', in *Ortus medicinae*, 645: 'serio rogavi Dominum, dignaretur alium me digniorem deligere'.

[4] Ch. 3, 'Causae et initiae naturalium', §1, 27: 'In domum desertam venio, sordes everrendas mihi reliquere. Plurima indaganda sunt, & docenda quae ignota; subvertenda quae male tradita; abstergenda quae immunda, & quae falsa, abjicienda erunt'.

[5] *Opuscula medica inaudita*, 'Tumulus pestis', ch. 1, 'Cujus generis sit Pestis', 830–31.

[6] For instance, ch. 1, 'Promissa authoris', II, §5, 9.

[7] Ch. 1, 'Promissa authoris', II, §3, 9: 'Itaque donata cognitionis charismata, sine invidia, fuco, lucri spe, aut vana ambitionis gloriola'. The term *charismata* suggests a divine gift.

[8] *Opuscula medica inaudita*, 'Philiatro lectori', 644.

work 'Ad judicem neutrum causam appellat suam et suorum philadelphus' (c.1621–1626), and an unfinished manuscript of about 150 pages tentatively called *Speculum philosophicoiatron*. The late period is the best-known one, comprising *De febrium doctrina inaudita* (1642), *Opuscula medica inaudita* (1644), most of the *Ortus medicinae* (1648) and the condensed Dutch-language version of the *Ortus*, *Dageraad, ofte Nieuwe Opkomst der Geneeskonst* (1661). Henceforth this monograph will focus on Van Helmont's late work in the *Ortus*, but will also highlight the major changes that occurred in Van Helmont's thought from his earliest to his last work.

Van Helmont's Childhood and Youth (1579–1608)

Van Helmont originated from a Flemish noble family called Berthout of Mechelen, sires of Helmont and Keerbergen.[9] In 1314, Duke Jan of Brabant gave Jan II Berthout van Berlaer (1280–1328) the Helmont land in exchange for certain domains in Lierre and thereabouts. The Berthout name continued throughout the 1400s and 1500s, although the genealogy is not very clear during this period.[10] In any case, the Berthout continued to rank amongst the oldest and richest Flemish nobility up to and beyond Jan Baptist's time.

Christian Van Helmont (1540–1580), Jan Baptist's father, was a high-ranking nobleman who acted first as counsellor, then as master of Brabant's Chamber of Accounts. In 1567, he married Marie de Stassart, who bore him at least seven children, of which Jan Baptist was the youngest.[11] The eldest son, Dierick, became mayor of Vilvoorde (a small town near Brussels).

Jan Baptist was born in Brussels on 12 January 1579 old style and was baptised at the church of Saint Gudula in the Catholic faith.[12] These were hard times for the Low Countries: the late sixteenth and early seventeenth centuries represented a period of unprecedented religious and political turmoil following the Reformation. In the mid-1500s, Protestantism, particularly of Calvinist persuasion, spread rapidly into the northern Low Countries and some urban areas in the South, leading to the fragmentation of the country along religious lines. The Low Countries ultimately broke up into the Protestant Netherlands and Catholic Belgium, still called the southern

---

9 Louis Stroobant, 'Les origins du Docteur van Helmont', in *Le Folklore brabançon: Numéro spécialement consacré à J. B. van Helmont, 1579–1644* (Brussels, 1933), 140. Helmont is a town of the Meyerij, situated over the Aa, not far from Eindhoven.

10 Stroobant, 'Les origins', 142, records that there were many illegitimate births in the family, which makes it hard to trace lineages.

11 Stroobant, 'Les origins', 148–9.

12 For a while there was confusion in regard to Van Helmont's actual date of birth, since Jan Baptist himself wrote that he was born in 1577. Desmarez discovered the birth certificate of Jan Baptist on page 135 of the baptismal register of Saint Gudula, with the date 12 January 1579; G. Desmarez, 'L'État civil de J.-B. Van Helmont', *Annales de la société d'archéologie de Bruxelles* 21 (1907): 108.

Netherlands during this period.[13] The Catholic lands remained under the rule of the Habsburg of Spain, who had inherited the Low Countries from the dukes of Brabant at the beginning of the 1500s.

When Jan Baptist was born, the Protestants and the Catholics were fighting over the hegemony of the southern Netherlands. Frequent military interventions degenerated into wanton violence and excess.[14] During this period, the cost of living rose, and the quality of food became dire, causing numerous epidemics. The highways and roads were often pillaged by bands of thieves called *gueux* who kidnapped travellers and merchants for ransom.[15]

Apart from the general misfortune befalling the country, Jan Baptist suffered a personal tragedy in his early childhood: in 1580, his father died, leaving his mother to tend a large family.[16] Presumably due to this hardship, Jan Baptist was sent at a tender age to the prestigious University of Louvain (Leuven), completing his arts studies there at the age of seventeen.

The University of Louvain, celebrated in the period for its humanist learning, offered a stimulating environment for the young Flemish gentleman. It was perhaps the fame of Cornelius Gemma (1535–1578) that initially attracted Van Helmont to the study of geometry and astronomy. At some point, however, he experienced a crisis of faith in the value of the education he was receiving. He began questioning not only astronomy, but the entire university curriculum.

Van Helmont soon became interested in an alternative course of study altogether. The Jesuits had arrived in Louvain a few decades earlier, but their attempt to teach natural philosophy and metaphysics there had been ill-received, particularly by university theologians. Louvain theology, although firmly Catholic, was different from the Thomist variety propounded by the Jesuits. Its guiding light was not Thomas Aquinas, but Augustine. Indeed, Louvain theology had evolved in an idiosyncratic direction during the period of Van Helmont's life. Michel de Bay or Baius (1513–1589), a Louvain professor of theology who became the Dean of the school in 1570, condemned Scholasticism and proposed a doctrinal return to the theology of Augustine and other Church Fathers.[17] Unfortunately, Baius's attack on Counter-Reformation Scholasticism was not well received in the Roman curia. Many of his propositions were condemned by Pope Pius V (1504–1572) in his Bull *Ex omnibus afflictionibus* of 1567 and reconfirmed by Gregory XIII (1502–1585).[18]

---

13 The southern Netherlands became formally known as Belgium in 1831. In the period, 'Belga' meant 'Dutch'.

14 Henri Pirenne, *Histoire de Belgique*, 7 vols (Brussels: Henri Lamertin, 1920), II, 359.

15 Alexandre Henne and Alphonse Guillaume Ghislain Wauters, *Histoire de la ville de Bruxelles* (Brussels: Librairie Encyclopedique de Perichon, 1845), 8–9.

16 Ch. 3, 'Studia authoris', §1, 14.

17 For a presentation of Baius's ideas from an orthodox Catholic point of view, see F.X. Jansen, *Baius et le Baianisme: Essai théologique* (Brussels: Dewit, 1927), 127–45.

18 E.J.M. Van Eijl, 'La controverse louvaniste autour de la Grace et du libre arbitre à la fin du XVIe siècle', in *L'Augustinisme à l'ancienne faculté de théologie de Louvain*, ed. M. Lamberigts and L. Kenis (Leuven: Leuven University Press, 1994), 215.

Despite these Papal interventions, Baianist doctrines did not disappear: on the contrary, they bred a new movement centred on the Bishop of Ypres, Cornelius Jansenius (1585–1638), another graduate and professor of Louvain. Jansenius, who wrote his masterpiece *Augustinus* (1640) as a defence of de Bay, was condemned by the papacy in 1642, despite protest by the university.[19] Henceforth Jansenist doctrine was considered heretical.

The influence of Louvain theology can be felt in the writings of Van Helmont. Not only did he embrace the teachings of Augustine, but he was sympathetic to the Baianist doctrine which held that the Original Sin was concupiscence; this doctrine drew on the concept of *libido* (desire) in Augustine.[20] Moreover, he inherited Louvain's reticence toward Jesuit teaching on religious matters.

Indeed, for Louvain theologians of Van Helmont's time, it was Jesuit doctrine that was heretical. When Leonard Lessius (1554–1623) taught a course at Louvain in 1587, the Theological Faculty protested that his doctrine of predestination and grace was in conflict with the teachings of Augustine, and issued the *Doctrina Lovaniensis*, which condemned Lessius's ideas as semi-Pelagian.[21] Tension between the Louvain theologians and the Jesuits continued to flare in the first half of the seventeenth century and throughout Van Helmont's life.

Despite the theological controversy, Van Helmont was curious about the 'new learning' promised by the Jesuits, particularly the courses on the nature of magic offered by the famous Jesuit scholar Martin del Rio (1558–1608). Del Rio's treatise, *Disquisitione magicarum libri sex* (1599) was recognised as one of the most reputed contributions to the study of witchcraft.[22] Van Helmont's own speculations on the subject evince the influence of del Rio's ideas.

Besides magic, the Jesuits helped Van Helmont discover alchemy. Around 1596, he attended, together with a friend called Jan Rubens,[23] the alchemical experiments carried out by François d'Aguillon (1567–1617), a Jesuit professor of mathematics at the College of Antwerp.[24] These experiments were not very fruitful, but they piqued

19 Pirenne, *Histoire de Belgique*, III, 46.

20 Nigel Abercrombie, *The Origins of Jansenism* (Oxford: Clarendon Press, 1936), 87–93. On concupiscence in Augustine, see Peter Burnell, 'Conscupiscence', in *Augustine throughout the Ages: an Encyclopedia*, ed. Allan Fitzgerald and John C. Cavadini (Grand Rapids, MI: Wm B. Eerdmans, 1999), 224–7.

21 Van Eijl, 'La controverse louvaniste', 210–11, 219–22.

22 On Del Rio, see Jan Machielsen, *Martin Delrio: Demonology and Scholarship in the Counter-Reformation* (Oxford: Oxford University Press, 2015); P.G. Maxwell-Stuart, introduction to *Martin Del Rio: Investigations into Magic*, ed. and trans. P.G. Maxwell-Stuart (Manchester: Manchester University Press, 2000), 8.

23 This Jan Rubens must have been a close kin of Peter Paul Rubens (1577–1640). It could not have been Peter Paul's father, Jan Rubens, who died in 1587. It would be interesting to explore the possible relationship between Van Helmont and the Rubens family. One must not forget that Peter Paul's mentor, the painter Otto van Veen (c.1556–1629), was also a Paracelsian thinker; see Halleux, 'Helmontiana I', 47–8.

24 Corneille Broeckx, 'J. B. Van Helmont: Ad Judicem neutrum causam appellat suam et suorum philadelphus', *Annales de l'Académie d'archéologie de Belgique* 22 (1869): 132. On d'Aguillon, see G.H.W. Vanpaemel, 'Jesuit Science in the Spanish Netherlands', in *Jesuit Science and the Republic of Letters*, ed. Mordechai Feingold (London: MIT Press, 2003), 395–96.

the young man's curiosity. We can presume he may have encountered Paracelsian writings around this time.

Van Helmont eventually felt unsatisfied with Jesuit learning.[25] Convinced at this point that Louvain could not offer him the truth he was seeking, Van Helmont refused the title of Master of Arts from the University and withdrew from his studies. Consequently, he fell into a period of introspection, as he pondered his life's calling. He first contemplated pursuing theology, as he was promised a canonship if he did so. However, he soon turned down this opportunity; he claimed that this was from a fear that he should 'eat the sins of the people' as Bernard of Clairvaux had stated. Instead, he was drawn to the study of Christian Stoicism, undoubtedly due to the strong influence of Justus Lipsius (1547–1606) in Louvain. Enthused by his reading of Seneca and Epictetus, Van Helmont contemplated becoming a Capuchin monk, identifying this order with the ideals of Christian Stoicism. Yet the rigours of an austere life seem to have taken a toll on his health, as he himself admitted. During this period of spiritual and physical weariness he had his first symbolic dream. According to his own words,

> I seemed to have become an empty bubble, whose diameter stretched from the Earth right up to Heaven: above it overhang a sarcophagus; but below, instead of the earth, lay an abyss of darkness, a true inferno.[26]

Van Helmont woke up frightened and concluded that 'in Christ Jesus, we live, move, and have our being. No one can even call on the name of Jesus for Salvation, without the special grace of God'.[27] In light of the dream, he now perceived Stoicism to be lacking substance and humility, and embraced a stronger Augustinian stance with its emphasis on Grace and predestination.

Dream-visions were essential in Van Helmont's life, thought and practice. According to his testimony, they occurred at crucial moments, whenever he was strongly questioning himself or his learning. The pattern that led to his visions is remarkably similar: it began with an internal crisis, triggered by a real-life conflict of some sort. The crisis was accompanied by strong negative emotions, such as feelings of worthlessness, inadequacy, falsehood or confusion. Van Helmont's reaction in these crises was to dedicate himself to ascetic practices. The combination of moral and physical exhaustion led to intense dreams that display strong visual imagery and are not straightforward at the first glance. The interpretation or understanding of the dream usually coincided with the waking moment.[28]

---

25 Ch. 3, 'Studia authoris', §6, 14.

26 Ch. 3, 'Studia authoris', §8, 15: 'Videbar factus Bulla inanis, cujus diameter, à terra, in coelum usque vergeret: superne enim immineret sarcophagus, inferne vero, loco terrae, abyssus erat obscuritatis'.

27 Ch. 3, 'Studia authoris', §8, 15: 'Quod in Christo Iesu vivimus, movemur & sumus. Quod nemo, vel nomen Iesu, ad salute dicere, sine peculiari gratia Dei, possit'. The quotation refers to Acts 17:28.

28 See, for instance, ch. 2, 'Confessio authoris', §13–14, 13.

We cannot, of course, exclude the idea that at least some of these dreams or visions were contrived as part of Van Helmont's autobiographical rhetoric. This is particularly so, as in the alchemical tradition fact and fiction were sometimes blurred. Alchemists often described 'dreams' or 'visions' that were really allegories of their science.[29] Nevertheless, the existence of this alchemical tradition does not exclude the possibility that Van Helmont actually had some of the dreams he is describing. While pointing out that Van Helmont's dreams could be integrated within the alchemical topos of trial before initiation, Halleux was also of the opinion that most were real.[30] What is certain is that Van Helmont explicitly mentions dreams and visions as essential for the acquisition of supreme mystical experience.[31]

Moving away from the practice of Christian Stoicism, Van Helmont took up plant medicine and the study of Greek herbalist Dioscorides (c. 40–90 AD) as a 'recreation'.[32] He considered that the knowledge of plants had not really progressed since ancient times.[33] He soon found himself enticed by the art of medicine, which he discovered to be a real 'science' (*scientia*) and a 'gift' (*donum*) of God. According to Van Helmont's confession, he voraciously read 600 books on the subject, including Galen, Hippocrates, Avicenna, and the more recent books of Leonhard Fuchs (1501–1566) and Jean Fernel.[34]

Undoubtedly Jan Baptist must have distinguished himself in his medical studies, as he was soon invited by the Medical College of Louvain to teach surgery. His professors Thomas Fyenus (1567–1631), Gerard de Villers (d. 1634) and Jean Stormius (1559–1650) strongly encouraged him on this path.[35] Van Helmont agreed to teach some courses, but eventually came to realise that it was presumptuous of him to teach what he had read, but not practised.

Discontented with his lack of practical medical knowledge, Van Helmont decided to take an apprenticeship with a physician; this, he hoped, would give him some medical experience. He was shocked to discover that, although he learnt how

29 One of the most famous is the vision of Arisleus in *Turba philosophorum*, see Julius Ruska, 'Die Vision des Arislei', in *Historische Studien und Skizzen zu Natur- und Heilwissenschaft*, ed. G. Sticker (Berlin: Julius Springer, 1930), 22–6.

30 Halleux, 'Helmontiana II', 25–31.

31 See below, Chapter 6, under heading 'Mystical Knowledge of One's Mind'.

32 Ch. 3, 'Studia authoris', §9, 15.

33 This is an exaggeration; we must keep in mind that the sixteenth and seventeenth centuries were a time of revival for herbalism. See K. Meier Reeds, 'Renaissance Humanism and Botany', *Annals of Science* 33 (1976): 519–42.

34 Ch.3, 'Studia authoris', §14–15, 15–16. Boasting about the number of books read seemed to be a current practice in this humanist period. Justus Lipsius maintained that Martin Del Rio had read no fewer than 1,100 books. See *Biographie Nationale de Belgique*, 27 vols (Brussels: Thiry & Van Buggenhout, 1866–1938), V, 475.

35 'Tumulus pestis', ch. 1, 833. On Thomas Fyenus (Feyens), see Jan Papy, 'The Attitude towards Aristotelian Biological Thought in the Louvain Medical Treatises during the Sixteenth and Early Seventeenth Century: The Case of Embryology', in *Aristotle's Animals in the Middle Ages and the Renaissance* (Leuven: Leuven University Press, 1999), 318–37.

to 'debate' (or converse about) a disease, he was never taught any healing methods.[36] As he bitterly confessed, at the end of his apprenticeship he did not even know how to cure a simple toothache. Disappointed, he perceived medicine as a vain art, which did more harm than good to a patient.

Unsatisfied by his medical studies, and uncertain how to proceed, Van Helmont fell into another period of introspection, questioning and prayer. Finally, his toil was rewarded by a new, defining dream-vision. He dreamt of the confused matter at the beginning of the world:

> I saw the whole universe, in the sight of truth, as an unformed *Chaos*, which was close to nothing. And I drew from it the conception of one word, which indicated the following: *Behold that you, and the things that you see, are nothing; whatever you desire is less than nothing in the sight of the Highest. He knows the ends of all things to be done; you may at most apply yourself to your own salvation.* In that conception there was an intrinsic precept, that I should become a physician, and that sometime *Raphael* himself should be given to me.[37]

This dream-vision can be placed sometime around 1599, since it seems to fit another dream he recounted in the preface of *Eisagoge* (1607).[38] According to this dream, on 24 September 1599, he fell asleep on the shore of the river Escaut and dreamt he entered in a large shining palace where he found Hermes Trismegistus, Paracelsus and Adepts of all nations. Guided by the ghost (*Evestrum*) of Paracelsus, he was shown a great ineffable light, before falling from a great height to the dark world below. This overly coherent dream is more likely than others to have been contrived, since the motive of the gathered Adepts was a popular alchemical topos due to the influential treatise *Turba philosophorum*.

In any case, it is clear that, by 1599, the young Van Helmont had become a fervent supporter of Paracelsian medicine. He determined henceforth to become a physician, but he faced one more hurdle. His family was not particularly supportive of this decision: his mother did not want him to go into medical practice, and Van Helmont apparently tactfully avoided mentioning it to his father's kin. There was no tradition in his noble family of anyone pursuing medicine, which may have seemed an odd and perhaps even dubious practice for a nobleman. Van Helmont himself felt a sense of inner conflict in pursuing his calling rather than fulfilling his family's expectations:

---

36 Ch. 3, 'Studia authoris', §17, 16.

37 Ch. 3, 'Studia authoris', §19, 16: 'vidi totum universum, in conspectu veritatis, tanquam aliquod informe chaos, quod merum pene nihil esset. Et hausi inde conceptum unius verbi; Qui significabat mihi, quae sequuntur. Ecce tu, & qua vides, sunt nil. Quidquid urges, minus quam ipsum nihilum, in conspectu Altissimi. Ipse scit fines omnes rerum agendarum; tu, tua saluti saltem intendas. Imo in isto conceptu, erat praeceptum intrinsecum, quod fierem Medicus, & quod mihi daretur quandoque ipsum Raphael'. The angel Raphael was traditionally seen as the patron of physicians.

38 The dream is transcribed and translated in French by Halleux, 'Helmontiana II', 20–25.

'I long lamented the sin of disobedience', he confesses.[39] He eventually deferred to God for the career decision:

> I often humbly entreated the Lord with a sorrowful heart that he may lead me to my calling, not to the one that I was carried to voluntarily, but to the one that might please Him most.[40]

In the context of Van Helmont's worldview, it is important to note that he did not pursue the physician's path until he had obtained divinely inspired sanction by a dream-vision. This gave him the courage to go against his family's tradition and specialise in a field that was not necessarily highly regarded at the time.

After having come to terms with his choice of career path, Van Helmont decided to accept the medical degree from Louvain. It was a decision that ran somewhat counter to his beliefs in the worthlessness of university education; however, Van Helmont's character seemed to display, together with his strong idealism, a practical streak. As shall be further shown, his largesse did not preclude a desire to provide for his family.

Even after receiving his medical degree, Van Helmont claimed to have still harboured doubts about his career path. In search of his destiny, he turned to travelling: 'I... commended myself to God, with an intention of travelling abroad, of forsaking medicine, and of never returning to my country'.[41] We may suspect Van Helmont was over-dramatising his travelling years, which may well have been part of a *peregrinatio academica* traditional in the epoch. Moreover, he was undoubtedly following the injunction of Petrus Severinus to leave everything behind and travel in order to accumulate medical experience.[42]

Before he left, Van Helmont bequeathed his whole inheritance to his widowed sister. He travelled to Switzerland and Italy between 1600 and 1602, returning home briefly before proceeding to France, Spain and England from 1602 to 1605.[43] In contrast with Paracelsus, Van Helmont seemed to have been well received wherever he went, a fact that undoubtedly had more to do with his noble descent than his medical reputation. He was even invited to parties organised by the English Court at Whitehall, one in the presence of the Queen herself.[44] In England, he became acquainted with two prominent alchemical philosophers, Sir Hugh Platt (1552–1608) and

---

39 'Tumulus Pestis', ch. 1, 834: 'Deflevi dudum peccatum inobedientiae'.

40 'Tumulus Pestis', ch. 1, 834: 'Ac moesto corde saepe supplex oravi Dominum, dignaretur me ducere ad vocationem, non qua mea sponte ferebar; sed in qua illi complacerem potissimum'.

41 'Tumulus Pestis', ch. 1, 834: 'Deo me commendavi, cum intentione proficiscendi peregre, Medicinam deferendi, atque in patriam redeundi nunquam'.

42 See Debus, *The Chemical Philosophy*, 70 for Severinus's paragraph.

43 J. Nauwelaers, 'La Maison de Van Helmont à Vilvorde', in *Le Folklore brabançon: Numéro spécialement consacré à J. B. van Helmont, 1579–1644* (Brussels, 1933), 171.

44 *Opuscula medica inaudita*, 'De lithiasi', ch. 2, 667. He may refer here to Queen Elizabeth I, which would situate Van Helmont's visit to January 1602 or 1603. However, it is possible that Van Helmont refers to James I's wife.

Robert Fludd (1574–1637).[45] Of Fludd his judgement was later negative: 'I have known the man in his country for a poor physician and an even worse alchemist'.[46] Van Helmont seems to have returned to England one more time in 1607, because he mentions in a letter to Mersenne that he saw a comet there on 28 September 1607.[47]

Although the record is not exactly clear, it was probably during these travelling years that he met a practising alchemist. As he recalled,

> an *Idiot* associated himself with me, who knew the art of the fire (*pyrotechnia*), or at least its practice. As soon as I beheld the inward part of bodies by the fire and understood many separations, which had not yet been set in books, being unknown even today, an avidity of knowing and operating increased in me day by day.[48]

In autumn 1605, Van Helmont went to Antwerp to help with an outbreak of plague: 'Returning from England to Antwerp in the autumn of 1605, I found hundreds of people falling into dropsy, after a malignant and ravaging fever. I cured many, while many under the inauspicious experiments of others perished'.[49]

The 'art of the fire' provided Van Helmont with the knowledge and tools he needed to dedicate himself completely to the practice of alchemical medicine. By his own admission, he started to practise his medicine for free, gaining a reputation mainly, but not only, amongst the poor he treated.[50] He claimed to have received flattering employment offers from Prince-Archbishop Ernest of Bavaria, as well as from his cousin, Emperor Rudolf II. Much has been written about Rudolf II's interest in the 'occult sciences', particularly alchemy, yet much less about Ernest, who entertained

---

45 I have conjectured the information on Sir Hugh Platt from two pieces of evidence. One is a mention Van Helmont made in his interrogatory of 21 March 1634, when he stated that one of his confiscated papers was a treatise by 'Hugo Place' (most likely Hugh Platt) originating from 1607–1608, see Corneille Broeckx, 'Interrogatoires du docteur J. B. van Helmont sur le magnétisme animal, publiés pour la première fois', *Annales de l'Académie d'archéologie de Belgique*, 13 (1856): 345; the other evidence is offered by the physician Robert Child in Samuel Hartlib's 'Ephemerides' for 1650: 'Mr Hinshaw hath Sir Hugh Plats MS. Amongst the rest hee saw one with an inscription Helmont's Althahest. It seemes Helmont was acquainted *with* Sir Hugh Plat's when hee was in England. Dr Child', HP 28/1/60B, 'Ephemerides'—1650, in *Hartlib Papers* (Ann Arbor, MI, 1995; enlarged ed., Sheffield, 2002) [on CD].

46 '19 Décembre 1630', II, 584: 'j'ay cognu l'homme en sa patrie pour un pauvre medicin et encore moindre alchymiste, sed erat garrulus, stentor, superficialiter doctus, parum qui constans'. At the same time, Van Helmont admits to having read Fludd's works.

47 '19 Décembre 1630', II, 583.

48 'Tumulus Pestis', ch. 1, 834: 'Idiota namque se mihi associabat, qui pyrotechniae, manualia saltem, noverat. Mox ego, ut penetrale quorundam corporum per ignem inspexi; comprehendi multorum separationes, tum libris nondum traditas, ac hodie aliquot incognitas. Dein indies crevit sciendi ac operandi aviditas'. For Van Helmont's use of the term 'idiot', see Chapter 6, under heading 'Mysticism and Empirical Investigation'.

49 Ch. 64, 'Ignotus hydrops', §11, 408: 'In Autumno anni 1605 ex Anglia Antverpiam rediens, reperi aliquot centenos, post malignam atque popularem febrim, Hydropicos: multos sanavi, multique sub infaustis aliorum tentaminibus interim perierunt'.

50 'Tumulus Pestis', ch. 1, 834.

similar interests in Cologne and Liège.[51] Ernest was a protector of alchemists and astrologers, and supported the publication of the Huser edition of Paracelsus's works in Cologne.[52] The Prince apparently dabbled in alchemy himself. Under the supervision of his personal physician-alchemist Philippe Gerinx (1549–1604), he took upon himself to analyse the waters of the Fountain of Pliny near Tongrès.[53] It was perhaps in the aftermath of Gerinx's death that Ernest invited Van Helmont to Liège to take up employment as his personal physician. However, Van Helmont realised that offering medical services to princes was not his true vocation, and rejected the offers. He was in the privileged position of not having to depend on the financial support of rulers.

Around 1607, the Flemish physician wrote his first known treatise, *Eisagoge in artem medicam à Paracelso restitutam*. This tentative Paracelsian apology was never published and was eventually confiscated by the Archbishopric of Mechelen from Van Helmont's house. This is the only work that can be attributed to Van Helmont's 'early' intellectual stage, and is a summary of Paracelsian medicine seen through the filter of Petrus Severinus's interpretation.

Maturity and Persecution (1609–1636)

In 1609, Van Helmont decided to retire to the town of Vilvoorde, part of his family's domains.[54] By this time, he had already married Marguerite van Ranst (n. 1590), daughter of Willem-Charles van Ranst and Isabelle van Halmale. The van Ranst family also originated from the Berthout, so the marriage was made within the same aristocratic line. It was also a practical and lucrative arrangement, because Van Helmont's wife was very wealthy. Jan Baptist apparently inherited from her the titles of Mérode, Royenborch, Oirschot, Pellaines and other domains.[55]

Isabelle van Halmale, Van Helmont's mother-in-law, seems to have been a well-connected and dynamic figure in Van Helmont's life, forming a close and affectionate relationship with her son-in-law. In her will, she recalled how, since 1609, Jan Baptist

---

51 On Ernest of Bavaria, see *Biographie Nationale de Belgique*, V, 632–45. On his court, see Robert Halleux, 'Nouveaux matériaux sur la cour savante d'Ernest de Bavière', *Cinquième Congrès de L'Association des Cercles Francophones d'Histoire et d'Archéologie de Belgique*, 394–404 and Robert Halleux and Anne-Catherine Bernès, 'La cour savante d'Ernest de Bavière', *Archives Internationales d'Histoire des Sciences* 45 (1995): 4–29.

52 Halleux, 'Helmontiana I', 47.

53 *Biographie Nationale de Belgique*, V, 643. On Gerinx, famous for his distillation of spa waters, see *Biographie Nationale de Belgique*, VI, 670–73.

54 Nauwelaers, 'La Maison', 171.

55 Stroobant, 'Les origins', 149–50. Nevertheless, some of these titles have been disputed by modern scholars. Desmarez, for instance, argues that he could not have really been the lord of all the lands of Mérode, Royenborch, and Pellaines. Nauwelaers, however, makes a good case for the Mérode and Royenborch lands. The Mérode title is confirmed albeit indirectly by Desmarez himself who found a tombstone of a Van Helmont that bore the Mérode coat-of-arms.

had 'nourished and cared for me without hope for re-payment'.[56] Isabelle energetically and relentlessly dedicated herself to liberating Van Helmont from house arrest.

Van Helmont seems to have had eight children: Pélagie-Livina (1611–1662), Olympie-Claire (b.1612), Léandre (b.1613), Franciscus Mercurius (b.1614 or 1618), Elzear (b.1616), Elisabeth (b.1616), Marie (b.1620), Clementine (?).[57] Some confusion remains about his children, since discovered records describe Franciscus Mercurius as the second eldest son, while Van Helmont talks about the death of his two eldest sons in 1634.[58]

Jan Baptist took an active interest in the education of his children, as Franciscus Mercurius later recalled, instructing them from a young age in the 'art of the fire'.[59] He refused to let his sons attend university, for whose Aristotelian-Galenic teachings he had nothing but contempt. While Franciscus Mercurius was proud of having been raised in the 'select school of Hermes', he seemed to have regretted the lack of a proper Latin education.[60] Franciscus might have been the son Van Helmont bragged about to Mersenne as having extraordinary artistic talents.[61] He grew to become, like his father, a renowned physician and alchemist, but his major contribution lies in the promotion of a new form of Christian Kabbalah influenced by the speculations of Isaac Luria (1534–1572).[62]

Van Helmont lived in Vilvoorde for at least seven years, from 1609 to 1616, experimenting with alchemy. After 1616, Van Helmont probably moved to Brussels.[63] It was there that he set up his medical practice. He claimed to have treated the

---

56 Nauwelaers, 'La Maison', 175.

57 Stroobant, 'Les origins', 156; Henri De Waele, *J-B Van Helmont* (Brussels: Office de Publicité, 1947), 12; Desmarez, 'L'État civil', 16. There is confusion with regard to Van Helmont's offspring amongst the scholars: for instance, Stroobant does not mention Elisabeth and Marie and has 1618 for the year of Francis's birth, while Desmarez fails to mention Elzear. Francis Mercury's birthdate is given as 1614 by De Waele. Clementine Van Helmont is mentioned only in his interrogatory of 21 March 1634, when it is stated that she died at 4 years of age; see Broeckx, 'Interrogatoires', 346. It is true that Van Helmont does not explicitly say she was his daughter, but one can assume this.

58 'Tumulus pestis', ch. 14, 873.

59 Franciscus Mercurius Van Helmont, 'Amico lectori', in J.B. Van Helmont, *Ortus medicinae*, [3] (there are no page numbers).

60 Van Helmont, 'Amico lectori', [11]. It is surprising to hear Van Helmont would not have taught his children Latin, which still lay at the core of the written alchemical tradition.

61 'Jean-Baptiste Van Helmont, à Bruxelles, à Mersenne, à Paris: 15 Janvier 1631', 31; '30 Janvier 1531', 53: 'Mon petit eagé de 11 ans [or 12 years old in the previous letter] prend d'une main le ciseau et de l'aultre une loppe de papier ou parchemin sans estre deline ou portraict, coupe selon l'idee luy propose, soit une histoire ou aultre phantasie. Et n'est merveille que l'on ne le croid pas, veu que les painctres d'Anvers le sont venu voir, ne croyant pas possible: toutefois il leur couppoit l'histoire d'Acteon'. However, if it was Franciscus Mercurius, it must mean that he was born in 1618 or 1619, not in 1614. Incidentally, this passage is another indication of Van Helmont's relationship with the Antwerp painters, perhaps even the Rubens family; Rubens certainly painted two Acteons, see p. 53, n. 2.

62 Allison Coudert, *The Impact of the Kabbalah in the Seventeenth Century: The Life and Thought of Francis Mercury Van Helmont (1614–1698)* (Leiden: Brill, 1999), 118.

63 This, at least, is Walter Pagel's conjecture, *Joan Baptista Van Helmont*, 7.

poor for free; at the same time, he was the physician of several high-standing dignitaries, whom he presumably charged for services.[64]

It was probably in Brussels that he witnessed the beginning of the heated exchange that pitted Rudolf Goclenius the Younger (1572–1628) against Jean Roberti (1569–1651) on the subject of the weapon-salve cure.[65] This ointment, which was thought by contemporaries to have been invented by Paracelsus, cured a wound by application to the injuring weapon rather than to the wound itself.[66] This odd treatment became the subject of an incredibly heated debate that dragged on and off for more than a hundred years. The origin of the controversy must be sought in the charged atmosphere of the early seventeenth century, when intellectuals clashed over matters of theology and natural philosophy.

Goclenius, a newly anointed professor at the University of Marburg, chose to vindicate the weapon salve in his inaugural speech at the school. The discourse was published as *Oratio* in 1609, followed one year later by the *Tractatus de magnetica curatione vulneris*. A new and improved edition was issued in 1613 in Frankfurt, entitled *Tractatus novus de magnetica vulnerum curatione*.

It was not a coincidence that Goclenius had chosen the magical weapon-salve cure as the subject of his first address at Marburg. During this period, the Lutheran university had come under the rule of Prince Moritz of Hessel-Kassel, whose religious sympathies leaned toward the Calvinist faith.[67] Moritz was profoundly dedicated to an eclectic philosophy that combined elements of Neoplatonism, Hermeticism, Paracelsianism and Kabbalah. Consequently, Goclenius sought to please his patron by proclaiming his official allegiance to Paracelsian and Hermetic ideas.

The *Oratio* was a defence of Renaissance Neoplatonic-Hermetic philosophy that upheld such ideas as the soul of the world, correspondences, astral influences and the analogy between the microcosm and the macrocosm.[68] According to Goclenius,

---

64 For instance, in 'Butler', 468, Van Helmont talks about the viscount of Ghent, whom he presumably treated for podagra. In the dedication to *Opuscula medica inaudita*, he mentions trying to give medical advice to his friend Lord Werner Spies von Bullensheim, 'Illustri viro D. Caspari Uldarico', 641. He was sought out even in house arrest; on 10 December 1635, he was allowed to go and treat the Lord of Droogenbosch; see Craig Harline, *Miracles at the Jesus Oak* (New Haven, CT: Yale University Press, 2011), 302.

65 The most comprehensive treatment of the weapon-salve controversy is the work by Roberto Poma, *Magie et guérison: La rationalité de la médecine magique (XVIe–XVIIe)* (Paris: Orizons, 2009). For the early history of the debate and Van Helmont's role in it see also Mark A. Waddell, 'The Perversion of Nature: Johannes Baptista Van Helmont, the Society of Jesus, and the Magnetic Cure of Wounds', *Canadian Journal of History* 38 (2003): 182; and Carlos Ziller Camenietzki, 'Jesuits and Alchemy in the Early Seventeenth Century: Father Johannes Roberti and the Weapon Salve Controversy', *Ambix* 48 (2001): 87–96.

66 In reality, the recipe for the weapon salve can be found in the spurious *Archidoxis magica* of Paracelsus; *Aureoli Theophrasti Paracelsi de summis naturae mysteriis libri tres* (Basel, 1570), 127–9.

67 For an excellent analysis of Moritz and his court, see Bruce Moran, *The Alchemical World of the German Court: Occult Philosophy and Chemical Medicine in the Circle of Moritz of Hessen* (1572–1632) (Stuttgart: Franz Steiner, 1991), esp. chapter 2.

68 For an analysis of the *Oratio* see also Poma, *Magie et guérison*, 139–41.

the action of the salve was not supernatural or demonic, but entirely natural, being based on magnetic force.[69] The work also contained some remarks that reflected Goclenius's Calvinist sympathies and anti-Catholic bias.[70]

Jean Roberti, Jesuit professor at the University of Douai, Treves, Würzburg and Mainz, responded to Goclenius swiftly and harshly. In *Anatome magici libelli Rodolphi Goclenii* (1615), Roberti castigated the weapon salve for making use of illicit magic and condemned its use. He raised three chief accusations against Goclenius: idolatry, necromancy and blasphemy. Roberti thought that the weapon salve could only work through the intercession of the devil.[71]

Goclenius responded with a defence, *Synarthrosis magnetica* (1617), in which he sought to clearly differentiate the natural magic of the weapon salve from demonic magic.[72] Roberti promptly launched a new attack called *Goclenius Heautontimoroumenos* (Luxembourg, 1618), followed immediately by two more publications, one in 1618 and another in 1619. Goclenius also replied three times, in 1618, 1619 and 1625.[73]

In these subsequent treatises, the religious differences became even more prominent. In *Goclenius Heautontimoroumenos*, Roberti accused the weapon salve of being a 'Calvinist heresy', while Goclenius denounced the idolatry of the Catholics, including the worship of the Eucharist and the canonisation of saints.[74] Thus, while the subject of the weapon salve started as a debate on natural magic, the discussion quickly became coloured by religious factionalism. Roberti, in particular, tried to associate Paracelsian medicine with Calvinism, hence seeking to discredit it in the Catholic lands.

Van Helmont followed the exchange with clear concern; he disagreed with both Goclenius's interpretation of the weapon salve and with Roberti's attempt to discredit Paracelsianism. He may have also been worried about the religious implications of the debate. Sometime around 1617, he wrote a defence of the weapon-salve doctrine called *De magnetica vulnerum curatione* in which he denounced both Goclenius's arguments and those of Roberti.[75] He upheld Goclenius's opinion that the salve cured by natural occult forces, but denied that these came from the heavens; instead, he argued that the ointment acted by the force of human imagination.

---

69 Rudolf Goclenius, *Oratio qua defenditur vulnus non applicato etiam remedio citra ulium dolorem curari naturaliter posse* (Marburg: Cattorum, 1609), 67.

70 Goclenius, *Oratio*, 8–9.

71 Jean Roberti, 'Anatome magici libelli Rodolphi Goclenii de curatione magnetica per unguentum armarium', in *Theatrum sympatheticum auctum* (Nuremberg: Sylvester Rattray, 1662): 228–9, 235–6.

72 Rudolf Goclenius, *Synarthrosis magnetica opposita infaustae anatomiae Joh. Roberti jesuitae pro defensione tractatus de magnetica vulnerum curatione* (Marburg: Jonas Saurium, 1617), 53–5.

73 De Waele, *J-B Van Helmont*, 28.

74 Goclenius, *Synarthrosis magnetica*, 132–50.

75 Halleux, 'Helmontiana I', 37 proposes 1615. However, in 1615 the Roberti treatise had just been published in Trèves, and only a year later in Liège; moreover, Van Helmont himself declared he read the treatises of Roberti and Goclenius in 1617; see also Broeckx, 'Interrogatoires', 318–19.

It is not clear whether Van Helmont wanted his treatise to be published; he later maintained that he intended it to be read only in private circles. This would partially explain why Van Helmont gave the treatise to Pierre Remacle Roberti to forward to his brother Jean. In turn, the Jesuit scholar went to Brussels to personally convince Van Helmont to publish the defence. Why would Van Helmont agree to such an idea, trusting someone who was railing so aggressively against the weapon salve? Did he think that his solution to the weapon salve would be acceptable to Catholic sensibilities? This is a possibility. Additionally, the gestures of Van Helmont seem to suggest that there was a direct or indirect relationship between the Robertis and Van Helmont that made the latter oblivious to the dangers to which he was exposing himself.[76] In any case, the details of the publication of the treatise are murky. Van Helmont sent it to his close friend Jean Gallé (fl. 1626), a mathematician of Liège, for editing and submission to the book censor of the town, Stevart.[77] The censor approved it in the first instance and gave it to the printer Hovius. At the last minute, however, the Jesuits intervened to revoke the licence. In fact, the censorship was only partially successful: the text had been copied and was subsequently published in Paris in 1621.

Van Helmont maintained that the Paris printing had been done against his will, but he did receive 23 copies of *De magnetica* from a certain abbot of Fontanes, which he subsequently distributed.[78] In turn, Roberti challenged Van Helmont's plea of innocence over the publication; he accused Van Helmont of intending to re-publish the work in Liège in 1623 or 1624.[79]

Roberti hurried to publish a refutation of Van Helmont called *Modesta responsio ad perniciosam disputationem JB ab Helmont Bruxellensis medici pyrotechnici contra eudem Roberti acerbe conscriptam* (Luxembourg, 1621). In it, he accused Van Helmont of corrupting sacred theology and Scripture, as well as of defending superstitious beliefs and demonic magic.[80] Roberti recommended that the newly installed Archbishop of Mechelen, Jacques Boonen (1573–1655), start formal proceedings against Van Helmont.[81]

However, Boonen, a fascinating figure in his own right, was not supportive of Jesuit influence in Flanders.[82] A graduate of Louvain, he had inherited both the

[76] This acquaintance with the Robertis, his studies with the Jesuits, particularly del Rio, and alchemical assistance to d'Aguillon raise interesting questions as to the nature of Van Helmont's relationship with the Jesuits and the reason for his fallout with them.

[77] Broeckx, 'Interrogatoires', 313–14; De Waele, *J-B Van Helmont*, 30. On Van Helmont's friendship with Jean Gallé and a brief presentation of the latter, see Halleux, 'Helmontiana I', 11.

[78] Broeckx, 'Interrogatoires', 25–6.

[79] Halleux, 'Le procès', 1064. Halleux's study is an excellent resource for the study of the Inquisition trial, including a handy timeline (Appendix 1). For an alternative 'lifelike' account of Van Helmont's trial see also Harline, *Miracles*, 179–241.

[80] Broeckx, 'J.B. Van Helmont: Ad Judicem', 76.

[81] Halleux, 'Le procès', 1064.

[82] On Jacques Boonen, see Felix Victor Goethals, *Lectures relatives à l'histoire de sciences, des arts, des letters, des moeurs et de la politique en Belgique*, 2 vols (Brussels: Vandooren, 1837), I, 119–31; also *Biographie Nationale de Belgique*, II, 700–5.

school's antipathy toward the Jesuits as well as its fascination for Baianist theology. A close friend of Cornelius Jansenius, Boonen tried to promote Baianism through the Privy Council of Brabant, and attempted to stave off the power of the Society of Jesus in the Mechelen area by inviting priests of the Augustinian Oratory of Pierre de Bérulle (1575–1629) in 1526.[83] Boonen's later support of Jansenism would bring him into open confrontation with the Jesuits.[84] In Van Helmont's case, Boonen did not stand against the Jesuits, but proved slow and ambiguous in his dealings with the case.

Since Boonen did not follow up on Roberti's initial request to investigate Van Helmont, the Jesuit scholar renewed his appeal to the Archbishop in a 1623 letter. This was accompanied by the censorious remarks of Francisco de Paz, the physician of the ruling archdukes Albert and Isabella, as well as of Thomas Fyenus and Gerard de Villers.[85] We have already encountered Fyenus and de Villers entrusting Van Helmont with a surgery course when the latter was around 17 years old. Yet time had passed, and the two medical professors may have borne some grudge against Van Helmont for abandoning the Galenic framework of medicine in favour of Paracelsianism. Moreover, as Halleux has pointed out, Fyenus had written a treatise explicitly denying the reality of natural magic and magnetism, except for the power a mother had over the foetus.[86] Despite their intervention, the letter did not compel Archbishop Boonen to action. Roberti, undoubtedly incensed by Boonen's silence, sent three more letters at the beginning of 1624. In these he threatened to refer the matter to Rome.[87]

In the meantime, Van Helmont was busy writing a further weapon-salve treatise, called 'Ad judicem neutrum causam appellat suam et suorum philadelphus', which was never finalised or published.[88] The treatise was an in-depth analysis of Roberti's *Goclenius Heautontimoroumenos* of 1618, in which Van Helmont sought to refute all of Roberti's arguments against the weapon salve. He appears to have worked on it until 1626 or even later, as it contains a mention of the condemnation of the Jesuit theologian François Garasse (1585–1631) by the University of Paris.[89] The manuscript was found in 1634 amongst his papers and confiscated by the Inquisition.

Van Helmont had other concerns as well, as evidenced by his commentary *Supplementum de Spadanis fontibus* on the *Spadacrene* (1614) of Henri de Heer (1570–1636). The *Supplementum* disputed de Heer's claim on behalf of the wondrous properties of the waters of the Spa. The treatise was published in 1624, in Liège, being

83 Pirenne, *Histoire de Belgique*, III, 46.

84 Pirenne, *Histoire de Belgique*, III, 46–7 and Goethals, *Lectures relatives*, 130–32.

85 Halleux, 'Le procès', 1079.

86 Halleux, 'Helmontiana I', 39; also Halleux, 'Le procès', 1070.

87 Halleux, 'Le procès', 1079.

88 Van Helmont maintained this was written in 1619, in response to the *Heautontimoroumenos* of Roberti. Halleux, 'Le procès', 1071, n.63 points out that 1621 is more plausible as a date, as Van Helmont makes reference to the *Modesta reponsio* of Roberti.

89 Broeckx, 'J.B. Van Helmont: Ad Judicem', 132.

dedicated to the Prince-Elector Ferdinand, the archbishop of Cologne. De Heer, who prided himself on 25 years of experience with the Spa waters, was greatly exercised by Van Helmont's comments on his work and responded acidly in two treatises, one in 1624 and another in 1630. In these, he appealed to an *ad hominem* attack, accusing Van Helmont of being a poor physician and having caused the death of a patient.[90] In the *Ortus*, Van Helmont claimed to have proven the correctness of his assertions by an experimental demonstration in the presence of de Heer and other persons.[91]

In the meantime, having had his letters to Archbishop Boonen gone unanswered, Roberti appears to have resorted to a cunning plan. In 1624, a new edition of the weapon-salve treatise suddenly appeared in Cologne, under the title *Joannis Baptista Helmontii medici et philosophi per ignem propositiones notatu dignae, depromptae ex ejus disputatione de magnetica vulnerum curatione Parisiis edita.*[92] This version was a summary of Van Helmont's ideas in 24 propositions, plus three Paracelsian propositions to which Van Helmont supposedly subscribed.[93] The publication was a product of Van Helmont's enemies, as it contained at the end a refutation of his ideas. The censure was signed by the five Louvain theologians, two Louvain physicians, Thomas Fyenus and Gerard de Villers, as well as six more physicians from Douai, Cologne and Luxembourg.[94]

It was this second publication that determined the Spanish Inquisition to begin legal proceedings against the Brussels doctor. The Inquisition's inquest was spurred on by the further censure of the Helmontian propositions by several Liège theologians, on 1 March, 10 April and 18 April 1625.[95] On 6 October 1625, the Inquisition officially condemned the 27 propositions of the Cologne tractate and demanded the arrest of Van Helmont on suspicion of heresy. Yet the Inquisition did not have a mandate to act directly in the southern Netherlands. On 23 February 1626, Sebastianus de Huerta, the secretary of the Inquisition, requested the Flemish tribunals to open a legal case against Van Helmont.[96]

This demand was not immediately acted upon. It was only a year and a half later, on 3 September 1627, that the ecclesiastical curia of Mechelen finally summoned the physician to appear in court. Van Helmont complied, but not before proclaiming his faith in the Catholic church and requesting the annulment of the trial.

Three days later, on 6 September, the prosecutor decreed that the books and responses of Van Helmont should be submitted to the judgment and censure of the

---

90 Henri de Heer, *Observationes medicae oppido rarae in Spa et Leodii animadversae* (Liege: Van Der Aa, 1630), 287.

91 Ch. 12, 'Aer', §11–13, 51–53.

92 'Joannis Baptistae Helmontii medici et philosophi per ignem Propositiones notatu dignae, depromptae ex ejus disputatione de magicae vulnerum curatione', in Hubert Dethier, Ivan Hermans and Sylvette Sue, *Johan-Baptist Van Helmont (1579–1644)* (Brussels: Vrije Universiteit Brussels, 1985), annex I.

93 De Waele, *J-B Van Helmont*, 34.

94 'Joannis Baptistae Helmontii medici', annex I.

95 Halleux, 'Le procès', 1080.

96 De Waele, *J-B Van Helmont*, 34.

professors of theology and medicine in Louvain. These spent the following three years analysing the documents. In the meantime, in 1628 the theologians of Douai, Cologne and Dillingen condemned Van Helmont's writings, followed in 1629 by several Lyon theologians and physicians.[97]

During this unsettled period (1625–1630), other events were taking place in Van Helmont's life. In 1625, or thereabouts, he claimed to have met an alchemical Adept, an Irishman called Butler, who had been imprisoned in the castle of Vilvoorde.[98] According to Van Helmont's story, Butler had cured a fellow inmate, a Franciscan monk, of *erysipelas* or swelling of the arm. Astonished by this, the town officials asked Van Helmont to talk to Butler and witness his cure.[99] Van Helmont became friends with the Irishman, and apparently the better for it, as the latter cured him of a slow poison given by an enemy. He later procured Butler's release.[100] According to Van Helmont's later testimony, the Irishman was in the possession of the universal medicine, but did not share its composition with him. Van Helmont pondered and meditated upon Butler's stone for a long time. Eventually, he grasped Butler's secret and claimed to have made the medicine (*Drif*) himself.[101]

Was this Butler real, and if so, who was he? Stories of mysterious Adepts visiting younger alchemists or being unjustly persecuted were popular in the period. In any case, attempts to discover his identity have failed thus far.[102] In 1652, the *Ephemerides* of Samuel Hartlib show that Frederick Clodius (1625–c.1661), Hartlib's alchemist son-in-law, was trying to obtain the manuscripts of Butler from a certain Higgins of Limerick, who claimed to have been Butler's servant.[103] Unfortunately, Clodius did not provide any further update on the matter. The fact that two other persons besides Higgins claimed to have manuscripts raises the question of whether any of those documents were genuine.

Incidentally, the Sloane collection at the British Library holds a manuscript by 'Buthler, philosophe anglais', which has obviously been transcribed in a neat hand in the French language.[104] It is difficult to ascertain the provenance of this manuscript, entitled 'Traité de l'usage de pierre', but the content is well in line with Van Helmont's description of Butler's stone as a universal medicine. This was probably

97 Halleux, 'Le procès', 1080–81.

98 Although Van Helmont's 'Butler' treatise did not mention the year of their acquaintance, Halleux, 'Helmontiana I', inferred from several pieces of information that it happened in 1625. Indeed, Van Helmont talks about Butler curing his wife of oedema contracted out of stress caused by his misfortune, ch. 79, 'Butler', 469.

99 Ch. 79, 'Butler', 467–8.

100 'Tumulus pestis', ch. 17, 'Zenexton', 884.

101 Ch. 79, 'Butler', 474–5. For more on this topic, please refer to Chapter 7, section on *Drif* or *Lapillus*.

102 Halleux, 'Helmontiana I', 40, thinks 'Butler' was an adventurer who assumed the identity of William Butler of Ipswich, a physician who died in 1617 or 1618. Although that is a possibility, this William Butler was an Englishman while Van Helmont's Butler is identified as an Irishman.

103 HP 28/2/28A-28B, 'Ephemerides', 1652.

104 London, British Library, MS Sloane 2879, fols 50^r—74^v.

written by a forger after Van Helmont's work had propelled the legend of Butler into alchemical consciousness.

In 1629, Van Helmont also began a legal action, demanding the division of the revenues originating from his wife's inheritance, the estate of Pellaines, which had previously remained undivided in the family of Mérode. He hence obtained for his wife 3,511 florins. The court's decision was disputed even after Van Helmont's death; in 1647, Jean de Mérode opened proceedings against Van Helmont's widow demanding the annulment of the decision. Mérode lost, and was condemned to leave to the van Ranst family the majority of the Pellaines estates. The trial only ended in 1663.[105] Van Helmont's involvement in this legal action reveals, on one hand, that his private concerns were not thwarted by the Inquisition's case at this point, and on the other, that the physician was concerned with ensuring his wife and children's financial future.

In May of the same year, Van Helmont was visited by the famous French philosopher Pierre Gassendi (1592–1655). Gassendi, fascinated by the topic of the weapon salve, had first met Jean Roberti, theologian Libert Froidmont (1587–1653) and Thomas Fyenus to discuss the subject. He then paid a visit to Van Helmont to get his opinion on the remedy and on magnetism. Gassendi was convinced of the reality of the salve, but proposed a corpuscularian explanation for its action.[106] Gassendi and Van Helmont conversed for several days and then corresponded with each other; unfortunately, only one letter is extant.[107] The debated issue was whether prelapsarian man was naturally carnivorous or vegetarian. Van Helmont considered that man, as a microcosm, contained within himself all animals and as such had to be carnivorous, while Gassendi took the position that meat-eating was naturally unsuited for humans.[108] Despite their formal disagreement, the letter leaves the impression of a profound esteem between the two natural philosophers. In fact, captivated by the Flemish physician, Gassendi recommended him to Mersenne, who first came into contact with him in May 1630.[109]

Consequently, Van Helmont conducted a bulky correspondence with Mersenne between June 1630 and July 1631. As the exchange ends abruptly, it is probable that the correspondence went on after this date, but further letters have been lost. Van Helmont offered his (negative) opinions on the writings of Jacques Gaffarel (1601–1681), who had favourably quoted from his *De magnetica vulnerum curatione* in his *Curiosites inouyes* (1629), as well as on Robert Fludd, whom he had met in England. Van Helmont also answered a large number of questions from Mersenne in regard

105 Stroobant, 'Les origins', 152–3.

106 Pierre Gassendi, *Opera omnia*, 6 vols (Lyon: Annisson, 1658), I, 456.

107 Gassendi, *Opera omnia*, VI, 19–24.

108 Guido Giglioni, 'Life and its Animal Boundaries: Ethical Implications in Early Modern Theories of Universal Animation', in *Ethical Perspectives on Animals in the Renaissance and Early Modern Period*, ed. C. Muratori and B. Dohm (*Micrologus Library* 55) (Florence: SISMEL, 2013), 111–37; Joseph Bougerel, *Vie de Pierre Gassendi* (Paris: J. Vincent, 1737), 48–57.

109 'Jean-Baptiste Van Helmont, à Malines, à Mersenne, à Bruxelles: Juin 1630', II, 496–8.

to medicine, alchemy, theology and philosophy.[110] His letters reveal that he was well immersed in the political realities of his day, as he sent Mersenne information about the military situation of Belgium, and the actions of French and Flemish nobles in Brussels.[111] He seemed to have had a good standing at the Flemish court, as he told Mersenne that he was trying to place one of his sons as page to the Ambassador in Constantinople.[112]

On 23 October 1630, the University of Louvain finally published a decision censuring Van Helmont's writings. It was signed by five theologians and two physicians, the same ones that had censured the 1624 Cologne *Propositiones*. Eight other physicians, including Henri de Heer, supported the action in general terms. The widespread support of the indictment by local physicians made Halleux wonder if, in fact, the case was fuelled by the medical establishment rather than the Jesuits.[113] However, Mark Waddell argues that the Jesuits still played a major role in Van Helmont's persecution, as they had a strong motivation to counteract his attack on their authority.[114]

Acting on the Louvain response, the Mechelen prosecutor interrogated and then asked Van Helmont to publicly retract his weapon-salve treatise. The Louvain theologian Johannes Schwenckelius provided a list of statements that Van Helmont had to retract and recommended the banishment of the physician from the southern Netherlands.[115] Van Helmont agreed to admit his guilt and withdraw his statements. The case was suspended without taking action on Schwenckelius's recommendations of expulsion.

Van Helmont enjoyed four years of respite. He continued to correspond with Marin Mersenne, who unsuccessfully tried to get the physician to come to France in 1631. Mersenne, who was close to the French Court, had asked for Jan Baptist's advice in regard to the infertility of the marriage between King Louis XIII (1601–1643) and Anne of Austria (1601–1666).[116] Van Helmont replied that he could make the Queen become pregnant by magnetism and asked for Mersenne's aid in travelling to France in order to disclose his secret to the King. Alas, the rest of this intriguing correspondence has been lost. However, an anonymous treatise beginning with the words 'Exterior homo' was found amongst Van Helmont's seized papers. In his interrogatory of 24 March 1634, the physician explained that this short work was part of a

110 On Jacques Gaffarel's work, see Van Helmont's letter of 26 September 1630, '26 Septembre 1630', II, 530–41; on Robert Fludd, letter of 19 December 1630, II, 582–94; Van Helmont's answers to Mersenne's questions extend to several letters, see III, 10–20, 31–51, 52–151.

111 'Jean-Baptiste Van Helmont, à Bruxelles, à Mersenne, à Paris: 21 Février 1631', III, 124; 'Jean-Baptiste Van Helmont, à Bruxelles, à Mersenne, à Paris: 1 Mars 1631', III, 139 on the duke of Vendôme and duke of Niebourg.

112 '21 Février 1631', III, 123.

113 Halleux, 'Helmontiana I', 54–7.

114 Waddell, 'The Perversion of Nature', 190–92.

115 As this statement was never signed, Harline wonders if it was ever shown to Van Helmont; *Miracles*, 219.

116 '1 Mars 1631', III, 138–40; '29 Mars 1631', 151–5. Anne of Austria managed to have a son, the future Louis XIV, but only in 1638, when she was 37.

consilium (medical consultation) for a certain prince, but denied that the treatise was his. He maintained it had been sent to him by Mersenne with the permission to make annotations on it, which he did.[117] The treatise can confidently be linked with the discussion of Louis XIII's childless marriage. In a recent article, I have undertaken the task of analysing this intriguing treatise, whose ideas are remarkably similar to those of Van Helmont himself. I have concluded that it is very likely Van Helmont wrote the piece as an introduction to the King, but lack of direct evidence prevents certain attribution of authorship.[118]

This was apparently not the only encounter of Jan Baptist with French politics. He took an active interest in the relations between France and the southern Netherlands, informing Mersenne of issues arising at the Brussels court.[119] Even more captivating is the fact that the horoscope of the Cardinal de Richelieu (1585–1642) was discovered amongst Van Helmont's papers confiscated by the Inquisition. Questioned about this, Van Helmont affirmed that it was given to him by Count Luca Fabroni, a favourite of France's Queen Mother Maria de Medici (1573–1642), to find out about Richelieu's political intentions.[120] Fabroni, a Florentine nobleman, was not only Maria de Medici's confidant but a famed astrologer, who predicted Louis XIII's early death.[121] It is captivating that an inveterate astrologer like Fabroni would approach Van Helmont, particularly considering that the latter emphatically rejected judicial astrology in his works.

The involvement of Van Helmont in French politics is indirectly confirmed by the testimony of the French Galenic physician Guy Patin (1601–1672). Patin, who disliked Van Helmont intensely, affirmed that the Fleming was only saved from 'being burnt at the stake' by the intervention of the Queen Mother.[122] In light of the trial records, the possibility that Van Helmont might have been condemned to death was remote, but it is clear that Patin was voicing some rumours that implicated the Flemish philosopher with the French faction that opposed Richelieu. It is notable that the Flemish and Spanish authorities took an active position in the quarrel between the Queen Mother and the anti-Spanish Cardinal de Richelieu by supporting her party and even welcoming her for a visit that took place in August 1631.[123] It was most likely during this visit that the Queen's protégé Fabroni came into contact with Van Helmont to obtain advice on her adversary Richelieu. Patin's affirmation strengthens the impression that the Flemish physician developed

---

117 Broeckx, 'Interrogatoires', 348–9.

118 Georgiana D. Hedesan, 'Paracelsian Medicine and Theory of Generation in "Exterior Homo", a Manuscript Probably Authored by Jan Baptist Van Helmont (1579–1644)', *Medical History* 58 (2014): 375–96.

119 For instance, '1 Mars 1631', III, 139–40, '8 Mars 1631', 141–2, 146, '29 Mars 1631', 151–4.

120 Broeckx, 'Interrogatoires', 45.

121 Julia Pardoe, *The Life of Marie de Medicis, Queen of France*, 3 vols (London: Colburn, 1852), III, 482; Herve Drevillon, *Lire et écrire l'avenir: l'astrologie dans la France du Grand Siècle, 1610–1715* (Paris: Champ Vallon, 1996), particularly 118, 165, 218.

122 Guy Patin, *Lettres* (Paris: J.H. Reveille-Parise, 1846), 16 December 1649.

123 Henne and Wauters, *Histoire de la ville*, 48–9.

a close, if slightly nebulous, relationship with the Queen Mother's party. What is clear is that Van Helmont was not a recluse and withdrawn person, as some scholars have portrayed him to be, but an intellectual that was actively involved not only in medicine but also in the life of his country.

Van Helmont may have thought the unfortunate matter of the weapon salve closed, but the tide turned in 1634, when Van Helmont's weapon-salve treatise was published again, this time at Liège, in the heartland of the southern Netherlands. The court was not prepared to be as lenient as it had been in previous years. On 3 March, Van Helmont was arrested, his house was searched and all his papers confiscated. On 6 March, the Flemish physician signed a declaration whereby he submitted to the judgment of the ecclesiastical court. He was thereby confined in a Franciscan convent against a bail amount of 6,000 florins. This was promptly paid by Willem van Ranst, Jan Baptist's step-brother.[124]

On 17, 21 and 24 March, Van Helmont presented himself at the Mechelen tribunal to answer several detailed questions. On 18 March, he wrote a letter requesting permission to stay at home rather than at the convent; this was granted against another surety of 6,000 florins if he left the house without permission. Henceforth, Van Helmont was to be held under house arrest for almost three years (until March 1637), without being tried or charged.

Van Helmont's mother-in-law, Isabelle van Halmale, sprang into action. She wrote several letters to the Chancellor of Brabant requesting Van Helmont's freedom. The petitions of Halmale were based on the arguments that, on one hand, Van Helmont could not be accused of heresy when he had openly declared his orthodoxy and brought many references in this respect, and on the other, that the ecclesiastical case did not respect the laws and customs of the land.[125] The Chancellor demanded the advice of three lawyers, the Archbishop of Mechelen and the official of the curia. The lawyers and the Archbishop gave him an ambiguous answer, while the curia official produced 79 court rulings without providing his own views on the matter. On 23 and 24 October 1634, the Van Helmont family came back with another request for freedom and new proofs of orthodoxy.

During this time, to compound Van Helmont's misery, a plague broke out in Brussels, which carried on for three years (1635–1638).[126] Confined to his house, the physician could do nothing to alleviate the sufferings of the sick; he did, however, request and receive an authorisation to tend an ill nobleman for one day. When his two eldest sons caught the disease, Van Helmont apparently helped them recover and then sent them to a convent to convalesce under strict supervision. However, as he later claimed, the strict medicine and diet he recommended was not kept up and the boys died away from their father. Although the stoic Jan Baptist did not comment in

[124] De Waele, *J-B Van Helmont*, 36. Most information presented henceforth is based on the in-depth study made by De Waele on the trial records, as well as from Broeckx, 'Interrogatoires', 330–49.

[125] Halleux, 'Le procès', 1074–5.

[126] Henne and Wauters, *Histoire de la ville*, 61.

depth on this tragic chapter of his life, he confessed in the plague treatise 'Tumulus pestis' that its writing was prompted by the loss of his two sons to the disease.[127]

Old Age (1637–1644)

On 16 March 1637, the Brabant court finally acceded to van Halmale's requests, demanding that the ecclesiastical court release Van Helmont against a bail of 6,000 florins. No decision in regards to the case was ever taken. However, the formal proceedings were only concluded in 1642, when Van Helmont was finally allowed to publish again.

In December 1638, Van Helmont wrote a letter of complaint to the Archbishopric of Mechelen in which he deplored the injustice done to him and demanded his confiscated papers back. Unfortunately, these were never returned to him, and are still in the possession of the Archbishopric today. He undoubtedly desired the manuscripts in preparation for his later writings.

In the latter part of his life, Van Helmont became better known and respected. An important factor in his rise to fame was the publication of the treatise *De febrium doctrina inaudita* (*De febribus*) in Antwerp in 1642. Van Helmont later maintained that its publication was occasioned by the death of a friend, Lord Werner Spies of Bullensheim, of tertian fever; according to the Flemish philosopher's testimony, he was not allowed to treat him because of the opposition of Bullensheim's Galenic physicians.[128] Following the publication of *De febribus*, Van Helmont confesses to have received many letters from alchemical enthusiasts across the continent demanding further illumination of his remedies.[129] Consequently, the Flemish physician published the *Opuscula medica inaudita* and *Tumulus pestis* (Cologne, 1644). These were dedicated to the German Baron Caspar Uldarick, described as a friend and protector of the 'art of the fire'.

Another patron of alchemists, Duke Frederick III of Schleswig-Holstein-Gottorp (1597–1659), seems to have contacted Van Helmont to obtain secrets, particularly the coveted and mysterious Liquor *Alkahest*.[130] According to a handwritten note by Robert Boyle to Samuel Hartlib, found in the Royal Society, the Duke sent a great

127 'Tumulus pestis', ch. 14, 'Proprietas pestis', 873: 'Horumque jacturam velut praesentem, frequenter intueor, hujusque tractatus initium dedisse putaveris'.

128 'Illustri viro D. Caspari Uldarico', 641.

129 'De lithiasi', ch. 8, §1, 704.

130 The prince was clearly passionate about alchemy, since he had also briefly sponsored the alchemist Michael Maier (1569–1622), while Pierre-Jean Fabre and Otto Tachenius dedicated treatises to him. On Maier, see Karin Figala and Ulrich Neumann, '"Author Cui Nomen Hermes Malavici": New Light on the Bio-bibliography of Michael Maier (1569–1622)', in *Alchemy and Chemistry in the 16th and 17th Centuries*, ed. Piyo Rattansi and Antonio Clericuzio (Dordrecht: Kluwer, 1994), 136. Pierre-Jean Fabre's *Manuscriptum ad Fridericum* has been republished by Bernard Joly, *La rationalité d'alchimie au XVIIe siècle* (Paris: Vrin, 1992), 132–218. See also Otto Tachenius, *Epistola de famoso liquore Alcahest* ([Hamburg], 1652).

gift to Van Helmont in an attempt to entice him to share his secret.[131] Unfortunately, the Flemish physician died before receiving it, and all Frederick could get was a version of the *Alkahest* from his widow—not a very good one according to Otto Tachenius (1610–1680)—as well as a number of manuscripts. These manuscripts, referred to as 'Canon Alkahesticus', were apparently brought to London by an unknown German alchemist around 1659–1660.[132] Boyle reports that he had managed to convince this alchemist to publish the manuscripts, but before this happened, they were burnt in the 1666 fire of London, save for one single page.

Van Helmont died on 30 December 1644, probably in Brussels. He was almost 65 and worn out by his life's tribulations, as he himself confessed. Nevertheless, his death was relatively abrupt, catching him in the middle of editing and finishing his most important works. As his son Franciscus Mercurius bears witness, in autumn 1644 his father caught a disease of the lungs (pleurisy), which lasted for seven weeks.[133] Franciscus insisted that Jan Baptist was not only fully conscious and in possession of his wits till the last moment, but that he even predicted the time of his own death. Franciscus's commentary was undoubtedly intended to counteract malevolent rumours according to which Van Helmont had gone mad. Thus, Patin had this to say: 'Helmont was an evil Flemish scoundrel who died mad several months ago . . . This man supported only a medicine based on chymical and empirical secrets, and . . . since he strongly condemned the practice of bleeding patients, he died enraged'.[134] It was precisely this idea that Franciscus sought to refute by his in-depth description of his father's last moments. Equally, he was keen to assert his rights as editor of Van Helmont's manuscripts, claiming that his father had given the papers to him to do as he saw fit.

In 1646, Van Helmont's widow obtained a posthumous rehabilitation from the Archbishop of Mechelen.[135] The place of interment of the physician is not known. Desmarez found a tombstone in the St. Gudula church in Brussels bearing the Van Helmont coat-of-arms, which he attributed to Jan Baptist.[136] However, his claim has not been endorsed by other scholars.

This chapter has attempted to show that Van Helmont's speculations were rooted in the environment of his time and place. The complexity of the period reveals many facets of Van Helmont's intellectual persona. He was a member of the landed gentry of the southern Netherlands, in a tumultuous period when the future state of Belgium was only beginning to take shape based on religious identity. He was a staunch supporter of the Catholic faith; at the same time, as a Louvain alumnus and a member

131 London, Royal Society, MS 187, fols. 35v-37r; also Robert Boyle, 'An Account of the Two Sorts of the Helmontian Laudanum, Communicated to the Publisher by the Honourable Robert Boyle', *The Philosophical Transactions* 9 (1674): 147–9.

132 The date is surmised by Clericuzio, 'From Van Helmont to Boyle', 312.

133 Franciscus Mercurius Van Helmont, 'Amico lectori', [1–2].

134 Patin, *Lettres,* 7 April 1645.

135 Halleux, 'Le procès', 1083.

136 Desmarez, 'L'État civil', 14–15.

of the Flemish landed nobility, he appeared lukewarm to Counter-Reformation doctrine. Like many of his intellectual contemporaries in Flanders, he leaned toward Augustinian and Patristic thought rather than toward Thomist Scholasticism. This was a religious tendency that coloured his worldview significantly.

Despite this conservative background, we will find that Van Helmont was one of the most radical reformers of natural philosophy in the period. He was a staunch anti-Aristotelian and anti-Scholastic, as well as a keen supporter of the alchemical philosophy. His worldview combined an alchemical perspective with a profoundly Christian faith influenced by Patristic thought.

Chapter 3

Van Helmont's *Ortus Medicinae* and Some of its Problems

Ortus medicinae was first published in 1648 by Van Helmont's son, Franciscus Mercurius. The unitary publication of the work made many scholars treat the *Ortus* as a coherent body of writing designed as such by Van Helmont himself. This view is perpetuated by the lack of a surviving manuscript of the *Ortus*. In fact, the external interventions of the editor and printer, and Van Helmont's own in-text evidence suggest that the assumption of the unity of the *Ortus* is questionable.

Let us first examine the external evidence for this. Franciscus Mercurius accompanied the publication of the *Ortus* with a lengthy introduction intended to show that he was the rightful possessor and editor of his father's papers. Franciscus stated that they were given to him by Van Helmont when the latter felt death was impending:

> A few days preceding his death, he called for me, [saying]: Take all my writings, those crude and unrevised, as well as those that have been thoroughly corrected, and join them together: I now commit them to your care; complete all things according to your own judgement. Thus it has pleased the Lord Almighty, who carries all things powerfully and directs them benevolently.[1]

This statement is important in understanding the nature of the *Ortus*. Franciscus here clarifies that the work is a compilation of revised, unrevised, and unfinished parts. He is implying that he cannot guarantee that all the parts belong together.

The possibility that the *Ortus* is not a unitary *magnum opus* is further raised by editorial interventions. First of all, the complete title of the book: *Ortus medicinae, id est, initia Physicae inaudita. Progressus medicinae novus, in morborum ultionem, ad vitam longam* (*The Birth of Medicine, that is, the Unheard-of Beginnings of Physics. The Progress of New Medicine against the Vengeance of Diseases and for Long Life*) suggests that the editor may have considered the separation of the *Ortus* into two. The first part was a book of medicine and natural philosophy; the second one of medicine and extension of life. Yet the separation was clearly tentative.

---

1 Franciscus Mercurius Van Helmont, 'Amico lectori', [2]: 'Paucis diebus obitum ejus praecedentibus, inquiebat mihi: *Cape omnia mea scripta, tam cruda & incorrecta, quam penitus expurgata, eaque conjunge; tuae curae nunc illa committo; omnia ad arbitrium tuum peragito. Omnipotenti Domino ita placuit, qui omnia fortiter aggreditur, omniaque benigne dirigit*'.

The table of contents gives us further clues in this matter (Figure 3.1). Here we encounter a four-section division of the *Ortus*. The first section has no headline, the second is called 'Tractatus de Morbis' (Treatise on Diseases), the third 'Supplementum' and the fourth 'Opuscula medica inaudita'. However, except for the fourth section, which is a reprint of the 1644 publication by the same title, the others do not match the internal structure of the *Ortus* very well. Chapters that should have been part of 'Tractatus de Morbis' (e.g. 'Asthma & Tussis', 'Ignotus hospes morbus') fall outside of it, while 'Supplementum' is a muddle of new chapters and old treatises from Van Helmont's middle period (that is, 'Supplementum de Spadanis fontibus' and 'De magnetica vulnerum curatione'). The numbering of the chapters is equally confusing and inaccurate. A dedication to the author (obviously not written by Van Helmont) is bafflingly set as chapter 1 of his work, thus derailing the entire numbering system until chapter 67 (which should be 66). After this, every time an additional section begins, the counting of the chapters resets to 1. All this further suggests that the editor was uncertain of how the different parts of the *Ortus* fitted together.

An additional piece of external evidence shows that the editor and/or printer sought to evade the issue of ordering the contents of the *Ortus*. They called all chapters 'treatises' ('*tractatuum*') to suggest that they could also be viewed separately from each other. In other words, the sequence of chapters and implicitly the sections themselves should only be regarded as relative. This clever strategy only works partially: a chapter like 'Gas aquae' ('The Gas of Water') could be described as a 'treatise', but one that is called 'Preface' (there are two chapters thus called) cannot. To complicate things further, some treatises are actually entitled 'tractatus' (e.g. 'Tractatus de anima'), but they actually contain several chapters.

In addition to external evidence, there are compelling internal clues in the *Ortus* itself that suggest Van Helmont did not conceive it as a unitary product. In chapter 105 (7 in the 'Supplementum' section), 'Aura vitalis' ('The Vital Aura'), Van Helmont points out that he has composed a volume called 'de ortu medicinae'.[2] This can be corroborated with declarations in the previously published *De febribus* (1642) and the *Opuscula medicina inaudita* (1644), that a 'large volume' on the knowledge and curing of disease, called 'Ortus medicinae. Id est, initia Physica inaudita' was being prepared.[3] Clearly, in all these instances Van Helmont was talking about the same work. Yet it makes no sense that Van Helmont would talk about a projected volume

[2] Ch. 105, 'Aura vitalis', 576: '. . . ac lectorem alio remittam ad volumen de ortu medicinae'.

[3] *Opuscula medica inaudita*, 'De febribus', 'Lectori scientiam', 738: 'Hic tractatus laceratus è magno volumine, quod inscribitur, Initia Physica nova & inaudita . . . Magnum enim volumen de morborum cognitionibus & sanationibus, ingens certè atque inauditum conscripsi, ab ipsis verae Philosophiae incunabilis' ; 'Philiatro lectori', 645: 'Ideoque in plena persecutionum tempestate, volume conscripsi; cujus titulus, Ortus Medicinae. Id est, initia Physica inaudita. In quo detexi errores Scholarum, medendo solitos'. See also 'Scholarum humoristarum passiva deceptio, atque ignorantia', ch. 1, 790. These statements do not necessarily mean that Van Helmont had already written the final version of the *Ortus* by 1642; it is more likely that he drew up the structure, or wrote drafts of it.

INDEX TRACTATUUM,

à JOANNE BAPTISTA VAN HELMONT editorum.

Figure 3.1 *Ortus medicinae* Table of Contents (1652 edition).

64 Igno-

Figure 3.1 (Continued)

TRACTATVS DE MORBIS.

***** 2 SVP-

Figure 3.1 (Continued)

SVPPLEMENTVM.

OPVSCVLA MEDICA INAVDITA.

VATI-

Figure 3.1 (Continued)

in a treatise that belonged in the volume itself. This suggests that at least the chapter 'Aura vitalis' was not meant to belong in the *Ortus* at all.

In fact, there are further clues that other treatises were not supposed to be part of the *Ortus* either. 'De lithiasi' and 'Tumulus pestis' in the *Opuscula medica inaudita* contain mentions of a 'Liber de vita longa' ('Book of Long Life').[4] Allusions to *Liber de vita longa* appear in other treatises of the *Ortus*.[5] However, no such book of Van Helmont has been published or remains in manuscript.

Instead, most, if not all of what should have been *Liber de vita longa* was incorporated by Franciscus within the *Ortus medicinae* itself. This is obliquely indicated in the long title of the *Ortus* ('. . . in morborum ultionem, ad vitam longam'). The 'Book' begins with a Preface ('Prefatio', chapter 88) and covers around 23 chapters. Amongst its most important treatises are 'De tempore' ('On Time'); 'Mortis introitus in naturam humanam decus virginum' ('The Entrance of Death in Human Nature. In Praise of Virgins'); 'Thesis' ('Thesis'); 'Demonstratur thesis' ('The Thesis is Demonstrated'); 'Imago Dei' ('The Image of God'); 'Arcana Paracelsi' ('The Arcana of Paracelsus'); and 'Arbor Vitae' ('The Tree of Life'). Given its position in the *Ortus* and topic, it is probable that 'Aura vitalis' belonged to the *Liber de vita longa* as well. All such internal evidence points to the work being a separate entity from *Ortus medicinae*.[6]

Of course, we should factor in this analysis Van Helmont's obvious intention of setting his *Liber* as a response to Paracelsus's long life treatises, one of which is actually called *Liber de vita longa*, and to Marsilio Ficino's *De vita libri tres* (1489). In this sense, Van Helmont's decision of calling his work *Liber* is clearly meant to be read as a reference to Paracelsus and Ficino. Yet Van Helmont's intention does not refute the theory; in fact, his clear ambition of 'competing' with these authorities makes the separation of the *Liber* from the *Ortus* even more necessary.

Incidentally, the influential German edition of the *Ortus*, *Aufgang der Artzney Kunst* (1683), supports these conclusions on the *Liber de vita longa*.[7] The editor, Christian Knorr von Rosenroth (1631–1689), spent a great deal of time revising the

---

4 'De lithiasi', ch. 9, §30, 715: 'Quae enim de vita jam dixi, toto libro De vita longa (ad quem propediem emersurum te remitto) dixi'; §35, 715: 'Prout libro De vita longa'; §54, 721: '. . . qui falsus est, ut ostendi libro de vita longa'; 'Tumulus pestis', ch. 4, 'Peregrina lues nova', 877: 'Quapropter libro de vita longa . . .'

5 E.g. ch. 6 'Archeus faber', §9, 34: 'libro de vita longa, ad longum elucidantur'; ch. 39 'Demens idea', §26, 224 'lib. de vita longa de mortis introitu'; ch. 48, 'Distinctio mentis à sensitiva anima', §1, 279: 'Inprimis Libro de vita longa, ad longum demonstravi, mortis introitum in naturam humanam . . .'; ch. 50, 'Nexus sensitivae et mentis', §6, 285: 'prout de vita longa, tractatu de introitu mortis in naturam humanam probavi' and others. It is true that Ch. 24 'Natura contrariorum nescia' and ch. 27 'Spiritus vitae' he refers to 'De vita longa' as a 'treatise' (§37, 139, §16, 159), and that within *De vita longa* Van Helmont refers to it as 'tractatus de vita longa' (e.g. 'Imago Dei', 563, 'Humidum radicale', 572, 'Lunare tributum', 582), but this does not mean he is not thinking of the *tractatus* as a separate entity from the *Ortus*.

6 The issue of whether Van Helmont intended for the *Ortus* and *Liber* to be published together is irrelevant to this argument.

7 Johann Baptista von Helmont, *Aufgang der Artzney Kunst, das ist: Noch nie erhörte Grund-Lehren von der Natur* (Sulzbach: Johann Andreae Endters, 1683), 'Verzeichnuss der Tractaten'.

contents of the *Ortus*, seeking to offer not only an accurate translation, but also a better-ordered version. He separated the last part of the work under the heading 'Vom langen Leben. Tractatus de Vita longa', which included 22 chapters plus a preface. Yet Rosenroth did not go as far as to clearly separate *Liber de vita longa* from the *Ortus*. Dividing the *Ortus* would have made little sense at a time when Van Helmont's fame was based on the publication of the unitary volume.

Finally, another piece of evidence, Van Helmont's Dutch-language *Dageraad*, strengthens the suspicion that *Liber de vita longa* was a separate entity from the *Ortus medicinae*. *Dageraad* contains treatises corresponding to the *Ortus*'s chapters on physics and diseases; it can even be called a digest of the Latin volume. However, *Dageraad* does not contain the treatises related to long life.

The case of the *Liber de vita longa* suggests that we need to approach the *Ortus* carefully. The safest approach to the *Ortus* is to view it as a publication rather than a unitary work. This strategy allows us to investigate the *Ortus* in its entirety, including the previously published *Opuscula medica inaudita*. In this sense it is worth mentioning that, when he first published *De febribus* (1642), Van Helmont presented it as an 'extract' from the *Ortus* rather than a stand-alone treatise. This was, obviously, a clever bid at pitching the future *Ortus* to the audience, but it also seems that at this time Van Helmont genuinely considered it as a part of the larger volume. For one, he apologised that he was presenting the treatise without the theoretical apparatus of the *Ortus*, thus making it appear slightly void of judgment.[8] Secondly, the first chapter of *De febribus* opens with a justification on why fevers have been allocated their own treatise within the much larger domain of 'diseases'. This is probably a reference to the section 'Tractatus de morbis' in the *Ortus*.[9] In any case, there are strong links between the *Ortus* and *Opuscula*, which suggest that they were conceived at roughly the same time and as part of the same project of the reformation of knowledge.

The strategy of treating the *Ortus* as a publication further allows us to examine it without shaky presuppositions. For instance, the assumption of a chronological coherence to the *Ortus* led Pagel to claim that the 'theological' treatises at the end of the volume were written after the natural philosophical and medical treatises. Pagel argued that the former were signs of Van Helmont's increasing pessimism caused by old age.[10] Yet most of the theological treatises he is referring to belong to the projected *Liber de vita longa*, and we have no proof that this was written later than the *Ortus*. In fact, the cross-references suggest they were probably written in parallel. The fact that these treatises are found at the end of the *Ortus* may have two causes: one is the editor's or printer's intention of setting them there, and the second, their position in the manuscript delivered by Van Helmont to Franciscus Mercurius.

---

8 'De febribus', 'Lectori scientiam', 738: 'Ex hoc opere, laceravi hunc tractatum de febribus.... Edidi ergo hunc tractatum, absque cognatis morborum doctrinis ... etsi hoc studis genus quandam pigritiam, & penuriam judicii, attestetur'.

9 'De febribus', ch. 1, §1, 739: 'Febri itaque non minus quam lithiasi, peculiarem tractatum debui. Quippe quae licet familiarissima, potissimum tamen nostra saginat coemiteria, & castra depopulatur'.

10 Pagel, *Joan Baptista Van Helmont*, 115.

Indeed, it is not difficult to surmise that Van Helmont's manuscript was in a more or less chaotic state. Why else would the middle treatises *De magnetica vulnerum curatione* and *De Spadanis fontibus* appear interspersed in the body of the projected *Liber de vita longa*? It is possible that Franciscus Mercurius failed to realise the logical coherence between the different chapters of *Liber de vita longa*, or even identify it as a stand-alone work. This suggests that he only paid cursory attention to this part of the manuscript.

Should we then dismiss Franciscus Mercurius as an essentially careless editor who printed the manuscript exactly as he found it? This is quite possible, given the fact that he claimed to have done little to the papers. In fact, he states in his introduction that his editorial work was done at the request of other scholars.[11] By portraying himself as a reluctant and harried editor, Franciscus Mercurius could exonerate himself from having to properly edit the manuscript, while reaping the benefits of being perceived as his father's intellectual heir.

In Franciscus Mercurius's defence, it is clear that the work of properly editing the manuscript would have been a huge task for anyone to undertake. Franciscus was more interested in advancing his own philosophy than spending a long time making sense of his father's writings. Consequently, it was left to future scholars, such as Rosenroth, to undertake this task.

Thankfully, the chaos of the *Ortus* is not absolute. The careful reader can uncover an inner logic to many of the chapter-treatises, particularly those at the beginning. Furthermore, there are thematic divisions that are often rendered explicit. For instance, the 'Tractatus de anima' is an explicit attempt at a grouping of chapters, which probably contained the subsequent 'Distinctio Mentis à sensitiva Anima' ('The Distinction of the Mind from the Sensitive Soul'); 'De Anima nostra immortalitate' ('On the Immortality of Our Soul'); and 'Nexus sensitivae & mentis' ('The Intertwining of the Sensitive Soul and the Mind'). Possibly the chapters 'Sedes Animae' ('The Seat of the Soul') and 'A sede Animae ad morbos' ('From the Seat of the Soul to Diseases'), which occur prior to the 'Tractatus', may be part of it. Similarly, 'Progymnasma meteori' ('Essay on Meteors') should be seen as a headline for several chapters, including 'Gas aquae' ('The Gas of Water'); 'Blas meteoron' ('The Blas of Meteors'); 'Vacuum naturae' ('Nature's Vacuum'); 'Meteoron anomalum' ('An Anomalous Meteor'); and 'Terra tremor' ('Earthquake').

Yet most thematic separations can only remain approximate.[12] I have attempted a general division in the *Appendix*, in order to clarify some of the author's intent in the *Ortus*. For instance, chapters 3–26 treat of natural philosophy, and form the core of the 'Physics' part of the *Ortus*. Chapters 27–31 may also belong in this group, but

---

11 Franciscus Mercurius Van Helmont, 'Amico lectori', [1]: 'Iam vero coactus rationibus ac literis aequorum multorum Sapientum, diversis hinc inde ex Regnis Statibusque, qui Salutis pignore adhortabantur, me teneri cuncta piè defuncti parentis mei scripta prelo subjicere . . . Id quod mihi extorsit, ut caetera cuncta relinquerem, praedictaque Scripta perlustrarem'.

12 For an alternative ordering, see also von Helmont, *Aufgang der Artzney Kunst*, 'Verzeichnuss der Tractaten'.

less evidently so. About 10 chapters, often disparate, refer to the soul. Many others treat of medicine and different types of diseases, including eight that deal with occult action.

Such separation is tentative; indeed, clarifying the structure of the *Ortus* would require a work of a rather different kind from the present one. The purpose of my analysis is to expound Van Helmont's ideas, rather than to examine the *Ortus* as a published work. However, one cannot and should not extricate ideas from their presentation. Van Helmont's ideas are not context-free units; rather, they interlink dynamically across the chapters, and eventually coalesce into something that is greater than their sum. It is this 'something' that I call, in accordance with Van Helmont's own term, his 'Christian philosophy'.

We should not conceive this 'Christian philosophy' as a finished product, but rather as an ideal Van Helmont worked to achieve. It is clear that a rather unexpected death robbed Van Helmont of the opportunity to set his papers in order, to edit his writing and to transcribe all of his ideas. Occasionally we can suspect that he may have still been ruminating on some fine philosophical points. His 'Christian philosophy' remained in a project stage because of his death, and as such a re-construction must be attempted in order to understand it. The labour may sometimes be compared to repairing a broken puzzle: the pieces that make it up can be anywhere on the floor, but with tenacity the puzzle can be reconstituted sufficiently to understand the image it represents.

Part II
The Principles of Van Helmont's 'Christian Philosophy'

There is little of the clarity, structure and elegance of Scholastic works in the *Ortus*. Van Helmont is prone to digressing on subjects only remotely connected with the matter at hand. Every now and then, the temptation to polemicise obscures the argument. In addition, he does not always define the terms he is working with, and some important notions like *Gas* are not given a chapter of their own. As far as evidence is concerned, Van Helmont freely juxtaposes personal testimony and experimental proof with examples taken from popular books of the period. His preference for empirical evidence occasionally leads him to take dubious accounts as trustworthy. Although implicit reasoning is often visible in his thought, he relies little on logical arguments.

Blaming the draft-like state of the manuscript for these shortcomings cannot fully account for this style of exposition. Clearly, Van Helmont was genuinely and consciously experimenting with alternatives to Scholastic writing. He believed that a new Christian philosophy had to rely on different evidence, writing style and approaches from Scholasticism. Such principles included a focus on empirical knowledge and fact, polemical refutations of heathen philosophy, ironic criticism of the nefarious habits of his day, and the use of personal testimony. Impassioned confessions, autobiographical insertions and lengthy prayers further suggest that Van Helmont's religious and literary model was Augustine, rather than Aquinas and the Scholastics.

The subsequent analysis is divided into three main chapters: on God, Nature and Man.[1] It is reconstructive in nature, and as such the coherence of ideas takes precedence over their positioning in the *Ortus*. The assessment also takes into account the development of Van Helmont's thought over the three proposed stages (early, middle and late thought); on some occasions, I refer to Helmontian works that are not part of the published *Ortus*.

1 By 'Man' I refer to the general category of 'humankind' and not to a specific person or gender. I have also used capitalised 'Nature' whenever I referred to the specific entity of Van Helmont's speculation.

Chapter 4
God

Van Helmont's Christian philosophy can only be understood in conjunction with his fervent belief in God. Van Helmont does not view God as an abstract principle or a philosophical Demiurge. Instead, he perceives God as being the divinity chiefly revealed in the Bible and confirmed by Christian doctrine as the Holy Trinity.

Despite the centrality of God in Van Helmont's thought, there is no treatise dealing specifically with this topic. There may be several reasons behind this reluctance. One might be that Van Helmont believed that his views on God were orthodox and in line with dogma, so he felt no need to elaborate on them in detail. Another reason might have been that, on the contrary, he felt some of his ideas were at odds with or liable to be misinterpreted by Catholic authorities. Given his persecution by the Inquisition and the Mechelen ecclesiastical court, this reason remains plausible. Even so, as we shall further see, Van Helmont did not shy away from affirming ideas that were controversial.

Van Helmont's ideas on God often transcend the boundary of the religious factionalism of his age. This is achieved by consciously appealing to the accepted doctrines of the Church Fathers, chiefly Paul, Augustine and Pseudo-Dionysius. Indeed, he openly calls for a return to 'apostolical sincerity', which he contrasts with the Scholastic disputations of his age.[2] His advocation of the faith of the early Christian church resonated with the contemporary Louvain theology, particularly Baianism, which dissented from the prevalent Thomism of the period.[3]

At the same time, Van Helmont's ideas of God can also be related to those of other Paracelsians in his era. Chief amongst these is 'panentheism', or the view that God is simultaneously immanent and transcendent. Van Helmont also supported a Christological view of Creation and man, with an emphasis on the regeneration of the soul. He was particularly keen on advancing the complementarity between Christian doctrine and alchemical ideas, a complementarity that stands at the root of his vision of a reformed Christian philosophy.

---

[2] *Opuscula medica inaudita*, 'Philiatro lectori', 644: 'In Theologicis vero, quid non augustum, per quaestionum acribologias, sperant cathedra? Utinam his cunctis careat humana necessitas. Meum atque tuum, cuique absque fuco cedat! Fides etiam (ut in Mahumetismo) staret citra disceptationem. Cedat quaeque subtilitas, cujus in die novissimo non habebitur ratio: Rediret sic enim Apostolica sinceritas: *Sic accepi, que tradidi vobis*'.

[3] M.W.F. Stone, 'Michael Baius (1513–1589) and the Debate on "Pure Nature"': Grace and Moral Agency in Sixteenth-Century Scholasticism', in *Moral Philosophy on the Threshold of Modernity*, ed. Jill Kraye and Risto Saarinen (Dordrecht: Springer, 2005), 51–90.

Apophatic and Cataphatic Theology

For his understanding of God, Van Helmont draws on the tradition of both apophatic (negative) and cataphatic (positive) theology. Apophatic theology emphasises the fundamental unknowability of God, while cataphatic theology focuses on God's self-disclosing revelation.[4] The important distinction was made in Christian thought by Pseudo-Dionysius the Areopagite, who, as a supposed apostolic figure, had a major impact on Western medieval thought.[5]

The apophatic approach to theology was developed by Origen and the Cappadocian Fathers.[6] Popularised in the West by John Scotus Erigena (c.815–c.877), apophatic theology was vigorously upheld by such thinkers as Thomas Aquinas, Johann Tauler (c.1300–1361), Nicholas of Cusa (1401–1464) and the anonymous author of *Theologia Germanica*.[7] In the Renaissance and early modern period, apophatic thought was embraced by Pico della Mirandola (1463–1494), Johannes Reuchlin (1455–1522), John of the Cross (1542–1591) and others.[8] The view was also adopted by many mystically oriented Paracelsian thinkers, including Jacob Boehme (1575–1624) and Oswald Croll.[9]

As part of his project of rooting his philosophy in orthodox Christian doctrine, Van Helmont argues that the knowledge which we have of God should be negative, as we cannot conceive of Him either by sense or mind.[10] He also repeatedly refers to God's essence and Christ's Incarnation as 'incomprehensible' and 'ineffable'.[11] Van

---

4 *The Westminster Dictionary of Christian Theology*, ed. Alan Richardson and John Bowden (Philadelphia, PA: Westminster Press, 1983), 167.

5 Pseudo-Dionysius, *Pseudo-Dionysius—Complete Works*, trans. Colm Luibheid (New York, NY: Paulist Press, 1987), 138–40. The Neoplatonists had already drawn the outlines of the difference between negative and positive theology based on Plato's *Parmenides*, but Pseudo-Dionysius developed the dual concept in a Christian sense by emphasising its paradoxical expression; see Andrew Louth, *Denys the Areopagite* (London: Continuum Books, 1989), 87–8. For Pseudo-Dionysius's influence on Western thought, see Paul Rorem, *Pseudo-Dionysius: A Commentary on the Texts* (Oxford: Oxford University Press, 1993), 214. Van Helmont quotes from Pseudo-Dionysius directly in ch. 43, 'Mentis complementum', §12, 252.

6 See Vladimir Lossky, *The Mystical Theology of the Eastern Church* (Bucharest: Bonifaciu, 1998), particularly chapter 2.

7 Rorem, *Pseudo-Dionysius*, 225.

8 Giovanni Pico della Mirandola, *On the Dignity of Man, On Being and the One, Heptaplus*, trans. Charles Glenn Wallis, Paul J.W. Miller and Douglas Carmichael (Cambridge: Hackett, 1965), 49–50. For Johannes Reuchlin see Wilhelm Schmidt-Biggemann, *Philosophia Perennis: Historical Outlines of Western Spirituality in Ancient, Medieval and Early Modern Thought* (Dordrecht: Springer, 2004), 96.

9 For Boehme, see Schmidt-Biggemann, *Philosophia Perennis*, 117–19; Croll, 'Admonitory Preface', 70, 215.

10 Ch. 4, 'Venatio scientiarum', §57, 25.

11 Ch. 25, 'Blas humanum', §23, 147; ch. 38, 'Imago mentis', §12, 216; ch. 43, 'Mentis complementum', §20, 254; ch. 101, 'Intellectus Adamicus', 562; ch. 111, 'Vita aeterna', 590. Van Helmont's view of the incomprehensibility of Christ's Incarnation seems to also draw on Pseudo-Dionysius, who said that 'the sacred incarnation of Jesus for our sakes is something which cannot be enclosed in words nor grasped by the mind, not even by the leaders among the front ranks of angels'; 'The Divine Names' in *Pseudo-Dionysius—Complete Works*, 65.

Helmont observes that the inability to understand divinity stems from Man's position as created being.[12]

Yet Van Helmont agrees with Pseudo-Dionysius that apophatic theology must be accompanied by a cataphatic one, otherwise knowledge of God becomes impossible. In accordance with Dionysian teaching, God manifests Himself by means of the Divine Names. Most of these Names are drawn from the Bible, but some originate in Greek philosophical thought.

Van Helmont emphasises God's oneness, following the Pseudo-Dionysian tradition according to which the One is the chief Name of God.[13] As Van Helmont points out in the treatise 'De tempore' ('On Time'), belonging to the *Liber de vita longa*, 'unity is an image of Divinity, because from it all numbers are made, and into the same are resolved'.[14] Van Helmont further adduces the philosophical testimony of Plato's *Parmenides* to argue that 'there are neither accidents in God, nor duality, distinct from His Essence'.[15]

According to orthodox post-Nicene doctrine, the Unity of God manifests itself outside of his inaccessible Oneness by means of the Trinity. Van Helmont concurs with this, pointing out that God is the 'holy sacred Trinity' (*sacrosancta Trinitas*) or 'thrice-glorious Trinity' (*tergloriosa Trinitas*).[16] This was a point of agreement between Christian dogma and Paracelsian philosophy. Paracelsus had extended the Trinitarian model beyond metaphysics to the natural world, arguing that 'all creatures are created in the number of the Trinity' and famously giving the example of the triad of salt, sulphur and mercury.[17] This view was not heretical, as it relied on the authority of Augustine's famous treatise *De Trinitate* (*On the Trinity*), where the Trinity was seen to be mirrored in Nature. The Paracelsian tenet of the presence of the Trinity in Nature was particularly popular amongst alchemical philosophers seeking a Christian understanding of the natural world.[18]

---

12 Ch. 89, 'De tempore', §33, 505.

13 Pseudo-Dionysius, *Complete Works*, 66–7.

14 Ch. 89, 'De tempore', §22, 504: 'Unde ipsa unitas Divinitatis typus est: eo quod exin omnes fiant numeri, atque iterum in eandem resolvantur'.

15 Ch. 38, 'Imago mentis', §28, 217: 'in Deo non sint accidentia, nec sint dualitas, ab ejus esse distincta'. Plato's *Parmenides* was a watershed for speculation on the ideas of God as One and as many (triad / henad), influencing both Neoplatonic thinkers (Porphyry, Proclus) as well as Christian ones (the Cappadocian Fathers and Pseudo-Dionysius); see Louth, *Denys the Areopagite*, 81–8, and Sarah K. Wear and John M. Dillon, *Dionysius the Areopagite and the Neoplatonist Tradition* (Aldershot: Asghate, 2007), 31–6.

16 Ch. 38, 'Imago mentis', §12, 214, §20, 216.

17 Weeks, *Paracelsus: Speculative Theory*, 118. Trinitarian motifs had also been present in the alchemical works attributed to Arnau of Villanova, as well as in the anonymous *Book of the Holy Trinity;* on this see Thorndike, *History of Magic and Experimental Science*, III, 75–6. Specifically on imagery, see also Barbara Obrist, *Les débuts de l'imagerie alchimique* (Paris: Le Sycomore, 1982), 67–8, 117–18.

18 Bruce Janacek, *Alchemical Belief: Occultism in the Religious Culture of Early Modern England* (University Park: University of Pennsylvania Press, 2011), 124–5.

Another important Name that Van Helmont uses in regard to God is 'Goodness' (*bonitas*).[19] This divine attribute is crucial to Van Helmont's speculation, as it defines God's relationship to His creation.[20] Van Helmont finds 'divine goodness to be actually, perpetually, everywhere, and immediately ruler' of all things.[21] God's bountiful goodness is expressed outwardly in Nature through creative Love, which spreads throughout Nature and makes it unified and harmonious.[22] Van Helmont emphasises that the concord of Nature ultimately reveals its divine roots.

The attribute of 'Goodness' is particularly poignant in respect to human beings, and manifests itself in the form of the 'free gift of God' (*donum Dei*). Unsurprisingly, the concept of divine goodness is intertwined with that of medical alchemy: one of the greatest gifts is the provision of efficacious medicine in nature, which must be extracted by alchemical means.[23]

Voluntarist Theology

Outstanding research by scholars Amos Funkenstein and Margaret Osler has brought to the fore the importance of voluntarist theology in the creation of the modern scientific mentality.[24] Voluntarist theology focuses on the absolute freedom of the will of God. While Funkenstein has concentrated on the contributions of the late medieval Nominalists, Osler has addressed this issue with respect to the French philosopher Pierre Gassendi.

---

19 See for instance ch. 92, 'Thesis', §1, 515; ch. 93, 'Demonstratur thesis', §78, 538; ch. 89, 'De tempore', §34, 505. The idea of God as all Good had a long tradition in Christian thought influenced by Platonic and Neoplatonic speculation. For Augustine's formulation and its influence see John Bussanich, 'Goodness', in *Augustine throughout the Ages*, 390–91; for Aquinas's speculations on divine goodness, see Eleonore Stump and Norman Kretzmann, 'Being and Goodness', in *Being and Goodness: The Concept of the Good in Metaphysics and Philosophical Theology*, ed. Scott MacDonald (Ithaca, NY: Cornell University Press, 1991), 98–128.

20 Ch. 43, 'Mentis complementum', §18, 253; ch. 79 'Butler', 471; ch. 109, 'Vita', 586.

21 Ch. 17, 'Meteoron anomalum', §15, 73: 'video divinam Bonitatem actualiter, semper, ubique & immediate praesidem'.

22 Ch. 24, 'Natura contrariorum nescia', §1, 131, §33–4, 137, §37, 138.

23 Van Helmont constantly talks about divine goodness in ch. 79, 'Butler', 468–75 in relation to the universal medicine, *Drif*.

24 Margaret Osler, *Divine Will and Mechanical Philosophy: Gassendi and Descartes on Contingency and Necessity in the Created World* (Cambridge: Cambridge University Press, 1994); Amos Funkenstein, *Theology and the Scientific Imagination from the Middle Ages to the Seventeenth Century* (Princeton: Princeton University Press, 1986), 117–92. His focus was on Nominalist thought, which emphasised the omnipotence of God. The importance of voluntarism to science has been disputed recently by Peter Harrison, 'Voluntarism and Early Modern Science', *History of Science* 40 (2002): 63–89; see John Henry's response, 'Voluntarist Theology at the Origins of Modern Science: A Response to Peter Harrison', *History of Science* 47 (2009): 79–113.

No scholar until today has analysed in depth the voluntarism of Gassendi's contemporary and friend, Van Helmont.[25] Yet the Flemish physician's ideas of God were permeated with voluntarism, which was a cornerstone of his approach to the natural world and Man. Indeed, his emphasis on God's freedom and divine will was part of his project to 'Christianise' philosophy. As is well known, Greek philosophy was deterministic, with gods and Nature being equally bound by necessity. The early Fathers of the Christian Church, particularly Augustine, sought to counteract Greek philosophy by emphasising divine will and freedom.[26] In the Middle Ages, Thomas Aquinas followed in the footsteps of Augustine in rejecting Greek determinism, but his work evinces a milder form of voluntarism than the late medieval nominalist school led by William of Ockham (1288–1347).[27]

Van Helmont's voluntarist view may not come across as clearly and eloquently as Gassendi's, nevertheless, time and again the Flemish physician emphasises God's omnipotence.[28] He is a 'Creator of infinite power' who 'encloses the Universe in His fist, can shake the Earth at His will, and alone can perform marvels (*mirabilia*)'.[29] God is often simply described as the 'Almighty' (*Omnipotens*) or the 'Power' (*Potestas*).[30] Other times, God is expressed through such attributes as 'vast', 'immense' or 'infinite'.[31] His will is immersed into the very fabric of Nature: 'Divine will forcefully reaches all things from end to end'.[32]

God's power is reflected in His absolute freedom. Indeed, Van Helmont emphasises, God does not need Creation and could have chosen to remain hidden forever. Hence Creation was God's own pleasure, and so was His decision to make Man in His image.[33] Essentially, He created all things for His own glory. Subsequently, He distributed His gifts as He saw fit, being unconstrained by necessity.[34]

25 The only scholar to touch on the subject was Guido Giglioni, who referred to the 'voluntarist formulation' of the treatise on earthquakes, which he contrasts with the earlier *De magnetica vulnerum curatione; Immaginazione e malattia*, 47.

26 On this, see Lossky, *The Mystical Theology*, 82–3; Louth, *Denys*, 84–6.

27 Osler, *Divine Will*, 20–35.

28 Osler, *Divine Will*, 52–5 on Gassendi. However, we must not forget that Gassendi was a trained theologian, so must have felt more at ease discussing theological matters than Van Helmont.

29 Ch. 89, 'De tempore', §26, 504; ch. 18, 'Terra tremor', §30, 82: 'Stat nempe arbitrio ejus, qui universum pugno claudit, terram pro nutu suo quatere, & mirabilia solus facere potest'.

30 For instance, ch. 4, 'Venatio scientiarum', §1, 17; ch. 9, 'Elementa', §3, 42; ch. 22, 'Ortus formarum', §74, 118.

31 Ch. 92, 'Thesis', §1, 515; ch. 38, 'Imago mentis', §10, 214; ch. 89, 'De tempore', §26, 504; ch. 21, 'Astra necessitant, non inclinant, nec significant de vita, corpore, vel fortunatis nati', §12, 98. Van Helmont may have been influenced in his definition of God as 'Infinite' from Nicholas of Cusa, whom he most certainly read. For the notion of 'infinite' in the theology in and prior to Nicholas of Cusa, see Etienne Gilson, *The Christian Philosophy in the Middle Ages* (London: Sheed and Ward, 1955), 535–9.

32 Ch. 25, 'Blas humanum', §3, 144: 'Sed nutus divinus, à fine usque in finem, attinget omnia fortiter'.

33 Ch. 79, 'Butler', 470: 'Deus . . . qui sua creata dotavit pro suo lubito'; ch. 109, 'Vita', 585.

34 Ch. 110, 'Vita brevis', 588; ch. 79, 'Butler', 470.

Van Helmont goes even further with the affirmation of divine omnipotence. He claims that God does not need to be stationary but could move as He wishes, 'being absolutely free, in his will and motion, as in rest'.[35] In a strictly philosophical sense, the idea of a moving God makes little sense, as it would raise the question of what He could possibly move against. Yet Van Helmont's point is precisely to contradict logic and expose our limited understanding of the divine.

Having advanced the unsettling possibility of a perpetually moving divinity, Van Helmont wisely chooses to steer away from controversy. In fact, his argument is purely hypothetical, since he eventually comes to the traditional theological conclusion that God is indeed immobile, albeit only out of His own volition.[36] Van Helmont's intention here is to make a voluntarist argument in the late Scholastic tradition of the difference between *potentia Dei absoluta* (absolute power) and *potentia Dei ordinata* (ordained power).[37] This voluntarist distinction was a key argument whereby late medieval Scholastics showed that God is independent of any necessity. In Van Helmont's interpretation of the doctrine, God's absolute power meant that He could have moved if He wished to (*potentia Dei absoluta*); however, He chose to remain unmoved in relationship with his Creation (*potentia Dei ordinata*). This was an ordained power which He could easily contradict if desired, so was essentially subsumed in His absolute power.

The orthodox doctrine of divine immobility leads Van Helmont to contend that the Word *Fiat* was sent into the world by God to do His bidding.[38] Hence the dynamic Word is contrasted with the fundamentally static nature of the divine essence. As Van Helmont likes to point out, the 'small' Word of God (*verbulum*) is so powerful that it generated the universe out of nothing.

Van Helmont's voluntarism has strong repercussions on the way he sees God's relationship with Nature and with Man. In Nature's case, God's will stands supreme: creatures have no free will and must implement the divine command (*nutus*).[39] The main mechanism by which God exercises His will in Nature is through Ferments and seeds. These spiritual principles, ultimately drawn from Augustine's concept of *rationes seminales*, represent latent forces of generation that manifest God's will in history.[40]

Thus, Van Helmont's voluntarist theology is extended in Nature to become a voluntarist natural philosophy. This doctrine is remarkably similar to Aristotelian

---

35 Ch. 25, 'Blas humanum', §3, 144: 'Sed prorsus liber, tam suo nutu & motu, quam quiete, indifferenter, & aeque potenter movet cuncta'.

36 Ch. 25, 'Blas humanum', §3, 144.

37 This topic, including its related relevancy to history of science, can be reviewed in Edward Grant, *Science and Religion: 400 BC—AD 1500, From Aristotle to Copernicus* (Baltimore, MA: Johns Hopkins University Press, 2006), 54–90; Funkenstein, *Theology*, 115–21.

38 Ch. 25, 'Blas humanum', §3, 144.

39 Van Helmont uses the term *nutus* very often to refer to God's will and omnipotence.

40 See below, Chapter 5, 'Nature'.

teleology, except that in Van Helmont's case the ideas are clearly drawn from Christian voluntarism.[41] Everything in Nature is in fact moved by the will of God: hence even insensitive beings are pushed onwards toward their own final goals.[42]

An important outcome of Van Helmont's voluntarist understanding of God is that he accepts the fact that Creation behaves in peculiar ways as part of an inscrutable and undecipherable divine plan. Hence he is always concerned with what he calls the quiddity (*quidditas*), the specificity and essence of each body and phenomenon. This concept is essential in framing Van Helmont's programme of empirical research.[43]

In contrast to Nature, God endowed human beings with free will (*voluntas humana* or *voluntas viri*), in order to better express their own divine nature. This free will eventually caused the Fall. Precisely because of its nefarious role in bringing about Original Sin, the Flemish philosopher held a distinctively negative view of human will: 'there is no power more destructive to man than free will', he stated.[44] Indeed, Van Helmont believed that the mystical path required the abandonment of the will, and that in Paradise human volition will finally and felicitously be obliterated.

Van Helmont's voluntarist ideas primarily stem from Augustine, who had specifically emphasised the role of will in human and divine life.[45] He was also probably influenced by the voluntarist views of Nominalist Scholastics, medieval mystics and some Renaissance humanists.[46] An emphasis on God's will was also part of a strong tradition in alchemical philosophy since the Middle Ages.[47] Here it was intertwined with the motif of the special election of alchemists. Such emphasis on divine will is also present in the speculation of several Paracelsians, notably in Oswald Croll.[48]

---

41 On Aristotelian teleology, see John Cooper, 'Aristotle on Natural Teleology', in *Language and Logos*, ed. Malcolm Schofield and Martha Nussbaum (Cambridge: Cambridge University Press, 1981), 197–222; Susan Sauve Meyer, 'Aristotle, Teleology and Reduction', *The Philosophical Review* 101 (1992): 791–825. Teleology was, of course, embraced by Christian thought. See Thomas Kennedy, 'Teleology', in *The Encyclopedia of Christianity*, 5 vols (Grand Rapids, MI: Eerdmans & Brill, 2008), V, 326–8.

42 Ch. 25, 'Blas humanum', §9, 145.

43 See below, Chapter 6, sections on 'The Mystical Knowledge of Nature' and 'Mysticism and Empirical Investigation'.

44 Ch. 38, 'Imago mentis', §25, 217: 'nulla est homini potestas pernicior, voluntate libera'.

45 On Augustine's concept of will, see John M. Rist, *Augustine: Ancient Thought Baptized* (Cambridge: Cambridge University Press, 1995), 148–202.

46 Charles Trinkaus has shown the similarity between humanist thought and Nominalism, even though the Renaissance philosophers were generally realists; see *In Our Image and Likeness: Humanity and Divinity in Italian Humanist Thought*, 2 vols (London: Constable, 1970), I, 60.

47 For instance, 'Raymundi Lulli Theorica & Practica', in *Theatrum Chemicum*, 6 vols (Strasbourg: Lazarus Zetzner, 1613), IV, 6: 'Hanc naturam supremus deus primo de nihilo creavit sua pura liberalitate & voluntate'.

48 Croll, 'Admonitory Preface', 186.

God in Nature

Panentheism

The immanence of God in Creation is a chief doctrine of Christian theology.[49] The version that prevailed in the Middle Ages was the moderate Scholastic view, which expressed it by means of the doctrines of God's ubiquity and His role as ultimate efficient cause. The first view maintained that God was essentially present in all places and at all times. This view had a long Scholastic tradition and led to the Lutheran dogma of the ubiquity of Christ's body.[50] The latter doctrine, postulated by Thomas Aquinas and held by most Scholastics, affirmed that God was the cause of being (*causa essendi*); as Funkenstein explains this view, 'He is in all by essence insofar as He is present to all as *causa essendi*'.[51]

Van Helmont accepted both Scholastic doctrines on the subject. He emphasised God's ubiquity, supporting the Nominalist association between God, space and infinity, a theory Funkenstein referred to as 'the body of God'.[52] The *causa essendi* tradition allowed Van Helmont to state the following paradox: God is the essence of all things but the individual essence of each being is not God.[53] He further explains that God is 'the true, immediate and absolute essence, from which the being (*esse*) of things issues and depends, being inseparable in nature'.[54]

Yet Van Helmont found Scholastic views insufficient to express the immanence of God in Nature. He postulated a stronger form of divine immanence, 'panentheism', a Christian tradition distinctly influenced by Neoplatonic-Stoic thought.[55] In principle, this tradition was orthodox, as it figured prominently in the work of the Greek Fathers, particularly pseudo-Dionysius and Maximus the Confessor. However, some theologians were distrustful of it as possibly leading to heretical pantheism, or the belief that the world was divine and uncreated. Indeed, the chief panentheists of the Western Middle Ages, John Scotus Erigena, Meister Eckhart (c.1260–c.1328) and Nicholas of Cusa, were all suspected of this heresy at one time or another.

---

49 Amos Funkenstein shows that Christian theologians were always attempting to balance this concept against pantheistic connotations as well as theories of absolute transcendence; *Theology*, 49–55.

50 For the Scholastic formulation of the doctrine of ubiquity, see Funkenstein, *Theology*, 50–56. For Luther's formulation, see ibid., 71–2.

51 Funkenstein, *Theology*, 50, Thomas Aquinas, *Summa Theologiae* (Cambridge: Cambridge University Press, 2006), vol. 2 (1a.2-11), 1a. 8, 4.

52 Ch. 22, 'Ortus formarum', §2, 105: 'Deum esse & hodie rerum immediatum principium, ubique praesens, operans omnem omnium perfectionem'. For a good discussion of Scholastic distinctions, see also Funkenstein, *Theology*, 54–5, 61–3.

53 Ch. 22, 'Ortus formarum', §16, 107: 'Suppono itaque Deum esse veram, perfectam, & actualiter omnem omnium essentiam'; 'Esse autem quod res habent, est entis, sive ipsius creaturae: Non autem Deus'.

54 Ch. 56, 'Tria prima chymicorum', §76, 333, 'vera, immediata, ipsissimaque essentia, à qua esse rerum defluit, & dependet, inseparabile in natura'.

55 On this topic, see John W. Cooper, *Panentheism: The Other God of the Philosophers: from Plato to Present* (Grand Rapids, MI: Baker Academic, 2006), 39–63.

In the early modern period, the followers of Paracelsus were also accused of being pantheists, and particularly of denying the doctrine of *Creatio ex nihilo*. In the mid-sixteenth century, the Lutheran theologian Thomas Erastus (1524–1583) had indicted Paracelsus for postulating the eternity of the world. Erastus's attack was based mainly on his reading of the spurious *Philosophia ad Athenienses*.[56] He did not take into account the explicit statement made by Paracelsus in the authentic *Labyrinthus medicorum errantium*, where he stated that God made all things out of nothing.[57]

Following Erastus's indictment, Paracelsian followers were eager to dispel the idea that the new alchemical philosophy was at odds with mainstream Christian faith. Richard Bostocke (fl. 1585) claimed that Paracelsus had been perfectly orthodox in asserting an uncreated *Mysterium magnum*, as this was nothing other than Christ the Word and Wisdom of God.[58] Similarly, Van Helmont's contemporary Jacob Boehme affirmed that, while Creation could only occur *ex Deo* as the Abyss (*Ungrund*), the world was not divine, being created by the Word *Fiat*.[59]

Yet all such arguments were liable of being deemed equally unorthodox as *Philosophia ad Athenienses*. Rather than risking controversy, Van Helmont preferred to affirm, unequivocally, that God created all things *ex nihilo*.[60] Hence, God was, in his opinion, 'the Parent of all things, the Author and Governor of Nature'.[61] He was also deeply involved in His Creation: Van Helmont viewed God as directly bestowing life.

The Presence of God as Life-Giver

For Van Helmont, God was the direct bestower of all souls in the Universe. To express this view, the Flemish philosopher appealed to a doctrine that, ironically, was not Christian in conception: that of *dator formarum*. This theory first appeared in the speculations of Avicenna (980–1037) on the form-giving planetary intelligence he called 'Cholcodea'.[62] Associated with Platonic thought by Albertus Magnus (c.1200–1280),

---

56 On the *Philosophia* and its Paracelsian reception, see also my article, Georgiana Hedesan, 'The Mystery of *Mysterium Magnum*: Paracelsus's Interpretation of Creation in *Philosophia ad Athenienses* and its Early Modern Commentators', in *Hidden Truths from Eden: Esoteric Interpretations of Genesis 1–3*, ed. Caroline Vander Stichele and Susanne Scholz (Atlanta: Society of Biblical Literature Semeia Studies, 2014), 141–62.

57 Walter Pagel, 'Prima Materia of Paracelsus', *Ambix* 9 (1961): 124–5.

58 Richard Bostocke, *The Difference between the Auncient Physicke . . . and the Latter Phisicke* (London: Robert Walley, 1585), ch. 21 (there are no page numbers).

59 Jacob Boehme, *Mysterium Magnum, or an Exposition of the First Book of Moses called Genesis* (London: Lodowick Lloyd, 1656), 23.

60 Heinecke, 'The Mysticism and Science', 70 deemed Van Helmont's philosophy of nature as having 'pantheist' tendencies and claimed that his mysticism tries to 'compensate for the threatening consequences of the premises inherent in his philosophy of nature'. This chapter amply shows that Van Helmont steers away from pantheist associations by clearly pointing out God's transcendence and apophatic quality.

61 Ch. 22, 'Ortus formarum', §2, 105: 'rerum Parens, naturae Conditor & Rector, creando perseverans'.

62 Dag Nikolaus Hasse, 'Avicenna's "Giver of Forms" in Latin Philosophy, especially in the Works of Albertus Magnus', in *The Arabic, Hebrew and Latin Reception of Avicenna's Metaphysics*, ed. Dag Nikolaus Hasse and Amos Bertolacci (Berlin: Walter de Gruyter, 2012), 225–51; Antonio Bruers, 'La "Colcodea" d'Avicenna e T. Campanella', in Tommaso Campanella, *Del senso della cose e della magia* (Bari: Laterza & Figli, 1925), 339–41.

Avicenna's theory received a lukewarm reception in Scholastic circles. The concept implied the dissociation of form and matter, which went against the spirit of Aristotle's hylomorphism; hence Thomas Aquinas and other Scholastics rejected it.[63]

Nevertheless, the doctrine did find adherents that helped to recast this view in Christian terms. The thirteenth-century Franciscan theologians Jean de la Rochelle (d. 1245) and Alexander of Hales (c.1185–1245) associated Avicenna's *dator formarum* with God.[64] Albertus Magnus took an ambiguous position towards this idea, but in some cases he equated the *dator formarum* with Augustine's 'Intelligible Sun' and James's notion of the 'Father of Lights'.[65] The late Middle Ages and Renaissance saw the further revival of the theory: John Buridan (c.1295–1358), Marsilio Ficino (1433–1499) and Tiberio Russiliano (fl. 1519) concurred with this view.[66] It was through such filiation that Van Helmont discovered and embraced this unusual theory.

For Van Helmont, as for Thomas Aquinas, God was essentially life.[67] Yet Van Helmont went further than Thomas to argue that life is bestowed directly by God on each individual being. Hence God became the 'Giver of Life', 'the Prince of Life' and Life itself. [68]

Van Helmont also viewed life and form as 'light'. He did not, of course, mean visible light, but an abstract type of light. Hence Van Helmont calls God 'the infinite, substantial, and thrice-glorious light'.[69] By associating form, life and light, Van Helmont connects the doctrine of *dator formarum* with the Christian view of God as 'Father of Lights', thus carrying Albertus's tentative association to its logical conclusion. In Van Helmont's speculation, the term 'Father of Lights' comes to denote the doctrine of God as the maker of all vital forms.[70]

This theory of God as direct bestower of form and life allows Van Helmont to develop an essentially Platonic vision of the Universe made up of juxtaposed matter and form. According to Van Helmont, both form and matter are created by God, and

---

63 Hasse, 'Avicenna's "Giver of Forms"', 238–44.

64 Dag Nikolaus Hasse, 'Plato Arabico-Latinus', in *The Platonic Tradition in the Middle Ages: A Doxographic Approach*, ed. Stephen Gersh and J.F.M. Hoenen (Berlin: Walter de Gruyter, 2002), 43.

65 Hasse, 'Avicenna's "Giver of Forms"', 238–44, Gilson, *Christian Philosophy in the Middle Ages*, 431. The term 'Father of Lights' occurs in James 1:17.

66 Hasse, 'Avicenna's "Giver of Forms"', 236.

67 Ch. 25, 'Blas humanum', §23, 147: 'Cum vita in abstracto, sit Deus ipse incomprehensibilis'. Thomas Aquinas had affirmed that God is *viventissimus*, 'most alive'; see Denis Des Chene, *Life's Form: Late Aristotelian Conceptions of the Soul* (Ithaca, NY: Cornell University Press, 2000), 63, 65. The association of higher forms with life was also a Scholastic tenet; see Alexander W. Hall, *Thomas Aquinas and John Duns Scotus: Natural Theology in the High Middle Ages* (King's Lynn: Continuum, 2007), 50.

68 Ch. 93, 'Demonstratur thesis', 522–3. The term 'Prince of Life' originates from Acts 3:15.

69 Ch. 22, 'Ortus formarum', §77, 118: 'lumen infinitum, substantiale, ter gloriosus'. It must be noted that Van Helmont sometimes uses the term *lux* and sometimes *lumen* for light, but does not clearly distinguish between the two.

70 See below, Chapter 5, 'Nature'.

originate *ex nihilo*. Matter and form are separate in a dualist manner and juxtaposed at a precise moment in the process of generation.

In turn, matter-form dualism enables Van Helmont to show that God is not bound to Nature; rather, God freely chooses to be continuously present in the universe working 'the perfection of all things', daily creating and ruling nature.[71] In fact, to press his point, Van Helmont emphasises God's 'jealousy' as a key attribute; the Creator, he asserts in a statement closely resembling Oswald Croll's, 'will give His honour of Creator to no creature'.[72] Hence God alone is the 'Creator of all souls'.

The Essential Presence of God: The Divine Word and Christ

Paracelsian philosophers sought to express the essential presence of God in matter in various ways. A favourite doctrine was that of the presence of the divine Word in the creation and maintenance of the Universe. In their desire to emphasise divine presence, Paracelsians sometimes conflated the Word in Genesis (which they saw as not only the *Fiat* but also the command 'increase and multiply', *crescite et multiplicamini*) with the *Logos*-Word (Christ). Such a view was bound to generate controversy, as theology traditionally separated the created Word of Genesis from the uncreated *Logos* of Christ.

Paracelsus had emphasised the role of the Word in the universe, but his approach on the subject appears orthodox. Thus, he argued in 'Opus Paramirum' that the Word created all the *semina* by separating all things from the primordial *yliaster*.[73] Moreover, he affirmed that the Word *Fiat* was the prime matter of all things. Yet Paracelsian supporter Richard Bostocke moved into heterodox realm when he stated that Christ, as the divine Word *Fiat* and the Wisdom of God, was the spiritual *Mysterium magnum* out of which the entire universe was made.[74] Similarly, for Adam Haslmayr, the divine Word was the centre and life of all things, and 'all creatures . . . the living incarnate word FIAT of God'.[75] Such views, inspired by the pseudo-Paracelsian *Philosophia ad Athenienses*, could indeed be suspect of pantheism.

In turn, Van Helmont also emphasised the important role in nature played by the divine Word.[76] Yet his view of the Word was sensitive to Catholic doctrine. According to him, Creation began with God's Word, *Fiat* or 'Let it be done', which figures in the first chapter of Genesis. As Van Helmont expresses it, 'The thrice-glorious

---

71 Ch. 22, 'Ortus formarum', §1, 105: 'quoque cunctorum . . . perfectionem'.

72 Ch. 22, 'Ortus formarum', §2, 105; ch. 93, 'Demonstratur thesis', 523. Croll had stated that God, 'who alone to be praised and blessed in all, and over all, will not give his Honour to another'; Croll, 'Admonitory Preface', 89.

73 Paracelsus, 'Opus Paramirum', in *Paracelsus: Essential Theoretical Writings*, ed. and trans. Andrew Weeks (Leiden: Brill, 2007), I, ii, 323. Martin Ruland defines Iliaster as 'the first chaos of the matter of all things, constituted of Sulphur, Salt and Mercury'; *Lexicon alchemiae* (Frankfurt: Johann Andrea & Wolfgang Endter, 1612), 263–4.

74 Bostocke, *The Difference*, ch. 21.

75 Gilly, '*Theophrastia sancta*', 175–6.

76 Broeckx, 'J.B. Van Helmont: Ad judicem', 110.

Omnipotence, by the simple and pure command of his cogitation and conceived Word [*Fiat*], made all creatures out of nothing (*ex nihilo*); and He put seminal virtues into them, durable through the ages'.[77] Here Van Helmont clearly distinguishes the Word 'Fiat' as being 'conceived' from the uncreated *Logos*.

In a letter to Mersenne, Van Helmont also noted that the properties of things are a 'gift infused by the word Fiat'.[78] In the *Ortus*, he described *Fiat* as the prime mover that set things in motion and impressed universal activity for all time. This Word was so powerful that the Universe would last forever should it be so commanded.[79]

In line with other Paracelsians, the command 'increase and multiply' also held great importance in Van Helmont's thought. This Word was responsible for bestowing fruitfulness to Nature: 'He, who by a small Word, made the stars out of nothing, constituted a similar power of the Word (*Increase and Multiply*) within the innermost part of seeds, which is to endure through the ages'.[80] Thus, this second pronouncement lends permanence to the world by allowing it to generate and increase. The doctrine of the generating Word is closely linked with Van Helmont's theory of seeds and ferments, which I will discuss further below.[81]

Van Helmont also upheld the role of the *Logos*-Word in the creation and maintenance of the universe. Christ had already been seen as the Incarnate Word by John, Paul and the Nicene creed, according to which the Universe was made through Christ.[82] A related doctrine Van Helmont appealed to was that of Christ Pantocrator, the Lord of all things, eloquently expressed by Paul in Colossians 1:13–23.[83] The concept of the Christ-Anthropos as creator, maintainer and saviour had been adopted by many early Christian Fathers such as Irenaeus, Clement of Alexandria, Origen and Tertullian. In the Middle Ages it appeared in the writings of Maximus the Confessor, John Scotus Erigena and Nicholas of Cusa.[84]

In turn, Van Helmont paraphrases both John and Paul when he states that 'Lord Jesus is the light (*lux*), life, beginning, way, truth, and the all of all things in an

---

77 Ch. 69, 'Progreditur ad morborum cognitionem', §13, 430: 'Tergloriosus omnipotens, suae cogitationis nudo puroque mandato, conceptoque verbo (fiat) totam creaturam ex nihilo fecit: illique vires seminales, in seculum durabiles, indidit'.

78 '30 Janvier 1631', 56: 'Proprietates rerum sunt dona infusa per verbum Fiat'.

79 Ch. 25, 'Blas humanum', §3, 144, ch. 45, 'Ignota actio regiminis', §4, 262.

80 Ch.110, 'Vita brevis', 588: 'Is enim, qui verbulo astra fecit de nihilo: consimilem Verbi potestatem (crescite & multiplicamini) in secula durabilem, intra seminum penetralia constituit'. See also, 'De lithiasi', ch. 1, 'Petrificatio', 650.

81 See 5.3.1 and 5.3.2.

82 John 1:1–3; 1 Corinthians 8:6; Colossians 1:15–20; Hebrews 1:2. See also Jaroslav Pelikan, *Jesus Through the Centuries* (New Haven, CT: Yale University Press, 1985), 57–70.

83 Paul himself drew on a rich heavenly Anthropos literature which had Jewish, Gnostic, Hermetic and Greek expressions; Dragos Giulea, 'Noetic Paschal Anthropos: Genesis 1:27 and the Theology of the Divine Image in Early Paschal Literature' (unpublished doctoral thesis, Marquette University, 2010), 41–82; Schmidt-Biggemann, *Philosophia Perennis*, 129–42.

84 Giulea, 'Noetic Paschal Anthropos', 135–70, David Albertson, 'That He Might Fill All Things: Creation and Christology in Two Treatises by Nicholas of Cusa', *International Journal of Systematic Theology*, 8 (2006): 184–205; and in Matthieu van der Meer, 'Nicholas of Cusa's View of Time and Eternity', in *Christian Humanism: Essays in Honours of Arjo Vanderjagt* (Leiden: Brill, 2009), 319.

incomprehensible manner'.[85] Van Helmont also asserts that Christ, the 'Lord of the Universe' (*Dominus Universi*), is 'alone administering the monarchy of Heaven and Earth'.[86] Even more poignantly, he argues that 'in Christ Jesus we live, move, and have our being'.[87] This statement equates the God of Paul's Sermon to the Athenians with Christ.

Indeed, the Flemish physician often extends God's attributes to Christ: for instance, he refers to Him as Infinite Goodness. He also explicitly exalts Jesus as the 'Prince of Life', underlining His role as life-giver and maintainer of long life.[88] This approach serves to re-affirm the identity of Christ with God and, implicitly, to emphasise the miracle of the Incarnation.

God in Man

God as Father of Mankind

According to standard Christian doctrine, God is the Creator of all things, but His relationship with humanity is much closer than with any other creature. Van Helmont observes that God decided to use His own Image as the 'archetype' for humanity.[89] The desire of God to create a being that would be like Him originates from the purest and highest love: 'The Almighty, out of his immense and voluntary goodness of love, prized and erected Man peculiarly for this purpose, that he might intimately and as closely as possible express His own image'.[90]

So powerful is God's love for mankind that Van Helmont denies that He would ever punish or curse it. Thus, commenting on the Book of Genesis, Van Helmont forcefully repudiates the views that God gave a law to Adam not to eat from the Tree of Knowledge of Good and Evil, or that He cursed him in the aftermath for breaking this law. Instead, God's words to Adam should be understood as a loving admonition that warned him of the consequences of eating from the Tree.[91] This interpretation leads Van Helmont to maintain that God did not create death, nor did He impose it as a punishment for the Original Sin: death is something that man caused himself.[92]

---

85 Ch. 114, 'In Sole tabernaculum' 622, 'Quia Dominus Jesus est incomprehensibili modo lux, vita, initium, via, veritas, & totum omnium'.

86 Ch. 22, 'Ortus formarum', §96, 120; ch. 82, 'De conceptis', 487: 'jam coeli terraeque Monarchatum solus administrans'.

87 Ch. 3, 'Studia authoris', §8, 15: 'in Christo Iesu vivimus, movemur, & sumus'.

88 Ch. 79, 'Butler', 466; ch. 63, 'Ignotus hospes morbus', §48, 396; ch. 69, 'Progreditur ad morborum cognitionem', §16, 431; ch. 103, 'Externorum proprietas', 571; 'Mons Domini', 629.

89 Ch. 38, 'Imago mentis', §2, 213.

90 Ch. 92, 'Thesis', §1, 515: 'Omnipotens ex sui immensa & spontanea amoris bonitate, hominem ad hoc peculiariter dilexit & erexit, ut sui imaginem intime & quamproxime exprimeret'.

91 Ch. 111, 'Vita aeterna', 590.

92 Ch. 92, 'Thesis', §2-§7, 515–16; see chapter 4 Man. The source of this idea lies in the deuterocanonical Wisdom of Solomon 1:13: 'For God made not death: neither hath he pleasure in the destruction of the living'.

Hence the Fall is exclusively Man's responsibility, and must not be blamed on anyone else.

Yet the Fatherhood of God means much more to Van Helmont than it does in traditional views. Thus, in the chapters belonging to *Liber de vita longa*, he holds the peculiar opinion that human beings were meant to be born from an earthly mother and a heavenly Father, through the agency of the Holy Spirit. The race of men, Van Helmont argues, was to be generated by the Holy Spirit through Eve. God intended to 'incarnate' his own Image in each human being by imprinting the Holy Spirit onto the flesh of perpetually virginal women:

> It is agreeable to reason, that if God wanted to make His own Image in flesh, and bless it in posterity, this had to be done in the mother being a Virgin, not in a woman defiled by Adam, as God could not have man to be His competitor in the intended Incarnation of His own image.[93]

According to this view, God would have hence assumed direct paternity of human beings, with no involvement of the male gender.

The implications of Van Helmont's doctrine are radical. Human beings are naturally and literally 'sons of God'.[94] Christ's birth from a Virgin was not the exception, but the prototype of all human births, if the Fall had not taken place. Indeed, as Van Helmont puts it, 'we may in some way conjecture by the most glorious Incarnation of our Lord the quality of human generation in virgin Eve, before the Fall'.[95]

The idea whereby God had originally planned to imprint His divine image on female wombs is unusual, and may have been Van Helmont's answer to the theological controversies on 'pure nature'. According to the late Scholastic formulation of this doctrine, Adam possessed a natural side ('human nature') and a juxtaposed supernatural side ('Grace'). Baius, on the other hand, rejected this idea and argued on Augustinian grounds that Adam's original state was a special union of nature and Grace.[96] Van Helmont seems to have inclined toward the concept of a unified Adam, but at the same time concurred with the Scholastics in envisioning God's creation of Man and his planned participation in human generation as essentially supernatural, a supreme form of divine gift (*donum Dei*).[97]

---

93 Ch. 93, 'Demonstratur thesis', §6, 522: 'Inprimis est rationi consentaneum: Si Deus volebat in carne facere sui imaginem, eamque per posteros beare, id debuisse fieri in Virgine matre: non autem in muliere per Adamum stuprata, ne Deus virum competitorem haberet in intenta suae imagines incarnatione'.

94 Ch. 92, 'Thesis', §32, 519.

95 Ch. 93, 'Demonstratur thesis', §23, 525: 'Qualitatem autem generationis humanae ante lapsum in Eva virgine quodammodo conjicere licet ex gloriosissima Incarnatione Domini'.

96 Henri de Lubac, *Augustinianism and Modern Theology*, trans. Lancelot Sheppard (London: Geoffrey Chapman, London, 1969,) 1–14.

97 For more on this topic, see below, Chapter 6, sections on 'Van Helmont's Views on Human Nature' and 'Van Helmont's Final System: the Duumvirate'.

Since God had intended to be the author of human generation, no carnal copulation was allowed in paradise. In fact, Van Helmont makes the startling affirmation that Adam and Eve were not created to be consorts, but companions. However, the devil interfered with God's plan, encouraging them to eat of the Tree of Knowledge of Good and Evil, which Van Helmont understood to be the embodiment of the seminal faculty.[98] The Original Sin, in his view, is the sexual act: 'the apple being eaten, their eyes were opened, and Adam began lustfully to covet after the naked virgin, and consequently defiled her'.[99] There is of course no such passage in Genesis, but Van Helmont argues that the Bible purposefully avoided mentioning intercourse out of shame.

To support his argument that the Original Sin was the sexual act, Van Helmont refers to Augustine and his doctrine of concupiscence.[100] The influence of Baius is evident in Van Helmont's identification of concupiscence as the source of the Original Sin.[101] Yet the most obvious source of Van Helmont's speculation is the Renaissance philosopher and magus Heinrich Cornelius Agrippa (1486–1535). Amongst Van Helmont's papers confiscated by the ecclesiastical court in 1634, the prosecutors found an extract in Van Helmont's own handwriting, which contains the statements that the Original Sin is the sexual act between Adam and Eve, and consequently that coitus is the source of death.[102] The court inquired about the source of these statements, which Van Helmont attributed to Agrippa's declamation 'On the Original Sin' (*De peccato originale*).[103] This work does indeed contain the idea of sex as being the Original Sin.[104] However, Agrippa does not go as far as Van Helmont to claim that the Holy Spirit was supposed to be the 'father' of prelapsarian man; this view may very well be Van Helmont's own elaboration.

Christ as Saviour

In line with his views of God as loving Father, Van Helmont further believes that God's affection for human beings did not alter because of the Original Sin. It was

[98] Ch. 92, 'Thesis', §15–16, 517. Again, the idea of 'Adamic' generation transmitted through the seed is Augustinian.

[99] Ch. 93, 'Demonstratur thesis', §29, 527: 'Manso autem pomo, statim aperti sunt oculi eorum, cepitque Adam libidinose concupiscere nudam virginem, eamque stupravit'.

[100] Ch. 93, 'Demonstratur thesis', §76, 538. He cites from Augustine's *City of God*, *Against Julian*, *The Flesh of Sin* and *On Marriages*, referring to Augustine's idea of the concupiscence of flesh. For most citations from Augustine I have used *The Works of Saint Augustine*, ed. J.E. Rotelle (New York, NY: New City Press, 1990–). Augustine postulated that Original Sin was transmitted through the concupiscence of flesh across generations, *Against Julian* in *The Works of Saint Augustine*, I.24, 4.4.3.4–4.8.44; see also Rist, *Augustine: Ancient Thought Baptized*, 319–27.

[101] See above, 2.1.

[102] MS 'Causa J.B. Helmontii Medici', II, 27–31.

[103] Broeckx, 'Interrogatoires', 346–7.

[104] Marc van der Poel, *Cornelius Agrippa: The Humanist Theologian and His Declamations* (Leiden: Brill, 1997), 225–46. Van der Poel claims that this declamation had no impact on theological speculation before the end of the seventeenth century; on the contrary, this study shows that at least Van Helmont was influenced by this declamation.

man who estranged himself from God, not vice-versa. In turn, God sought to remediate his fallen status. Divine love led Christ, the second Person of the Trinity, to embody Himself as the son of the Virgin Mary and provide a path not only to reverse the effects of Original Sin, but to be glorified as Sons of God by new Birth.[105]

Van Helmont advocates a central role for Christ as the Saviour of mankind. To redeem us, Christ had to be born in human shape, in 'the form of a servant'.[106] Christ took the image of 'Adamic Man', but He was not born of sinful flesh. His 'true' self, Van Helmont points out in a 1631 letter to Mersenne, is revealed to human beings only on mount Tabor.[107] The Taboric transfiguration of Christ represents Jesus's assumption of His true sinless, virginal and spiritual body in front of the Apostles.

In Van Helmont's view, Christ brought with him the message of regeneration through the mystery of the Baptism, which to the Flemish thinker means the restoration of the original 'Virginal generation'. As he emphasises,

> The joyful news was brought unto us, that Baptism would be given for the remission of sins, by which Man would be so renewed by water and the Holy Spirit, that his soul would be born again as it were by a new birth, and be made partaker of the unspotted humanity of Christ the Saviour, as framed by the Holy Spirit.[108]

The effects of Baptism are striking, changing the person into a living image of Christ.[109] Baptism brings about a profound transformation that is seen only by faith in one's lifetime, but will be visible in the afterlife, when the old body will be replaced by the new, glorified body. The Elect will then become true 'Sons of God'; in other words, they will assume the Taboric body of Christ. In all these views, Van Helmont is clearly drawing on the orthodox Paulinian tradition.[110]

The Flemish philosopher opines that Baptism is sufficient for children, but in adults it must be complemented by the Eucharist due to the frailty of our Adamic body.[111] The Eucharist, Van Helmont states, 'is wine which buds virgins'.[112] The Eucharist allows us to 'commune' and 'unify' ourselves to Christ's mystical body,

---

105 Ch. 93, 'Demonstratur thesis', §78, 539.

106 Philippians 2:7: 'But made himself of no reputation and took upon him the form of a servant and was made in the likeness of men'.

107 'Jean-Baptiste Van Helmont, à Bruxelles, à Mersenne, à Paris: 14 Février 1631', 103. On the transfiguration of Christ, see Matthew 17:1–17:2.

108 Ch. 111, 'Vita aeterna', 589: 'laetum nuncium nobis allatum fuit, quod unum baptisma datum esset in remissionem peccatorum, per quod homo ita renasceretur ex aqua & Spiritu sancto, quod ejus anima, tanquam nova nativitate renasceretur, & fieret particeps illibatae humanitatis Christi Salvatoris, fabricatae per Spiritum sanctum'.

109 Ch. 93, 'Demonstratur thesis', §47, 530.

110 On the Paulinian concept of participation in Christ and eschatology, see James D.G. Dunn, *The Theology of Paul the Apostle* (Edinburgh: T&T Clark, 1998), 390–412, 466–98.

111 Ch. 93, 'Demonstratur thesis', §53–4, 532.

112 Van Helmont refers to Zecharias 9:17 which in the Vulgata version reads 'vinum germinans virgines'.

thus 'participating in His incomprehensible and amorous Incarnation'.[113] By 'eating the Lord's Body', human beings experience a profound transformation whereby the mortal flesh is altered into that of the Lord, being readied for eternal Life after death.

Without Baptism and Holy Communion, Van Helmont further emphasises, there is no Eternal Life breathed into a person. Hence at the Last Day, even though all human beings will arise, only those granted Eternal Life will gain access to Paradise. The new body will still be made of flesh, but this will be a virginal and spiritual flesh.

Van Helmont's insistence on the Sacraments as a 'mystical union' between man and Christ is orthodox. However, he does not address the contentious medieval doctrine of transubstantiation, which had been upheld by the Catholic Council of Trent.[114] Even as he upholds the real presence of Christ in the Eucharist, he avoids expressing himself in Scholastic terms.

Alchemy, Paracelsianism and Van Helmont's Theology

We cannot be sure whether and how familiar Van Helmont was with Paracelsus's theological treatises, most of which remained in manuscript until the twentieth century but were widely circulated in Van Helmont's time.[115] Nevertheless, one cannot help noticing that their arguments concerning the two Sacraments and the spiritual body are similar. Paracelsus had maintained that both the body and the soul will enter heaven, except this is not the earthly, Adamic body, but the immortal, spiritual one.[116] The new body is obtained by sacramental participation in the 'divine-spiritual

113 Ch. 93, 'Demonstratur thesis', §56, 533. The whole passage runs: 'Communio itaque castissimi illius corporis, unit nos suo mystico corpori, reditque nos participes incomprehensibilis & amorosae suae Incarnationis, quatenus renascendo virginitatem ejus (in qua salvari debemus) participative induimus'. See also ch. 111, 'Vita aeterna', 589.

114 Carter Lindberg, *European Reformations* (Oxford: WileyBlackwell, 2009), chap 14, 'Catholic Renewal and Counter-Reformation'.

115 Dane Thor Daniel refers to Paracelsus's sacramental theology as 'the immortal philosophy' (*untötliche philosophei*); 'Paracelsus on Baptism and the Acquiring of the Eternal Body', in *Paracelsian Moments*, ed. Gerhild Scholz Williams and Charles D. Gunnoe, Jr. (Kirksville, MI: Sixteenth Century Journal Publishers, 2002): 117. See also Dane T. Daniel, 'Invisible Wombs: Rethinking Paracelsus's Concept of Body and Matter', *Ambix*, 53 (2006): 131. Paracelsus's theory of the sacraments has been analysed chiefly by Kurt Goldammer, *Paracelsus in neuen Horizonten: Gesammelte Aufsätze* (Vienna: Verband der wissenschaftlichen Gesellschaften Osterreichs, 1986), 87–122; Hartmut Rudolph, 'Hohenheim's Anthropology', 187–206; and Michael Bunners, *Die Abendmahlsschriften und das medizinisch-naturphilosophische Werk des Paracelsus* (Berlin: Humboldt University, 1961).

116 Rudolph, 'Hohenheim's Anthropology', 192–3, based on his study of Paracelsus's Eucharistic writings. It was a theological commonplace in the era that the body would accede to heaven in some form, but there were disagreements on how this body would look like and behave. See Caroline Walker Bynum, *The Resurrection of the Body in Western Christianity, 200–1336* (New York, NY: Columbia University Press, 1995), 236, 250. Both Paracelsus and Van Helmont might have been influenced by Paul's speculations of a spiritual body ascending to heaven in 1 Corinthians 15.

corporeality of Christ' as Kurt Goldammer has called it.[117] Indeed, both Paracelsus and Van Helmont emphasised that this spiritual body is obtained through Baptism and the Eucharist, which allow human beings to take the form of Christ.[118] Incidentally, both played down the other Catholic Sacraments, with the qualification that Van Helmont did accept the sacrament of marriage as essential in order to rein in the concupiscence of the flesh.[119]

There are other elements that draw Van Helmont closer to Paracelsus's theology. Like Van Helmont, Paracelsus's theological ideas were generally Christocentric.[120] He also upheld the idea that Christ was of a different constitution from ordinary human beings.[121] It was orthodox to state that Christ had a body that was free from Original Sin, hence in some sense 'different' from the other fallen human beings. Van Helmont would have deemed himself in accordance with the Council of Chalcedon's statement of faith in affirming that the true humanity was Christ's humanity, rather than ours.[122] In doing so, Van Helmont emphasises Christ's nature as the new, or second Adam, a topos originating in Paul's writings and adopted, amongst others, by Paracelsus himself.[123]

In order to show how it was possible for Christ to be born in a sinless body, both Van Helmont and Paracelsus upheld the idea of the Immaculate Conception of the Virgin Mary as a logical consequence.[124] Van Helmont believed Christ actually formed his Mother with His divine Hand. Thus, the Virgin Mother was 'sanctified in her mother's womb' before birth.[125] This concept differs from Paracelsus's idea of

---

117 Kurt Goldammer, 'III. Einleitungen zu den Schriften dieses Bandes, 2. De genealogia Christi', in Theophrastus von Hohenheim, gen. Paracelsus, *Sämtliche Werke II: Abteilung: Die theologischen und religionsphilosophischen Schriften*, ed. Kurt Goldammer (Wiesbaden: Steiner Verlag 1955), III, xxx–xxxi.

118 Gause, 'On Paracelsus's Epistemology', 142.

119 Ch. 93, 'Demonstratur thesis', §80, 539.

120 Arlene Miller Guinsburg, 'Paracelsian Magic and Theology', 135; on the other hand, Ute Gause, 'On Paracelsus's Epistemology', 214 argues that Paracelsus is not always Christocentric, at least in respect to the concept of 'heavenly magic'.

121 Paracelsus insisted that Christ was made of an 'angelic body, but in blood and flesh'; 'De genealogia Christi', in *Sämtliche Werke II: Abteilung: Die theologischen und religionsphilosophischen Schriften* (Wiesbaden: Steiner Verlag, 1955), III, 3.71. Dane T. Daniel deemed Paracelsus a heretic of the Docetist or Monophysitic type (these denied the full humanity of Christ); 'Medieval Alchemy', 130–31. Yet Paracelsus did state that Christ was present in flesh and blood on earth.

122 The Council of Chalcedon upheld the idea of Christ's two natures, 'true God and true man', but the 'true man' term implied that He is 'like us in all things apart from sin'; see Gerald O'Collins, *Christology: A Biblical Historical and Systematic Study of Jesus* (Oxford: Oxford University Press, 2009), 196.

123 Romans 5:12–21; see also Dunn, *The Theology of Paul*, 199–204, 241–2.

124 Paracelsus was particularly outspoken about Immaculate Conception as doctrine; in *De salve regina et magnificat* (Leiden, Codex Vossianus Chymicus 24, fols 367r–372v), he stated that 'Far be it from us to teach the heretical doctrine that she is conceived in original sin'. For a commentary on Paracelsus's Mariology, see Weeks, *Paracelsus: Speculative Theory*, 83–5, 88; Arlene Miller Guinsburg, 'The Counterthrust to Sixteenth Century Misogyny: The Work of Agrippa and Paracelsus', *Historical Reflections* 8 (1981), 18–22. Of course, Van Helmont's support of the Immaculate Conception did not have to derive from Paracelsus, as it was popular in the era and upheld, amongst others, by the Jesuits.

125 Ch. 93, 'Demonstratur Thesis', §55, 533.

a genealogy of Christ which had been preserved from Original Sin throughout history.[126] However, Van Helmont and Paracelsus were similar in their devotion to the Virgin Mary.

One can conclude that the approach of Van Helmont and Paracelsus in regard to the theological issues of the Sacraments, the Virgin and Christ are similar, but a direct influence cannot be proven at this point. One scholar, Maximilian Bergengruen, has affirmed that Van Helmont's theological stance is a reiteration of the most important ideas of Paracelsus:

> Van Helmont's theological position is a combination of the key topoi in the theological division of the Paracelsian discourses: the structure of salvation history as a direct line from Adam to Christ (the new Adam), the idea of an inner *Imitatio Christi* as the most important task of the believer, the New Birth or Rebirth in Christ, and the loss of individuality and the individual will.[127]

Yet many of the theological ideas common to Van Helmont and Paracelsus could be found in other tracts of the era, particularly those influenced by German mysticism.

If it cannot be proven that Van Helmont read Paracelsus's theology, it can, however, be shown that many of his theological ideas resonated with alchemical thought. This conclusion can be drawn from Van Helmont's own testimony. His middle treatise 'Ad judicem neutram causam appellat suam et suorum philadelphus' contains a long list of statements that Van Helmont claims can be demonstrated 'by the fire'. The Flemish physician seems to have believed that any alchemist would have agreed with these ideas as 'statements of faith'. At points 51–65 in the list, Van Helmont unexpectedly moves from an explanation of nature based on primordial seeds, *Archei* and spirits to theological doctrine. What is remarkable is that most of these latter statements reflect the same ideas later expounded in *Ortus medicinae* as theology, such as: Original Sin can only be removed through the regeneration of the whole man; this regeneration must be done through water and the eternal fiery spirit; Christ could only be born from a Mother who did not know menstruation or sin; the glorified body must be united with a body that is drawn from the Virgin.[128]

The inclusion of these theological statements in explicit connection with alchemy much chagrined the ecclesiastical court which was trying Van Helmont in 1634.[129]

126 Paracelsus's ideas about Christ's genealogy were primarily stated in *De genealogia Christi*. See Daniel, 'Medieval Alchemy', 130; also Bunners, *Die Abendmahlsschriften*, 94–96. Van Helmont's argument was in line with that of Thomas Aquinas, who had maintained that the Virgin's sanctification in the womb erased both Original Sin and all subsequent sins as well; see Aquinas, *Compendium of Theology*, trans. Richard J. Regan (Oxford: Oxford University Press, 2009), 181.

127 Maximilian Bergengruen, *Nachfolge Christi—Nachahmung der Natur: Himmlische und Natürliche Magie* (Hamburg: Meiner Verlag, 2007), 51 (my translation). Bergengruen himself points out that the themes of the old-new Adam, the New Birth and Imitatio Christi are present in Luther's thought as well; 57–8.

128 Broeckx, 'J.B. Van Helmont: Ad Judicem', 116–17.

129 Broeckx, 'Interrogatoires', 341–2.

They suspected that Van Helmont's sentences meant that alchemists could demonstrate in the laboratory the truths of the Christian religion, an idea utterly blasphemous to them. To such accusations, Van Helmont replied that all those statements were in fact *Decknamen*, or metaphors for abstruse alchemical processes. Forthwith he explained that Christ actually meant philosophical mercury; Original Sin: the sordid *materia prima* of the alchemical work; the regeneration of the whole man: the perfection of metals into gold; the perpetual fiery spirit: the alchemical *aqua permanens* or solvent; the birth of Christ from a sinless Mother: the purity of mercury and the matter wherefrom it is extracted; the glorified body: the union between sulphur and mercury.[130]

In offering this explanation, Van Helmont sought to avoid any suggestion that alchemists believed mysteries of faith could be elucidated in the laboratory. This idea had been vociferously denounced by Mersenne, Gassendi, and the Jesuits. Nevertheless, the answer did not fully satisfy the judge. He asked Van Helmont if, perchance, the use of theological terms for purely alchemical matters might be construed as blasphemy. Van Helmont replied that the alchemists use these terms so that 'pagans' and 'heretics' could be made to know and understand the mystery of faith by laboratory processes, implying that alchemy could be used to convert unbelievers to Christian (or even Catholic) faith. By affirming a conversion or educational role for alchemy, Van Helmont was implicitly maintaining the correspondence between the natural and supernatural planes, between the work of the laboratory and matters of faith. Such an argument did not theoretically blur the boundaries between the two planes, but could still appear suspect in the eyes of some clergy. In this case, the judge was apparently satisfied by the reply and did not pursue the matter further.

Perhaps because he was never reprimanded on this subject, the later Van Helmont advocated the didactic role of alchemy in the *Ortus medicinae* as well. He may have been encouraged in this respect by the recent writing of Pierre-Jean Fabre, *Alchymista Christianus* (1632), although there is no direct proof that he had read this work. The *Christian Alchemist* was dedicated to Pope Urban VIII and argued that alchemy and Christianity, particularly that of the Catholic persuasion, were not at odds with each other; on the contrary, alchemy helped illuminate Christian beliefs.[131]

In turn, Van Helmont used the analogical approach to illustrate his Christian eschatological views. He openly associated the regeneration of souls through the Eucharist with the projection of the philosophers' stone.[132] He hence makes an analogy between the transmutation of metals into gold by means of the elixir and the transformation of the soul of man glorified by Eucharist and Baptism. Whereas the metal is turned into gold by the heat of fire, the soul is changed by the heat of devotion.[133] Thus, just as the philosophers' stone has the power of projection or

---

130 Broeckx, 'Interrogatoires', 341–2.

131 Pierre Jean Fabre, *Alchymista Christianus* (Toulouse: Petrus Bosc, 1632), 130–36.

132 Ch. 93, 'Demonstratur thesis', §58, 534. For Van Helmont's views of *chrysopoeia*, see chapter 7.

133 Ch. 111, 'Vita aeterna', 590.

multiplication, 'a small quantity of this mystical and divine super-celestial bread regenerates, restores and renews a huge number of the Elect'.[134]

As in the interrogatory, Van Helmont is careful not to assert a direct relationship between the philosophers' stone and the Eucharist, except for the fact that they both occur in the sublunary world.[135] Apart from this, the relationship is that of analogy or similarity.[136] The use of the philosophers' stone as a didactic tool lies in its perceived capacity to make hidden divine forces visible by means of transmutation. Since the effects of Baptism and the Eucharist cannot be perceived in this world, Van Helmont contends that, by witnessing or understanding the transmutation of the stone, Christians may more easily believe regeneration.[137]

In his argument, Van Helmont cannot escape the implication that the philosophers' stone is a type of 'miracle' occurring in the sublunary world, similar in its supernatural status to the Eucharist. Petrus Bonus of Ferrara, who similarly advocated the didactic role of alchemy in the Middle Ages, clearly viewed the stone as supernatural. There are many signs that Van Helmont concurred with this view, albeit not so openly for fear of heterodoxy.

Christ as Wisdom

Besides His fundamental roles as 'ruler of the world' and 'Saviour' of humanity, Christ has another attribute that defines his relationship with human beings, which is Wisdom. In 1 Corinthians 1:24, Paul called Christ 'the wisdom of God'.[138] While there are disagreements in regard to Paul's intentions, subsequent Christian tradition associated the Wisdom (*Sophia*) of the Old Testament's Psalms and Proverbs with Christ.[139] Christ as Wisdom figured prominently amongst medieval writers like Bonaventure, Johann Tauler and Nicholas of Cusa.[140] It became a key trope for many

---

134 Ch. 93, 'Demonstratur thesis', §58, 534: '... tantillum hujus mystici & Divini panis supercoelestis regenerat, restaurant, atque innovat immensum electorum numerum'.

135 Ch. 111, 'Vita aeterna', 590.

136 Ch. 93, 'Demonstratur thesis', §57, 534.

137 Ch. 111, 'Vita aeterna', 590.

138 See also 1 Corinthians 1:30: 'But of him are ye in Christ Jesus who of God is made unto us wisdom and righteousness ...'

139 H. Windisch, 'Die göttliche Weisheit der Juden und die paulinische Christologie', *Neutestamentliche Studien für Georg Heinrici* (Leipzig: J.C. Hinrichs,1914), 232; A. Feuillet, *Le Christ Sagesse de Dieu d'après les Épitres pauliniennes* (Paris: Études Bibliques, 1966), 52, 193, 385 argued that Christ of the New Testament (especially Paul's) was the divine Wisdom of the Jews, while A. Van Roon negated such intention on the part of Paul, 'The Relationship between Christ and the Wisdom of God according to Paul', *Novum Testamentum* 16 (1974): 207.

140 Ilia Delio, *Simply Bonaventure: An Introduction to his Life, Thought and Writings* (New York, NY: New City Press, 2001), 139; Johann Tauler, *The Sermons and Conferences of John Tauler of the Order of Preachers*, trans. Walter Elliott (Washington, DC: Apostolic Mission House, 1910), 262; Eugene F. Rice, 'Nicholas of Cusa's Idea of Wisdom', *Traditio* 13 (1957): 348–51, 357.

Paracelsian writers, including Paracelsus, Heinrich Khunrath, Adam Haslmayr and Oswald Croll.[141]

In line with this tradition, Van Helmont affirms that Jesus is 'the only wisdom of the Father', or 'the principle of the Father's Wisdom, the unlimiting end, the Alpha and Omega'.[142] Thus Van Helmont clearly implicates Christ in the epistemological process: 'only Wisdom, the Son of the everlasting Father of Light, bestows knowledge'.[143] Hence, in Van Helmont's vision, true *scientia* can only be obtained from Christ, and not through one's own efforts.[144] Christ chooses those whom He illuminates:

> And whom the Lord Jesus shall call to Wisdom, he, and no one else shall come to it. He that has arrived at its summit shall still be able to do very little, unless the bountiful favour of the Lord shall shine upon him.[145]

The vision of Christ as giver of Wisdom is linked with His role as mediator between God and human beings. Van Helmont once again paraphrases John 14:6 to state that

> The Lord Jesus is hence the Way, Truth, and Life: the way to His Truth, and to the Life of the Father of Light . . . Where He is, the Kingdom of God is present with all its free gifts.[146]

The fact that Christ shows the Way to Truth, or knowledge, Van Helmont argues, is indicated by Jesus Himself through the Lord's Prayer. Hence the Flemish physician affirms that the Prayer contains an esoteric, mystical teaching that is not immediately apparent.[147] Van Helmont's analysis of the Lord's Prayer concentrates on what he calls the 'three wishes': 'Our Father, which art in Heaven, hallowed be thy Name, thy Kingdom come, thy Will be done'.[148] As the Prayer calls for the sanctification of the Name

---

141 For Paracelsus, see Schmidt-Biggemann, *Philosophia Perennis*, 183; for Heinrich Khunrath, *Amphitheatrum Sapientiae Aeternae* (Frankfurt: Tobias Gundermann, 1609), 6, 8, 17; for Haslmayr see Carlos Gilly, '*Theophrastia sancta*', 163–76.

142 Ch. 47, 'Nexus sensitiva & mentis', §1, 284: '. . . ad Dominum meum Iesum redeo, qui solus paternae sapientiae principium, interminans finis α & ω'.

143 Ch. 7, 'Logica inutilis', §24, 37: 'Scientias autem, sola dat sapientia, filius sempiterni Patris luminum'.

144 'Promissa authoris', §14, 8.

145 Ch. 3, 'Studia authoris', §19, 16: 'Et quem Dominus Iesus vocaverit ad sapientiam, ille, & non alius, venturus est. Imo qui ad fastigium pervenerit, minimum adhuc poterit, nisi Domini favor benignus affulserit'.

146 Ch. 43, 'Mentis complementum', §17, 253: 'Dominus Iesus itaque via est, veritas & vita. Via inquam ad se veritatem, & ad vitam Patris luminum . . . Ubi ipse est praesens Regnum Dei, cum omnibus suis charismatibus'.

147 Interpretations of 'Our Father' were given by numerous early Church Fathers such as Tertullian, Origen, and Gregory of Nyssa; see Frederic Henry Chase, *The Lord's Prayer in the Early Church* (Cambridge: Cambridge University Press, 2004), 25–36.

148 Matthew 6:9–10.

of God and the coming of His Kingdom, Van Helmont links these ideas with Matthew 6:33, 'But seek ye first the kingdom of God, and his righteousness'. The prayer, Van Helmont affirms, is in fact for the Kingdom of God within us to come closer to us; thus is God's Name hallowed. This is achieved by a mystical knowledge of the self, whereby 'the Kingdom of God comes to us and is reborn'.[149] The emphasis is on the presence of God within the soul. Based on the equivalence of the Kingdom of God with God and the soul, Van Helmont concludes that 'thy Will be done' means that we naturally desire to annihilate our will so that His Name is sanctified within.[150]

Van Helmont's understanding of the Lord's Prayer is most likely drawn from Johann Tauler. In his Second Sermon for the Fifteenth Sunday after Trinity, Tauler talks about the beginning of the Lord's Prayer, maintaining that:

> it is really a prayer to obtain possession of God's own self, for the kingdom of God is God Himself, reigning in all created things . . . If He finds our souls ready, then (as the *Pater Noster* proceeds) He makes His name known, hallowed and lovingly adored in our interior life.[151]

Conclusions: The Innerness of God

The image of God that pervades Van Helmont's writings is that of a divinity which, although fundamentally unknowable, is paradoxically present in each individual creature. God is the invisible 'centre' of things, deeply involved in His Creation and incessantly producing life. As this chapter has shown, such panentheistic views of the innerness of God had firm roots in an early Christian vision fuelled by Neoplatonic ideas, revived in the writings of medieval German mystics and further expressed by Renaissance thought.

In Van Helmont's times, this divine centrality was also promoted by early modern alchemical philosophy, which drew on the quintessence theme of late medieval alchemy. The quintessence was seen as a substance that had was related to, or even originated from, the heavens. Its 'otherworldly' character made it susceptible to religious interpretation. In one of the earliest elaborations, Pseudo-Lull had emphasised that the quintessence present in Nature was made by God Himself.[152] Some early modern alchemists went further to view the quintessence as divine, being the Spirit of God and even, in some more daring accounts, God Himself. Thomas Charnock (1516/1524–1581), for instance, talked about the 'invisible Godhead' as being the

---

149 Ch. 4, 'Venatio scientiarum', §48, 23, 'quod etiam Regnum Dei, nobis velut adventat, atque renascatur'.

150 Ch. 43, 'Mentis complementum', §18, 254.

151 Tauler, *The Sermons*, 523.

152 'Raymundi Lulli Theorica & Practica', 6. For more on the quintessence, see F. Sherwood Taylor, 'The Idea of Quintessence', in *Science, Medicine and History*, ed. E. Ashworth Underwood, 2 vols (London: Oxford University Press, 1953), I, 247–65.

quintessence of all things.[153] Heinrich Khunrath referred to the quintessence as being *Ruach Hochmah-El*, the Spirit of the Wisdom of God,[154] while others talked about it being the Eucharistic blood of Christ. This religious take on the quintessence resonates with the vision of Christ as *Logos* and Wisdom, cosmic Anthropos, the archetype of man and the essence of the universe.[155]

Van Helmont stayed away from most heterodox alchemical speculation on God in Nature, yet he also saw Christ as central to both the search for Wisdom and eternal salvation. For a philosopher thirsting for knowledge, Christ offered epistemological certainty. As Wisdom, He is the author of all knowledge, and as Wisdom-Archetype and *Logos*, He is also knowledge itself, the *scientia* which Van Helmont sought in nature. Thus, Van Helmont's Christocentrism provides a unitary understanding of the universe and Man. As we shall see in more detail in the next chapters, this belief in the ubiquity and immanence of God and Christ dominated Van Helmont's views on Nature and Man.

---

153 Thomas Charnock, 'The Camp of Philosophy', in *Theatrum Chemicum Britannicum*, edited by Elias Ashmole (London: Nathaniel Brooke, 1652), 306–7.

154 Khunrath, *Amphitheatrum Sapientiae Aeternae*, 19.

155 For the motifs of the Christ Anthropos in alchemy, see Urszula Szulakowska, *The Sacrificial Body and the Day of Doom* (Leiden: Brill, 2006), 15–36.

Chapter 5
Nature

Van Helmont views Nature as created by God in order to implement His will. Consequently, the Flemish philosopher does not think that Nature needs to be 'subdued' or 'wrestled with' by Man: she naturally obeys him as the image of God. At the same time, Van Helmont does not perceive Nature as inferior to Man, since it is at all times permeated by the divine Wisdom.

This chapter is separated into three major parts: Van Helmont's general views of Nature, on matter, and on spirit. This division further reflects Van Helmont's understanding of the world in dualist terms: passive matter and active spirit.

Pure matter, Van Helmont thinks, is inert. He describes it as 'empty, void, dead, and idle, unless it has been, or will be formed at some point by a seminal principle'.[1] As we shall see, Van Helmont bases his views on the Biblical account of Genesis. The Flemish philosopher condemns Aristotle for arguing that matter might originate anything, and that a principle of motion belonged to it.[2] On the contrary, matter in itself is incapable of change, which can only be caused by the intention of spiritual agents.

Like most Paracelsians, Van Helmont was particularly concerned with the spirit. Paracelsus had argued that medicine and natural philosophy ought to pay more attention to spirit than matter. His view was taken up by many alchemical philosophers at the turn of the seventeenth century. In turn, they sought to clarify the relationship between matter and spirit by associating them with the qualities of passivity and activity, respectively. Joseph Du Chesne classified the elements as being those things which 'more cleave and inclyne to Matter than to Forme', and are fundamentally 'passive'; by comparison, the seeds and the *tria prima* 'come néere to Forme, than to Matter' and are 'active'.[3] Du Chesne did not base this analysis on Genesis, but his contemporary Timothy Willis (c. 1551–1627) did. Willis contended that the Spirit of God hovering over the waters was fundamentally spiritual and active while the 'waters' below were passive.[4]

---

1 Ch. 5, 'Causae et initia naturalium', §3, 27: '. . . inanis, vacua est, mortua, ac deses, nisi vitali, aut seminali adesse principio, fuerit constituta'.

2 Ch. 23, 'Magnum oportet', §11, 122–3.

3 Du Chesne, *The Practise of Chymicall and Hermeticall Physicke*, ch. 4. It is remarkable that Du Chesne considers the *tria prima* as being more spiritual than material.

4 Timothy Willis, *The Search of Causes, containing a Theophysical Investigation of the Possibilitie of Transmutatorie Alchemie* (London: J. Legatt, 1616), 9–11.

Van Helmont followed the Paracelsian tradition to postulate a wide array of spiritual agents in the world, including the Ferment, the *Archeus*, light, *Gas* and the soul. Why so many, one may ask? There are several reasons for it. First, the multiplicity of spiritual agents suggested their importance in the economy of the universe, something Van Helmont thought to have been insufficiently recognised by Aristotelian philosophy. Secondly, the Flemish physician preferred particular agents to universal ones because they were more useful in clarifying his natural philosophy. For instance, Van Helmont's spiritual entities play specific roles in his theory of generation. Thirdly, the multiplicity of spiritual agents is related to Van Helmont's understanding of the interaction of matter and spirit. A comparison with the views of his contemporary Descartes may be instructive. Descartes also differentiated between matter and spirit, but he did so by reducing spirit to the incorporeal human soul, while implicitly multiplying materiality. Consequently, he had trouble accounting for the interaction between body and soul. By comparison, Van Helmont admitted other incorporeal entities besides the human soul, many of which were embedded in and acted through matter. He also imagined the corporeal and incorporeal to be in continual interaction with each other, generating composite entities that can best be described as liminal.

Van Helmont's matter-spirit model was primarily drawn from the alchemical tradition, which often viewed the world through a sexualised lens. All beings were generated by the interaction of the male principle (active) with the female principle (passive). The template for this may have been the *Tabula smaragdina* of Hermes Trismegistus, which posited that '[t]he Sun is its father, the Moon its mother, the wind hath carried it in its belly, the earth is its nurse'. Based on this tradition, Pseudo-Paracelsus's *De Pestilitate* posited that the generation of all things happened through the engendering of seed originating from the Sun and the Moon into the watery matter.[5] This sexualised scheme postulates that matter and spirit are not opposite; on the contrary, they have a natural relationship of attraction, or 'love' toward each other. Van Helmont thought this alchemical worldview was an important point of concordance with the Christian tradition, which also viewed the alliance of matter and soul as unbreakable and necessary.

In Van Helmont's view, the interaction of matter and spirit also creates a wide array of invisible beings that lie at the boundary between corporeality and incorporeality. In this sense, Van Helmont followed the Neoplatonic and Pseudo-Dionysian tradition to postulate an entire spectrum of spiritual forces acting on matter, from the finely corporeal to the incorporeal forms-lives. Some spiritual agents are described as very fine matter combined with incorporeal principles, such as the *Archeus* and *Gas*. Others, like the *magnale*, are described as liminal. Still others, such as the Ferment, the light and the soul, are plainly incorporeal.

[5] Philippus Theophrastus Bombast von Hohenheim Paracelsus gennant, 'De Pestilitate', in *Dritter Theil Der Bücher und Schrifften*, ed. Johannes Huser (Basel: Conrad Waldkirch, 1589), 30–36.

Van Helmont's Views of Nature

A Voluntarist Definition of Nature

Van Helmont's voluntarist understanding of God as the omnipotent architect of creation had a major repercussion in the way he viewed Nature. Nature, he says, is 'the command of God whereby a thing is that which it is, and acts as it is directed to act'.[6] This definition shows that Van Helmont did not understand Nature as matter. Instead, Nature is *physis*, the source of order and motion.[7]

The rejection of the notion of Nature as material is part of Van Helmont's polemic with the concept of the Aristotelian *prima materia*, *hyle*, which he rejects as fictional. Instead, Nature is a principle of organisation that makes the universe a 'cosmos' rather than 'chaos'.[8] This idea is also articulated in Aristotle, but Van Helmont attributes the concept to the father of medicine, Hippocrates, whose *Aphorisms* described *physis* as a dynamic, vital and regenerative force.[9] Following Hippocrates, Galen endowed *physis* with rationality and ordering intelligence in his treatise *De usu partium* (*On the Use of Parts*).[10]

The alchemical tradition eminently internalised the view of Nature as an active principle. The earliest known alchemical treatise, *Physika kai Mystika* by Pseudo-Democritus (c. 300 BC), contained the famous exhortation: 'Nature rejoices with Nature; Nature conquers Nature; Nature restrains Nature'.[11] This view of Nature also appears in medieval alchemical treatises like the *Turba philosophorum*.[12]

During the early modern period, Paracelsus perceived Nature as a vital invisible force that shaped and produced the visible universe.[13] Joseph Du Chesne defined Nature as 'the divine virtue, which God hath put and implanted in all creatures'. He emphasised the medieval notion of *Natura naturans* ('creating nature') over that

6 Ch. 24, 'Natura contrariorum nescia', §39, 140: 'Natura jussus est ille Dei, quo res est id, quod est, & agit, quod agere jussa est'.

7 Robin George Collingwood, *The Idea of Nature* (Oxford: Oxford University Press, 1960), 3. For a good overview of the concept of Nature in the Middle Ages and Renaissance, see also Kurt Goldammer, *Der Göttliche Magier und die Magierin Natur* (Stuttgart: Franz Steiner Verlag, 1991), 10–11, n. 3–4.

8 Ch. 5, 'Causae et initia naturalium', §2, 27. Aristotle stated that 'Nature is everywhere a cause of order', 'Physics', in *The Complete Works of Aristotle*, 2 vols, ed. Jonathan Barnes (Princeton: Princeton University Press, 1984), I, VIIII, I, 252a12, 17; see also Helen S. Lang, *The Order of Nature in Aristotle's Physics* (Cambridge: Cambridge University Press, 1998), especially chapters 1 and 8.

9 Ch. 66, 'Respondet author', §1, 419. On Hippocrates's notion of *physis*, see also Heinrich von Staden, 'Physis and Techne in Greek Medicine', *The Artificial and the Natural: an Evolving Polarity*, ed. Bernadette Bensaude-Vincent and William Newman (Cambridge, MA: MIT, 2007), 21–51.

10 Jacques Roger, *Les sciences de la vie dans la pensée française du XVIIIe siècle*, 2nd edition (Paris: Albin Michel, 1993), 58.

11 Matteo Martelli, ed., *The Four Books of Pseudo-Democritus* (Leeds: Maney Publishing, 2013), S95.

12 *Turba philosophorum*, in *Artis auriferae quam chemiae vocant*, 2 vols (Basel: Waldkirch, 1593), I, 1–62.

13 Paracelsus, 'Paragranum', 149.

of *Natura naturata* ('created world').[14] Fludd described infinite Nature as 'a Spirit immense, ineffable, unintelligible, above all imagination and essence'.[15] Perhaps closer to Van Helmont is Sendivogius's definition of Nature as a volatile spirit whose 'seat' is in the will of God.[16]

Van Helmont's definition of Nature must also be perceived in the context of an early modern reaction to what some intellectuals saw as a 'pagan' trend of 'divinising' Nature in the Renaissance. Such tendencies were opposed by a 'Christianising' view that sought to distinguish Nature from God as being created and inferior. This trend viewed Nature as divine order, but rejected any possibility that it was in any way comparable to or independent from God. This argument was present in Jacob Schegk's writings, became prominent with Van Helmont's friend Gassendi and culminated with Robert Boyle's 'Free Enquiry in the Vulgarily Conceived Notion of Nature' and the work of the Cambridge Platonists.[17] It was also present in alchemical philosophy, for instance in David de Planis Campy.[18]

In turn, Van Helmont's view of Nature reflects his own voluntarist tendencies and project of creating a truly Christian philosophy. For him, Nature is God's subordinate instrument, always performing the works that had been assigned to it. As it was created out of nothing, it could easily be reduced to nothing if God so pleased.[19] In an elaborate dream Van Helmont describes at the beginning of the treatise 'Potestas medicaminum' ('The Power of Medicine'), he portrays Nature as the 'footstool' on which God rests His feet: a powerful image of subordination.[20] Van Helmont emphasises that Nature by itself is unable to confer form, life or individuation; all these come from God.[21]

Moreover, the Flemish philosopher affirms that Nature is possessed by a desire to obey its Maker. This is in fact a natural inclination that springs from within creatures, an aspect that the middle-period Van Helmont had already explained in *De magnetica vulnerum curatione* by the example of the magnet. As he had argued, it

---

14 Du Chesne, *The Practice of Chymicall and Hermeticall Physick*, [6]. On the medieval background of this distinction, see Olga Weijers, 'Contribution à l'histoire des termes 'natura naturans' et 'natura naturata' jusqu'à Spinoza', *Vivarium* 16 (1978): 70–80.

15 Robert Fludd, *Utriusque cosmi maioris, scilicet et minoris metaphysica, physica atque technica historia*, 2 vols (Oppenheim: Theodor de Bry, 1617), I, 19.

16 Michael Sendivogius (Divi Leschi Genus Amo), *De lapide philosophorum Tractatus duodecim* (Frankfurt/Prague, 1604), 15–16: 'Natura autem non est visibilis, licet visibiliter agat, nam est spirito volatilis, qui in corporibus suum agit officium, locum & sedem sua habet in voluntate Divina'.

17 The views of Pierre Gassendi and Robert Boyle were analysed by Margaret Osler, 'The Intellectual Sources of Robert Boyle's Philosophy of Nature: Gassendi's Voluntarism and Boyle's Physico-Theological Project', in *Philosophy, Science and Religion in England*, 1640–1700, ed. Richard Kroll, Richard Ashcraft and Perez Zagorin (Cambridge: Cambridge University Press, 1992), 178–98; for Schegk and the Cambridge Platonists, see Hiro Hirai, *Medical Humanism and Natural Philosophy: Renaissance Debates on Matter, Life and Soul* (Leiden: Brill, 2011), 86–7.

18 David de Planis Campy, *Traicté de la vraye, unique, grande et universelle medecine des anciens, dite des recens Or Potable* (Paris: François Targa, 1633), 47.

19 Ch. 16, 'Vacuum naturae', §28, 71.

20 Ch. 61, 'Potestas medicaminum', §3, 377.

21 Ch. 22, 'Ortus formarum', §96, 120.

was not the earth pole that coerced a magnet to point north; instead, the magnet was driven towards it by an inner desire of obedience.[22] This example remains valid for Van Helmont's later description of Nature. His God is a benevolent ruler that does not coerce, but inspires emulation: everything in Nature wants what God wants.

By virtue of its close relationship with God, Nature must be esteemed and respected. Van Helmont disagrees with views that seek to diminish its dignity. First, he denounces the arrogance of the Scholastics who force Nature to be what they wish it to be.[23] Such presumption is shown in the tendency towards mathematising Nature; this is criticised as forcing an artificial dimension on it.[24] He equally disagrees with overzealous medical alchemists who think Nature always needs perfecting. Instead, Van Helmont concurs with Hippocrates and Galen that Nature has an uncanny ability to regenerate itself with no external help.[25] Thus, medicine should only be given when Nature's powers are exhausted.

Universal Harmony in Nature

To understand the created world, Van Helmont goes back to Genesis 1:31 where it is stated: 'And God saw that whatsoever things he had made, were good'.[26] Armed with this Biblical authority, he similarly concludes that all Creation is 'good' in essence, and that there is 'unity and concord' which reflects the nature of God Himself.[27] Since God created Nature from nothing, He imprinted His own character of divine goodness on it.[28]

Consequently, Van Helmont rejects any idea of opposition in Nature. There can be no real hostility between things, as this would imply that this evil would come from God Himself. Van Helmont is particularly discontent with the idea of strife between the elements, as propounded by the Aristotelians and Galenists of his era. Although elements may be different in action, this does not mean they are contradictory or at war.[29] Since God created the universe in His own image, the result could only be an orderly and harmonious Nature: 'the Father of the Universe, loving concord, hates discord and quarrels, and particularly in the elements'.[30]

The Flemish philosopher further argues that the idea of contrariness in Nature has no ontological reality, being a projection of the human mind, which after the

---

22 Ch. 113, 'De magnetica vulnerum curatione', §68, 606.

23 There are numerous examples of his condemnation of the 'Schools', for instance ch. 25, 'Blas humanum', §1–2, 144, or 'De lithiasi', ch. 7, §1, 692.

24 Ch. 5, 'Causae et initia naturalium', §41, 12.

25 Ch. 59, 'Victus ratio', §39, 364.

26 Ch. 24, 'Natura contrariorum nescia', §47, 142; ch. 99, 'Paradoxum sextum', §2, 555; originally in Genesis 1:31.

27 Ch. 24, 'Natura contrariorum nescia', §38, 139.

28 Ch. 60, 'Pharmacopolion ac Dispensatorium modernum', §45, 373.

29 Ch. 14, 'Gas aquae', §44, 64: 'Non voluit enim benedictus Naturae Parens, quod elementa hostiliter essent opposita & apposita, mutuam perniciem, & voracitatem perpetuo anhelarent'.

30 Ch. 10, 'Terra', §12, 45: 'Ast Pater universi, concordiae amans, discordiam & rixas odit, maximeque in elementis'.

Fall was tainted by hostility and strife.[31] This happens both at an unconscious and a conscious level, as people seek to justify their own aggressiveness. Yet by doing so, men implicitly accuse God of being the creator of contrariety and adversity, which is equally impious and preposterous.[32] In reality, poison, the spittle of a mad dog, or the serpent bite do not harm human beings by contrariety, but by simply doing what God has commanded them to do from the beginning of time.

The idea of universal harmony was current in Van Helmont's time, being one of the strongest legacies of Renaissance thought. It was not only a mainstay of alchemical philosophers such as Severinus, Du Chesne and Fludd, but it was also upheld by Johannes Kepler (1571–1630) and Marin Mersenne.[33] In his beliefs, Van Helmont was in line with the philosophers of his time, except that he was not interested in mathematical and musical harmonies.

The Universal Force: Blas

Van Helmont revered Hippocrates as one of the greatest ancient philosophers.[34] Consequently, the Flemish philosopher was inspired by the ancient physician's concept of the *impetus faciens* (*to enormon*) to devise a comprehensive vision of the force determining movement within the Universe. This is the *blas*, also called *vis enormontica*.[35] The term *blas* probably originates from the Dutch word *blazen*, 'to blow' (thus conceiving this force as a type of wind).[36] Van Helmont justified the novelty of the name by arguing that this power (*vis*) lacked a name, even though, of course, he could have used the Hippocratic term for it.[37]

Although Van Helmont did not give a proper definition of *blas*, he had a rather well-contoured idea in his mind. In the treatise 'Ignotus hospes morbus' ('Disease is an Unknown Guest'), Van Helmont explained:

> [E]very natural agent is born to produce its like, except that which acts by *blas* . . . Thus, Heaven generates meteors, not Heavens; man, by a voluntary *blas* and the *Archeus*, by an ideal and seminal *blas*, can cause diverse alterations.[38]

---

31 Ch. 24, 'Natura contrariorum nescia', §47, 142.

32 Ch. 24, 'Natura contrariorum nescia', §37, 138, §48–9, 142–3.

33 Robert Westman, 'Nature, Art and Psyche: Jung, Pauli and the Kepler-Fludd Polemic', in *Occult and Scientific Mentalities in the Renaissance*, ed. Brian Vickers (Cambridge: Cambridge University Press, 1984), 177–229; Peter Dear, *Mersenne and the Learning of the Schools* (London: Cornell University, 1988), 109–16.

34 See my forthcoming article, Georgiana D. Hedesan, 'From Interpretation to Integration: Van Helmont and His Medical-Alchemical Hippocratism', in *Alchemy and Medicine from Antiquity to the Enlightenment*, ed. Jennifer M. Rampling and Peter M. Jones (Aldershot: Ashgate, forthcoming).

35 Ch. 15, 'Blas meteoron', §5, 65.

36 Albert Roodnat, 'J.B. Van Helmont', in *Dictionary of Gnosis and Western Esotericism*, ed. Wouter J. Hanegraaff (Leiden: Brill, 2006), 471.

37 Ch. 63, 'Ignotus hospes morbus', §62, 398.

38 Ch. 63, 'Ignotus hospes morbus', §62, 398: 'omne agens naturale, natum esse producere sui simile, excepto eo, quod agit per Blas . . . Sic Coelum meteora, non coelos generat. Et homo, per Blas voluntarium; itemque Archeus per Blas ideale, atque seminale, diversas suscitat alterationes'.

Thus, it appears that *blas* is a force of spiritual origin that agents can employ to cause transitive change. Such changes can be simply of position (our ordinary understanding of motion), but can also be transformative. In this sense, *blas* can equally act on the surface of things and penetrate within them.

At first glance, *blas* resembles the late Scholastic notion of *motus* (from Aristotle's *kinesis*), which defined movement as the result of the action of an agent on a patient.[39] Yet Van Helmont views *blas* as more than simply change: it is a power (*vis*) and prerogative of the actor, similar to 'will'. Intentionality seems to be the key to Van Helmont's view. *Blas* can be described as the means by which superior things impose their will on inferior ones.[40] Here again we perceive his Christian philosophy in action, since he uses voluntarism as a unifying concept of natural philosophy.

Blas does not require direct contact; in fact, it often involves action at a distance or 'occult influence'. Van Helmont gives the power that stars have on the sublunary world as an example of how action at a distance works.[41] This can be a direct influence caused by the movement of the heavens, or a subtler and more important one, whereby sublunary things harmonise with or emulate the stars. In *De magnetica vulnerum curatione* (1621), Van Helmont had argued that 'there is a celestial and *enormontical* nature in things themselves, which excites and governs itself according to a superior rhythm'.[42]

Van Helmont further differentiated between 'motive' *blas* (also called 'local' or 'operative') and 'alterative' *blas*. Thus, as he states:

> There are natural agents which have a motive force that I call a motive *blas*, while the agents themselves I call strong movers. But other movers [have what] I call alterative *blas*, which evidently operates by the seminal force of Ferment.[43]

This distinction between 'motive'/'local' and 'alterative' *blas* mirrors the similar division of motion amongst Scholastics.[44] Yet, Van Helmont's *blas* distances itself from its Scholastic origins by describing an intentional impetus rather than simple movement.

In this sense, 'motive' *blas* is a force that determines local motion. The typical example of motive *blas* is again that caused by stars and planets, as well as that caused by

---

39 Dennis Des Chene, *Physiologia: Natural Philosophy in Late Aristotelian and Cartesian Thought* (Ithaca, NY: Cornell University Press, 1996), 17–52. This kinship is also indicated by the 'Explicatio aliquot verborum artis' in the *Opuscula medica inaudita* (Cologne: Jodocus Kalcoven, 1644): '*Blas*, Etymi defectu voco, vim motus, tam alterativi, quam localis.' This 'Explication' probably (but not undoubtedly) originated with Van Helmont.

40 Ch. 45, 'Ignota actio regiminis', §18–19, 266.

41 Ch. 15, 'Blas meteoron', §2–5, 65.

42 Ch. 113, 'De magnetica vulnerum curatione', §60, 604: 'Sat est itaque . . . naturam coelestem & enormonticam rebus ipsis inesse, quae tamen ad rhythmum superioris lationis sese concitet ac moderetur'.

43 Ch. 45, 'Ignota actio regiminis', §5, 262: 'Sunt autem agentium naturalium, quaedam habentia vim motivam, quam Blas motivum dixi: ipsa vero agentia, voco robora moventia. Reliqua autem moventia, Blas alterativum dico, quae videlicet operantur per fermenti vim seminalem'.

44 Cees Leijenhorst, *The Mechanisation of Aristotelianism: The Late Aristotelian Setting of Thomas Hobbes' Natural Philosophy* (Leiden: Brill, 2002), 179–80.

the *Archeus*, the pulse, and 'sympathetic' medicine.[45] The stars exert an 'enormontical, motive force' (*enormonticam motivam vim*), also called a 'pulsive virtue' (*virtus pulsiva*) or a 'masculine' force (*masculum*). This causes seasons, winds, storms and floods.[46]

The activity that produces 'alterations' within things belongs to the other type of *blas*, the alterative one. Examples of alterative *blas* include odours of the principiating Ferments, the Sun and the Moon. The alterative *blas* is particularly connected with the entire process of generation from seed to the adult being.[47]

The most elaborate example of *blas* Van Helmont gives is the one existing in man, the '*blas humanum*'. This *blas humanum* should not be understood as a third category of *blas* besides the motive or alterative, as Pagel presented it, but as an illustration of *blas* in general. To elucidate the nature of the human *blas*, Van Helmont first postulates that there is an internal heaven within us, which can initiate this force.[48] The concept is directly drawn from Paracelsus, who maintained that human bodies have an internal heaven that mirrors the movements of the external heaven, according to his theory of microcosmic-macrocosmic harmony.[49] Van Helmont agreed with the Swiss physician that the human body, created last in Genesis, follows the movements of the stars which were created beforehand. Thus, each organ within our body shadows the movements of a planet.[50]

Van Helmont points out that *blas humanum* becomes manifest in diseases, when the human body precedes and forecasts future tempests. Likewise, the *blas* of the healthy human follows the changes of seasons. Van Helmont classifies this type of unconscious *blas* as one occurring by natural motion (*naturali motu*), as opposed to the conscious *blas* of the will.[51] Yet the two are ultimately not that different: the movements of the inner *blas* also have implied intentionality.

If not the conscious will, what will does the *blas humanum* reflect? To explain, Van Helmont draws on the New Testament, specifically on John and Paul, to distinguish a 'will of the flesh' from the 'will of man' and the 'will of God'.[52] Van Helmont advances the idea that the 'will of the flesh' is an unconscious will which governs an animated body and acts by means of *blas*. It is a first mover that implements its will both in the seed and in the body grown out of the seed.[53]

Van Helmont is careful to explain that this 'will of the flesh' is not inherently evil, despite the negative connotations given to it by Paul. True to his strong voluntarist theology, the Flemish philosopher affirms that all will in Nature ultimately

45 Ch. 76, 'Recepta injecta', §5, 454, ch. 45, 'Ignota actio regiminis', §22, 267.

46 Ch. 15, 'Blas meteoron', §1, 65. I found no support for Pagel's contention that *blas* is a 'gravitational force', as he claims in *Joan Baptista Van Helmont*, 87.

47 Ch. 15, 'Blas meteoron', §6, 66.

48 Ch. 25, 'Blas humanum', §5, 145.

49 Paracelsus, 'Paragranum', 173–7; see also Weeks, *Paracelsus: Speculative Theory*, 120.

50 Ch. 25, 'Blas humanum', §5, 145.

51 Ch. 25, 'Blas humanum', §9, 145.

52 John 1:13, Ephesians 2:3. The 'will of the flesh' was also emphasised by Augustine, *Confessions*, in *The Works of Saint Augustine*, I, 7:9.

53 Ch. 25, 'Blas humanum', §9, 145.

belongs to God Himself. Hence, the 'will of the flesh' is absolutely natural, except in the case of Man. In human beings, the 'will of the flesh' is the origin of idols and envy.[54]

Van Helmont's *blas* can be seen as a universal force that extends, in modern terms, beyond the realm of physics into that of biology and even psychology. By introducing the term *blas*, Van Helmont sought to focus philosophical attention not only on spirit as an agent, but on its action as well. In this sense, it is important to note that, while the spirit appears under different guises (*Archeus*, *Gas*, *magnale*), the spiritual force is always one: the *blas*. Moreover, *blas* belongs not only to quasi-corporeal spirits, but to souls as well, thus clearly transcending the material realm. This is a fascinating development, as it does not clearly differentiate the acts of soul from those of spiritual agents.

The Criticism of Aristotelian Causes

Van Helmont's criticism of the Aristotelian causes, mainly contained in the treatise 'Causae et initia naturalium' ('The Causes and Beginnings of Natural Things'), is essential to an understanding of his natural philosophy. Here he clearly articulates his dualist views of matter and spirit, his understanding of form as life and light, and the voluntarist perspective of Nature.

In her analysis of Van Helmont's attack on Aristotle, Alice Browne wonders whether his criticism should be read as 'displaying knowledge of it in a parody way' or as a 'serious use of Aristotelian concepts against particular statements in Aristotelian science'.[55] In reality, these stances are not mutually exclusive. Overtly, Van Helmont is taking a polemical attitude against Aristotle; at the same time, an examination of his ideas shows that many are tributary to the Greek philosopher, often as interpreted by Thomist Scholastics. Van Helmont does not reject the Aristotelian philosophical framework in itself, but attacks several of its tenets, which in his mind are not suited for a Christian philosophy.[56] He hence attempts to reframe Aristotelian discourse in terms that reflect his own Christian and alchemical beliefs. In doing so, he appropriates those Aristotelian notions that have a strong explanatory power into his idiosyncratic system, with the inevitable consequence that an Aristotelian scaffolding often props up the new philosophy he is proposing.[57] This does not, of course, mean

54 Ch.113, 'De magnetica vulnerum curatione', §74, 607–8.

55 Alice Browne, 'J.B. van Helmont's Attack on Aristotle', *Annals of Science* 36 (1979): 584. Browne concludes by observing that Van Helmont's attack was rather on Aristotle's authority in the university, much more so than on the text of Aristotle, 591. In reality, Van Helmont's relationship with Aristotle is more complex and nuanced than Browne's analysis suggests.

56 Aristotle's theory of causation is enunciated in *Physics* II.3, *Posterior Analytics* II.11, and *Metaphysics* A.3 and Δ.2.

57 This is not unusual in the epoch. As noted in chapter 2.2, Aristotle was being challenged by numerous philosophers but remained influential in debates. See also Charles Schmitt, *Aristotle in the Renaissance* (Boston, MA: Harvard University Press, 1983), 27.

that Van Helmont is a 'closet' Aristotelian, as Pagel described him: his attempt to find alternatives to the Greek philosopher is sincere and often clearly contoured. It is more accurate to say that Van Helmont's philosophy formed in dialogue with, and in reaction to, the late Scholasticism of his age.

Van Helmont's particular approach to Aristotle is apparent in his critique of causation. His discussion begins by clarifying that his analysis only refers to those beings that are 'subject to change'.[58] This assertion should be understood in Aristotelian terms as referring to the domain of physics, and not of metaphysics, the philosophy of immutable things.

The philosophical framework is hence Aristotelian, but the content is not. Van Helmont asserts that his treatment will not apply to the elements, since these 'constantly remain the same as they were from the beginning'; hence they do not fall within the realm of physics.[59] This view of elements is evidently opposed to that of Aristotelians. The exclusion of the elements is part of his programme of affirming both their permanence and their unchanging nature, as will be shown further in the section below on Matter.

As befits a Christian philosopher, Van Helmont makes it clear from the beginning that God is the ultimate first cause of everything that exists. Although in many cases 'ordinary' causes can be found to explain a certain phenomenon, behind these there is always the fundamental cause which is God.[60] God is never very far, as He infuses 'the knowledge of ends and dispositions' into the natural efficient cause. Moreover, God is the efficient cause of life, the only one who can imprint a 'light' into the seed.[61] Or, as Van Helmont explains it, 'God concurs with the generation of beings as the universal, independent, total, essential, and efficiently efficient cause; any other created being concurs only as the dependent, partial, particular, and dispositively efficient cause'.[62]

The Flemish philosopher proceeds to analyse each of the four Aristotelian causes (material, formal, efficient and final). He begins with the material cause of individual beings. First, Van Helmont clarifies that Creation as a whole does not have a material cause, as it springs directly from God.[63] However, in ordinary beings, he accepts that the material cause is necessary to account for the generation of bodies.

---

58 Ch. 5, 'Causae et initia naturalium', §3, 27: 'Causas ergo rerum naturalium, intelligo vicissitudinem pressuponere'.

59 Ch. 5, 'Causae et initia naturalium', §2, 27: 'hodie constanter quoque maneant eadem, quae ab initio fuerunt'.

60 Ch. 17, 'Meteoron anomalum', 73, ch. 5, 'Causae et initia naturalium', §12, 28.

61 Ch. 5, 'Causae et initia naturalium', §20, 29.

62 Ch. 22, 'Ortus formarum', §16, 107–8: 'Deus ergo, ad generationem entis, concurrit ut causa universalis, independens, totalis, essentialis, & efficienter efficiens; sed ens creatum concurrit, ut causa dependens, partialis, particularis, & dispositive efficiens'. God as ultimate efficient cause was a medieval topos; see Nicholas H. Steneck, *Science and Creation in the Middle Ages: Henry of Langenstein (d. 1397) on Genesis* (London: University of Notre Dame Press, 1976), 92. However, the common understanding was that His will was done by means of the heavenly bodies, while Van Helmont favours direct divine intervention.

63 Ch. 5, 'Causae et initia naturalium', §2, 27.

Van Helmont rejects Aristotle's second cause, the formal cause. For the Flemish thinker, form is not an active agent—a function which is reserved for principles alone. He makes the following argument:

> Surely, the form of a composite thing cannot be the cause of a produced thing: rather, it is the ultimate entelechy of generation, the essence and perfection of the generated thing, for the attainment of which all other matters are directed. Therefore I meditate that the form is as an effect, rather than a cause of things.[64]

In other words, for Van Helmont the form is the outcome of the intervention of a principle in matter. The form is described as 'entelechy' or actuality, which for Van Helmont is a type of blueprint that is established prior to generation and actualised by the *Archeus*, the efficient internal agent. Since form is perfection, it is essentially 'passive'.

Van Helmont accepts the third Aristotelian cause, the efficient (moving) cause. However, this cause is criticised primarily because it proposes an external source as the origin of motion.[65] For Van Helmont, this is preposterous since essential movement in fact originates from within: 'the total efficient cause in Nature is inward and essential'.[66] In other words, the physical, procreating father is not the true efficient cause of an offspring, but a 'remote' cause: the direct cause is the seed and the *Archeus* contained in the body.

Finally, Van Helmont criticises Aristotle's fourth, or final cause for being distinguished from the efficient cause. This opinion should be correlated with Van Helmont's theory of the seeds, which contain all the information needed to frame an individual before birth.The purpose or finality of a body is already inscribed in the efficient cause. As Van Helmont explains: 'Since the efficient contains all ends within, as in instructions of things to be done by it, the final external cause of the Scholastics, which exists only in artificial things, cannot be found in Nature'.[67]

Van Helmont's criticism of the Aristotelian final cause is consonant with the documented rebellion of early modern philosophy against Aristotelian finality.[68] Yet his intention was fundamentally different from that of mechanical philosophers: he

---

64 Ch. 5, 'Causae et initia naturalium', §5, 27–8: 'certe, forma compositi, causa esse nequit producti; sed potius Entelechia ultima generationis, ipsissimaque generati essentia, atque perfectio: propter cujus affectionem, caetera cuncta diriguntur. Itaque formam, potius tanquam effectum, quam ut causam rei meditor'.

65 Aristotle, 'Generation of Animals', in *The Complete Works of Aristotle* I, 1.

66 Ch. 5, 'Causae et initia naturalium', §9, 28: 'Tota alioqui causa efficiens, est interna, essentialis, in natura'.

67 Ch. 5, 'Causae et initia naturalium', §12, 28: 'Fines denique cum omne efficiens in se contineat, tanquam rerum à se agendarum instructiones, ideo finalis causa Scholarum externa, quae duntaxat in artificialibus locum habet, vana prorsus est in natura'.

68 Jeffrey K. McDonough, 'The Heyday of Teleology and Early Modern Philosophy', *Midwest Studies in Philosophy* 35 (2011), 179–204.

was not rejecting teleology in itself. Instead, his criticism stemmed from his belief that Aristotelian thought did not emphasise teleology sufficiently. For Van Helmont, the *telos* is the fundamental reason of existence.[69]

Consequently, Van Helmont's critique of Aristotelian causation results in the reduction of the four causes to only two: the material and the efficient. This is done by merging the efficient and final causes, and by eliminating form as cause of change. Van Helmont's double causation is consonant with his theory of the duality of matter and spirit. Thus, it separates the passive matter from the active agent, with the latter being the sole container of the information needed to shape and order beings. Van Helmont could have eliminated the 'material cause' completely; indeed, there seems to be tension between his preservation of the material cause and his firm belief that matter cannot be active in any form. What he means is that, although matter does not act, it enters into the composition of other beings as part of the generative process.

Matter

Genesis and the Elements

Van Helmont deems matter to be synonymous with the elements.[70] His understanding of these is drawn on the written authority of the Bible, more specifically the first books of Genesis. Commentaries on Genesis were nothing new: 'hexameral' exegeses had begun in Late Antiquity and extended up to Van Helmont's day.[71] Yet Genesis commentaries held a special place in Paracelsian philosophy. Alchemical thinkers believed that the Book of Genesis, thought to be authored by Moses himself, was a source of natural knowledge, and sought to elucidate it in alchemical terms.[72] Their

---

69 It must be noted that Aristotle would have agreed that the final does not always differ from the efficient cause in practice; see for instance *Physics* 198a25. However, for Aristotle the distinction of final from efficient causes is discovered by reasoning, while for Van Helmont such inquiry makes no sense.

70 Ch. 13, 'Progymnasma meteori', §11, 56: 'Elementa vero, sunt corpora, non autem spiritus'.

71 For a review of the hexameral tradition in Late Antiquity and the early Middle Ages, see Grant, *Science and Religion: 400 BC—AD 1500,* 114–137; also P. Nautin, 'Genèse,1, 1–2, de Justin à Origène', in *In principio. Interpretations des premiers versets de la Genèse* (Paris: Centre d'Etudes des religions du livre, 1973), 61–94. In the late sixteenth century the most popular hexameral commentary was that of Benito Pereira; see William Poole, 'Francis Lodwick's Creation: Theology and Natural Philosophy in the Early Royal Society', *Journal of the History of Ideas* 66 (2005): 256.

72 This was first signalled by Allen Debus in his *The Chemical Philosophy*, 56. See also my article, Hedesan, 'The Mystery of *Mysterium Magnum*', as well as Michael T. Walton, 'Genesis and Chemistry in the Sixteenth Century', in *Reading the Book of Nature*, ed. Allen G. Debus and Michael T. Walton (Kirksville, IL: Truman State, 1998), 1–14; Peter J. Forshaw, 'Vitriolic Reactions: Orthodox Responses to the Alchemical Exegesis of Genesis', in *The Word and the World: Biblical Exegesis and Early Modern Science*, ed. Kevin Killeen and Peter J. Forshaw (Basingstoke: Palgrave Macmillan, 2007), 111–36; Didier Kahn's 'L'interprétation alchimique de la Genèse chez Joseph Du Chesne', 641–92; and Michael T. Walton's *Genesis and the Chemical Philosophy : True Christian Science in the Sixteenth and Seventeenth Centuries* (New York: AMS Press, 2011).

exegetical tradition was chiefly based on one of Paracelsus's writings, the spurious *Philosophia ad Athenienses.* This portrayed God as creating the world in an alchemical manner, through separation (*Scheidung*). Other writings of Paracelsus (considered authentic) further attempt to interpret Genesis alchemically.[73]

Yet it was the early Paracelsian alchemist Gerhard Dorn (1530–1584) who set the tone for the vogue of the alchemical interpretation of Creation. His lead was followed by numerous other Paracelsians, including Richard Bostocke, Joseph Du Chesne, Timothy Willis, and Robert Fludd. The interest in interpreting Genesis alchemically was a key part of the Paracelsian attempt to harmonise natural philosophy with religion.

Van Helmont also set great store by his Biblical interpretation, and particularly that of the Book of Genesis. Indeed, his physical interpretation of the account of Creation must be considered fundamental to his Christian philosophy: it was the cornerstone on which he based his theory of matter and spirit.

Genesis 1:2 gives Van Helmont the conviction that matter and spirit are fundamentally different. According to this verse, 'the earth was without form, and void (Heb. *tohu-va-bohu*); and darkness [was] upon the face of the deep (Heb. *tehom*). And the Spirit of God (Heb. *Ruach Elohim*) moved upon the face of the waters'. Van Helmont interpreted the 'deep' and 'the waters' to refer to the primeval material chaos. He considered this primordial matter to be fundamentally passive and activated only by the intervention of the Spirit of God.[74]

Importantly, Van Helmont does not think in Aristotelian terms that passivity and lack of movement are negative aspects. On the contrary, he believes that 'the desire for permanence is older, stronger, and more natural than the desire of permutability'.[75] The reason for this, from his perspective, is that 'being' is more basic than 'change', an idea which betrays a strong Platonic bias. Indeed, Van Helmont emphasises that all beings seek to conserve themselves, and change only brings about their destruction. He considers self-love (*philautia* or the principle of conservation) to be natural in matter.[76]

Van Helmont considers the Spirit of God as being the archetype for all spiritual agents. He further theorises that the world contains 'principles' (*initia* or *principia*) that bestow life, movement and order to matter.[77] Such principles cannot be material.

---

73 For instance, in Philippus Theophrastus Bombast von Hohenheim Paracelsus gennant, 'Philosophia de Generationibus et Fructibus quatuor Elementorum', 54–159, and 'Liber de mineralibus', 334–60, in *Achter Theil Der Bucher und Schriften*, ed. Johannes Huser (Basel: Conrad Walkirch, 1590).

74 'De lithiasi', ch. 1, §5, 650–51.

75 Ch. 23, 'Magnum oportet', §12, 122–3: '. . . desiderium permanendi, antiquius, fortius, & naturalius sit, appetitu permutabilitatis'.

76 See Giglioni, *Immaginazione e malattia*, 55–56; more generally Cesare Vasoli, 'L'amorosa filosofia: dall'amore platonico all' universale *philautia*', in *Francesco Patrizi da Cherso* (Rome: Bulzoni, 1989).

77 Ch. 5, 'Causae et initia naturalium', §3, 27. His views betray the influence of the Aristotelian concept of the form.

Van Helmont attributes this erroneous idea to Aristotle and contends that the latter confuses the principle with the material cause.

Having thus established the nature of matter and spirit, Van Helmont seeks further enlightenment on the elements from the Bible. He notes that, according to Genesis 1:1, the Heaven (*Coelum*) and the Earth (*terra*) were made *in principio*, prior to the first day.[78] He next argues that the term 'Heaven' (*shamayim*, in Hebrew) had an esoteric meaning that had to be decoded. This mysterious term was the subject of intensive speculation by Renaissance Christian Cabalists and alchemists. Van Helmont relates it with the mention of 'waters' in the same verse, as well as with Genesis 1:6.[79] He observes that the Biblical account does not discuss the creation of water, so he concludes that the element must have initially been indistinguishable from 'heaven'. He believes that this reading is further confirmed by the mention of the 'face of the deep' and 'face of the waters' in Genesis 1:2. He concludes that 'the deep' referred to the confused mixture of waters above and below Heaven.[80]

Van Helmont enhances his argument by translating the Hebrew word *shamayim* as 'where waters are (or can be found)'.[81] In doing so, he moves away from traditional Cabalist interpretations, which ultimately drew on the Jewish *Midrash Bereshith Rabba*, according to which *shamayim* was a composite word made of *esh* (fire) and *mayim* (water).[82] The opinion that the heaven was originally made up of fire-water was upheld in the early modern period by Johannes Reuchlin, Heinrich Khunrath and Johann Heinrich Alsted.[83] This view accorded well with those alchemists that viewed *shamayim* as a divine *quinta essentia*, or secret substance that could be harnessed in the laboratory.[84]

Yet Van Helmont refused to follow suit.[85] It is possible that his choice may have been inspired by a diffidence toward theosophical ideas, so harshly condemned by Marin Mersenne and the Jesuits. At the same time, Van Helmont's view of *shamayim*

---

78 According to Genesis 1:1, 'In the beginning God created the heaven and the earth'. In Latin, 'in principio creavit Deus caelum et terram' (*Biblia Sacra Vulgata*).

79 Genesis 1:6–1:8: 'And God said, Let there be a firmament in the midst of the waters, and let it divide the waters from the waters. And God made the firmament, and divided the waters which were under the firmament from the waters which were above the firmament: and it was so. And God called the firmament Heaven'.

80 Ch. 9, 'Elementa', §3, 42.

81 Ch. 9, 'Elementa', §4, 43: 'ideoque Coelum Ebraeis (ubi sunt aquae) sonat'.

82 *Midrash Bereshith Rabba in Hebrew and English* (Jerusalem: Shalom Publications, 1988), 1:31.

83 Johannes Reuchlin, *On the Art of the Kabbalah*, trans. Martin and Sarah Goodman (Lincoln, NA: Abaris Books, 1983), 99; Heinrich Khunrath, *Vom Hylealischen . . . Chaos* (Magdeburg: Johann Francken, 1616), 39, and *Amphitheatrum Sapientiae Aeternae*, 129; Johann Heinrich Alsted, *Lexicon theologicum* (Hanover: Eifridus, 1634), 108.

84 See, for instance, Heinrich Khunrath's elaborate description of *Schamaim* as *aether*, the quintessence of Heinrich Cornelius Agrippa, a divine fiery water, entelechy, form, and *Ruach Elohim*, the Spirit of God; *Vom Hylealischen . . . Chaos*, 39–40.

85 Did Van Helmont know Hebrew? He seems to imply this; however, he may have also extracted his ideas from Christian Cabalists, so no certain verdict can be given. Indeed, the term *shamayim* was rather common amongst alchemists and Christian Cabalists, and we can presume that not all knew Hebrew.

as water accords well with his entire philosophical project. Based on this interpretation, he concludes that the initial substance of creation was water. Water was the first element and the fundamental matter of all beings. This heavenly water lay both above and below, hence the lower waters of the earth are, and continue to be, akin to the heavenly ones.[86]

Having thus decoded the word *shamayim*, Van Helmont adds the surprising remark that the term also signifies air. Although he does not as clearly explain how he etymologically derived 'air' from *shamayim*, he may have been aware that the Hebrew version of Genesis 1:26 uses *shamayim* to indicate air.[87] Another possible source of inspiration may have been the *Turba philosophorum*, according to which heaven (*coelum*) was made up of water and air.[88]

Air and water are, then, the two primordial elements.[89] Van Helmont strives to prove that the other two elements, earth and fire, are secondary. Fire, he insists, is never mentioned in the Genesis account as being created; therefore, it cannot be a true element as Aristotle believed. Van Helmont was not unique in this assertion; many Paracelsians thought that fire did not naturally lie in the sublunary world, but was a force of celestial origin.[90] Nevertheless, Van Helmont equally rejects Paracelsus's explanation that fire appears in the Genesis account in the form of light and stars, and consequently can be deemed a superlunary element.[91] His refutation is linked with his belief that an element must have been 'first-born' (*primigenius*); that is, framed *ab initio*, before the first day.

Earth too is a secondary element. This opinion was harder to justify in terms of Genesis 1:1, where Heaven and earth are mentioned together. However, Van Helmont argues that the fact that earth is mentioned after Heaven means that the former is created later, and is hence inferior.[92] Implicitly, the element water is also superior to earth, being simpler and more heavenly.

Still, Van Helmont concedes that the Genesis account is insufficient to ascertain what earth is. Consequently, he appeals to his alchemical knowledge to demonstrate that earth is not elementary. Indeed, his 'art of the fire' (*pyrotechnia*) can prove that all things, including minerals, earth and metals can be transformed into water by the

---

86 Ch. 9, 'Elementa', §6, 43.

87 Genesis 1:26.

88 *Turba philosophorum*, I, 4: 'Creavit Deus coelum ex aqua & aere, ita coelum quoque ex duobus est compositum'. This saying is attributed to Pythagoras.

89 Van Helmont calls them *primigenia*, first born things, to emphasise their superiority to the earth, ch. 9, 'Elementa', §7, 43.

90 On the views of Paracelsus, see Pagel, *Paracelsus*, 92–3; on Heinrich Khunrath's concepts of divine fire and light, see Urszula Szulakowska, *The Alchemy of Light: Geometry and Optics in Late Renaissance Alchemical Illustration* (Leiden: Brill, 2000), 94–6; also Joseph Du Chesne, *The Practise of Chymicall and Hermeticall Physicke*, ch. 11.

91 Ch. 9, 'Elementa', §9, 43.

92 It appears that Van Helmont read 'the heavens and the earth' as meaning 'the heavens and then, the earth'.

means of the 'Circulate Salt' of Paracelsus or the *Alkahest*. Experiment can show that earth is nothing but an offspring of water.[93]

In his analysis of the quiddity of earth, Van Helmont clearly assumes that the knowledge acquired in the laboratory complements his reading of Genesis: 'Thus I have learned by the fire that God, before there was day, created water and air, and of the water an elementary earth, which is the *Quellem* sand'.[94] Knowledge given by the fire is thus taken to illuminate Scripture. We see here a prime example of how Van Helmont conceived the Book of Nature to be in harmony with the Book of Grace, the Bible.

However, there was one difficulty to this process of harmonisation, which he addresses in the treatise 'Complexionum atque mistionum elementalium figmentum' ('The Fiction of Elementary Complexions and Mixtures'). Here he observes that the Bible affirms 'Thou art Earth and into Earth thou shalt go', which could pose a serious objection to his theory of the constitution of matter.[95] He responds that this verse should not necessarily be read to mean that bodies are fundamentally made of earth. Instead, it simply means that they more closely resemble earth than water. This reading highlights the fact that Van Helmont is determined to affirm the unity of the Book of Scripture with that of Nature at all costs. To achieve this harmonisation, he is willing to interpret the Bible metaphorically, going beyond a simple and literal reading of it. We can conclude that, at least occasionally, alchemical theory and practice does influence Van Helmont's reading of Genesis.

The Flemish philosopher further interprets the primordial character of the elements to mean that they must be simple and indestructible, being appointed by God to bestow constancy upon the Universe.[96] This view of elements leads him to three important conclusions. First, he upholds the theory of the conservation of matter. Thus, he affirms that 'the water which existed from the beginning of the Universe is the same, undiminished, and shall remain so until the end of the world'.[97] Secondly, the theory leads him to maintain that any subtilisation of the element results in 'trituration', or fine division into particles. Corpuscularianism follows as a logical consequence; indeed, he employs the notion of the 'atom' of water to explain its behaviour.

Finally, the fundamental unchangeability of the elements also means that they cannot be converted into one another.[98] The concept of the mutability of the elements, formulated by Aristotle, was generally accepted by medieval alchemy but not

93 Ch. 13, 'Progymnasma meteori', §20, 57.

94 Ch. 9, 'Elementa', §12, 43: 'Didici ergo per ignem, quod Deus, ante diem, creaverit aquam, & aerem, & ex aqua terram elementalem, quae est arena *Quellem*'. Van Helmont describes '*Quellem* sand' as being the innermost layer of soil, which no mattock can penetrate.

95 Ch. 19, 'Complexionum atque mistionum', §43, 90. See Genesis 3:19: 'for dust thou [art], and unto dust shalt thou return'.

96 Ch. 13, 'Progymnasma meteori', §5, 55.

97 Ch. 14, 'Gas aquae', §29, 63: 'Sicque eadem, & non imminuta aqua quae ab initio universi, exstitit est; eritque in ejus finem'.

98 Ch. 10, 'Terra', §12, 45; ch. 12, 'Aer', §3, 51; ch. 14, 'Gas aquae', §44, 64. For Aristotle's views of the cyclical transformation of the elements, see *Metaphysics*, 1050b29–30, *Physics* 255b5–31, *On Generation and Corruption*, 331b4–11–34.

by Paracelsus, who viewed elements as passive matrices or wombs of things.[99] Yet, as medieval alchemy became more fashionable in the early modern period, so did the traditional view of the elements. It stood at the core of Sendivogius's philosophy, where the transformation of water into air and vice-versa was of particular importance.[100] Influenced by Sendivogius, other influential alchemical philosophers took up this idea, including Cornelis Drebbel (1572–1633) and Heinrich Nollius.[101]

Van Helmont strongly disagreed with the principle of the mutability of the elements.[102] However, instead of attacking fellow alchemists, he preferred to condemn the Scholastics for their belief that air could be changed into water through condensation. If such a theory were true, he argued, air compressed by a bullet in a pistol would turn into water, something which did not happen even in wintry conditions.[103] Clearly air and water were stable, and this was due to their lack of desire, appetite, appointment or necessity of changing one's nature.[104]

In Van Helmont's system, elements may not have degrees or proportions, but do have properties. Thus, he estimates that both air and water are cold in nature, contrary to Aristotelian doctrine, which viewed air as hot and water as cold.[105] Coldness originates from their reduced vitality; thus, despite his formal rejection of heat as a sign of life, Van Helmont seems to follow the Aristotelian tradition of the 'vital heat' when it comes to the elements.[106]

Temperature is the only point of similarity between water and air. Van Helmont stresses that each element is fundamentally different from the other. As always, his arguments are buttressed by Biblical quotations. The air must be dry rather than moist, since in the aftermath of the flood, God sent winds to dry the face of the Earth.[107] Moreover, since Creation, air has continuously acted as separator of the waters, a fact which Van Helmont understands to mean that the air must be dry and exceedingly cold.[108]

Finally, each element has its specific trait: water can become earth by means of seed, while the air's peculiarity is that it has *vacua*, or empty spaces into which water in the form of vapour can be insinuated.[109] This specific character is called an

99 See Daniel, 'Invisible Wombs', 134–6; Weeks, *Paracelsus*, 127.

100 Sendivogius, *De lapide philosophorum*, 76–81.

101 Heinrich Nollius, *Naturae sanctuarium quod est physica hermetica* (Frankfurt: Nicolas Hoffmann, 1619), 112; Cornelis Drebbel, *Ein kurtzer Tractat von der Natur der Elementen* (Leiden: Henrich von Haestens, 1608), [3]–[5].

102 Although the late Van Helmont did not acknowledge Paracelsus's influence in this theory, this is evident in 'De Spadanis fontibus' where he calls elements 'matrices'; ch. 95, 'Paradoxum alterum', §2, 547–8. Interestingly, Van Helmont accepted the (ultimately Aristotelian) idea of the transformation of water into earth but, in order to save his theory of elementary immutability, rejected earth's status as element.

103 Ch. 12, 'Aer', §3, 50.

104 Ch. 14, 'Gas aquae', §44, 64.

105 Ch. 12, 'Aer', §10, 51.

106 On vital heat in Aristotle, see Gad Freudenthal, *Aristotle's Theory of Material Substance: Heat and Pneuma, Form and Soul* (Oxford: Clarendon Press, 1995), 6–72.

107 Ch. 13, 'Progymnasma meteori', §1, 54, referring to Genesis 8.

108 Ch. 14, 'Gas aquae', §40, 64.

109 Ch. 12, 'Aer', §10, 51.

'elementary privilege' (*elementali privilegio*), or a particular appointment given by God.[110] Here again we may notice Van Helmont's emphasis on the specific, *quidditas*, and his voluntarist view of Nature: the elements act in very specific ways because they are so commanded by God.

The Element of Water

Van Helmont firmly believed that water is the origin of all bodies, or the material cause of all sublunary things.[111] Water gives birth to bodies by means of the seed, the efficient cause. The seed first changes water into a viscous substance called *Leffas*, the primordial juice of plants, *Bur*, the thick liquid of metals, or the sand *Quellem*, the original earth.[112] Conversely, all bodies can eventually be resolved into the element of water, whence they came.

Van Helmont adduces experimental proof of this contention. He describes, in detail, his famous 'willow tree experiment', heralded by many historians of science as one of the first quantitative experiments of chemistry.[113] This experiment arose out of Van Helmont's reading of Nicholas of Cusa's *The Idiot*, where a similar experiment is described.[114] However, it is unclear if Cusa himself undertook it, apparently considering it only theoretically.[115]

The belief in the primacy of water has essential implications for Van Helmont's alchemical practice. Although he acknowledges the importance of the fire in altering bodies, he believes that meaningful transmutation is achieved by means of dissolution. Van Helmont emphasises the alchemical maxim according to which 'bodies cannot be changed into each other, unless they are first reduced into their first, viscous matter'.[116] This liquefaction was primarily to be obtained through the action of

110 Ch. 14, 'Gas aquae', §44, 65.

111 Ch. 5, 'Causae et initia naturalium', §31, 30.

112 Ch. 9, 'Elementa', §12–13, 43; ch. 10, 'Terra', §3, 44.

113 Ch. 19, 'Complexionum atque mistionum', §30, 88: 'Caepi enim vas terreum, in quo posui terrae in clibano arefactae lb 200, quam madefeci aqua pluvia, illique implantavi truncum salicis, ponderantem lb 5 ac tandem exacto quinquennio, arbor inde prognata, pendebat 169 lb & circiter uncias tres. Vas autem terreum, sola aqua pluvia, vel distillata, semper (ubi opus erat) maduit, eratque amplum, & terrae implantatum, & ne pulvis obvolitans terrae commisceretur, lamina ferrea, stanno obducta, multoque foramine pervia, labrum vasis tegebat. Non computavi pondus foliorum quaterno autumno deciduorum. Tandem iterum siccavi terram vasis, & repertae sunt eadem librae 200 duabus circiter unciis minus. Librae ergo 164 ligni, corticum, & radicum, ex sola aqua surrexerant'.

114 Nicholas of Cusa, *The Idiot in Four Books: The First and Second Wisdome. The Third of the Minde. The Fourth of Statick Experiments, or Experiments of the Balance* (London: William Leake, 1650), 188–9.

115 For a description of Cusa's proposal, and antecedents, see Pagel, *Joan Baptista van Van Helmont*, 53–6.

116 Ch. 52, 'Latex humor neglectus', §45, 306: 'Quod corpora non mutentur in invicem, nisi prius reducta sint in primam sui & sequacem materiam'. This edict, also known as *Sciant artifices*, originates from the Latin translation (and modification) of Avicenna's *De congelatione et conglutinatione lapidum*, which was at the time considered to be authored by Aristotle. See William R. Newman, *Promethean Ambitions: Alchemy and the Quest to Perfect Nature* (Chicago, IL: University of Chicago Press, 2004), 43–44.

the liquor *Alkahest*.[117] Elsewhere he suggests other simpler ways of converting matter to a liquid state: for instance, he argues that all oils, sulphurs and fats can be neutralised by alkali salt, transformed into soap and then distilled into water.[118]

Since water lies at the root of all bodies, it is not surprising that Van Helmont believed it could be found everywhere. In a blatant rejection of the traditional Aristotelian-Scholastic system, water stands at the centre of the Earth.[119] Thus, Van Helmont argued that there was a

> . . . unique fountain and spring of waters, which [God] has placed in the middle and extremity of the Earth, and is afterwards diffused in a thousand veins, which almost everywhere pierces the Globe of the Earth, for far better uses.[120]

This view of the centre of the Earth also runs counter to the predominant version of alchemical philosophy at the time, according to which the centre of the Earth was hollow and fiery. This theory had been advanced by Sendivogius and was well received in the period, especially by Athanasius Kircher (1601–1680).[121]

Since water was the fundamental matter out of which all bodies were formed, the Flemish philosopher was keen to perform a complete analysis of the element. Van Helmont believed that water was found in the air in the form of vapour and *Gas*, and was the cause of 'meteors', or atmospheric phenomena. Under the influence of heat, water changed into vapour and rose to the higher region of the air. There, it rarefied into its smallest atoms; yet, under the impact of cold air, it finally condensed and fell upon the earth as water or dew.[122] This was obtained by the alteration of the water atom's inner parts (*tria prima*) outward or inward.[123]

It was commonplace knowledge that heated water turned to vapour; however, where Scholastics believed that this vapour was air, Van Helmont argued that it was still water in a different state. This was easily proven alchemically by condensing vapours back into water by the use of a still head; Van Helmont showed that the liquid thus obtained suffered no loss of weight.[124]

---

117 Ch. 13, 'Progymnasma meteori', §7, 55, ch. 61, 'Potestas medicaminum', §44, 384.

118 Ch. 19, 'Complexionum atque mistionum', §27, 88.

119 Ch. 10, 'Terra', §8–9, 45. The Scholastics followed Aristotle in imagining that the sphere of the earth lay at the centre of the Universe and that water formed a sphere surrounding it; Steneck, *Science and Creation*, 78–80.

120 Ch. 11, 'Aqua', §2, 46: 'Unicus demum fons, & aquarum scaturigo, quam in terrae meditullio atque cacumine collocaveras, est deinceps in venas mille diffusus, quae ubique fere, terrae globum terebrant, ad longe meliores usus'.

121 Debus, *The Chemical Philosophy*, 89–93.

122 Ch. 13, 'Progymnasma meteori', §7, 55. For Van Helmont's conception of atoms and corpuscles, see Newman, 'The Corpuscular Theory', 189–91.

123 Ch. 14, 'Gas aquae', §8–9, 60. For the role of the *tria prima* in Van Helmont's physical explanations see below, section 'The *Tria Prima*: Critique and "Corpuscularian" Use'.

124 Ch. 13, 'Progymnasma meteori', §1, 54.

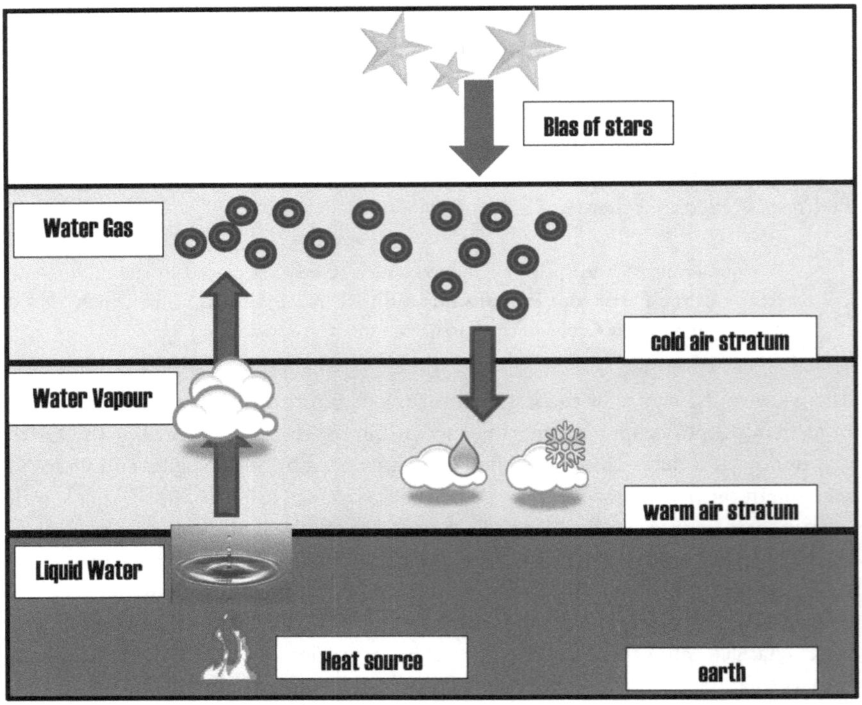

Figure 5.1 Van Helmont's view of the transformations of water in higher atmosphere

The Flemish philosopher further argued that, although water was naturally passive, it did not completely lack sensitivity. Van Helmont believed that 'water has a certain kind of sense and choice, for all beings participate in life in some manner. *Come let us worship the King by whom all things live*'.[125] The 'choice' of water was revealed by the fact that it preferred to be transformed into a vapour rather than a *Gas*. Moreover, water actively protected itself by creating an icy crust when it encountered cold air.

In his speculation on the 'sense' and 'choice' of water Van Helmont appears to have been influenced by the thought of Tommaso Campanella (1568–1639), who had maintained that all things possess sense.[126] Indeed, in his middle-period *De magnetica de vulnerum curatione*, Van Helmont had claimed that most things have sense

125 Ch. 14, 'Gas aquae', §36, 63: 'Quod aqua aliqualem sensum habeat, & electionem, adeoque omnia entia quodammodo vitam participent. Regem cui omnia vivunt, venite adoremus'. This expression originates from the Catholic Matins Office of the Dead.

126 Tommaso Campanella, *De sensu rerum et magia* (Frankfurt: Egenolphus Emmelius, 1620).

by virtue of their tendency for sympathy and antipathy.[127] This is repeated in the *Ortus*, while 'Tumulus pestis' connects this concept with *philautia* (self-love) and the ability to project or receive images, which Van Helmont saw as the basis of sense.[128]

Yet Van Helmont did not equate this sensitivity with life. Indeed, in his system, water cannot be alive because it is pure matter without form. Instead, it might be more accurate to say that water fosters life, and perhaps also 'participates' in life by virtue of having been directly created by God, the vital source.

Van Helmont's Elementary Water, Paracelsianism and Alchemy

Although Van Helmont primarily attributes his understanding of elementary water to his close reading of the Book of Genesis, alchemy clearly contributed to his views. A source of Van Helmont's idea may have been Paracelsus, particularly the Pseudo-Paracelsian text *De Pestilitate*.[129] According to this treatise, all things sprang from 'the mother of all things', water, corresponding to the 'waters' of Genesis.[130] However, water cannot give birth to anything without a seed, which is described as 'fire' or the alchemical principle of 'sulphur'. This quasi-sexual view of the procreation of all things closely fits Van Helmont's theory of generation. The result of the impregnation of the water matrix by the seed also resembles Van Helmont: *De Pestilitate* maintains that the water turns into a mucillaginous substance, which it calls 'the sperm of frogs'.[131] This viscous first matter recalls the *Leffas*, *Bur* and *Quellem* of Van Helmont.

The idea of water as the matrix of all things occurred in authentic Paracelsian treatises. In 'De mineralibus', which Van Helmont briefly mentions, Paracelsus describes all minerals as created from the element of water.[132] In 'Paragranum', he further commented on the Genesis account of the Spirit of God hovering over the waters to state that 'water was matrix, for in water heaven and earth were created'.[133]

These Paracelsian statements fit well with those of several authoritative medieval treatises, notably the *Rosarium philosophorum*. According to this anonymous work, Genesis showed that 'all things are created out of water'.[134] Based on such authorities, the idea that water is at the root of all matter was widespread amongst the Paracelsian alchemists of the late sixteenth and early seventeenth century.[135] One can find this

127 Ch 113, 'De magnetica vulnerum curatione', §32, 599.

128 Ch. 69, 'Progreditur ad morborum cognitione, §13, 430, 'Tumulus pestis', ch. 9, 'Imaginantur Mineralia & Herbae suo modo anomalo', 855.

129 Sudhoff has classified this work as spurious; Karl Sudhoff, 'Vorwort zum 14. Band', in Theophrastus von Hohenheim gen. Paracelsus, *Sämtliche Werke*, I, 14 vols (München and Berlin: Oldenbourg, 1933), 33.

130 Paracelsus, 'De Pestilitate', 30.

131 Paracelsus, 'De Pestilitate', 35.

132 Paracelsus, 'Liber de mineralibus', 336.

133 Paracelsus, 'Paragranum', 159.

134 *Rosarium philosophorum*, in *Artis auriferae quam chemiae vocant*, II, 281.

135 Charles Webster, 'Water as the Ultimate Principle of Nature', 98 points out that this tenet was popular amongst sixteenth-century alchemists.

contention, for example, in Michael Sendivogius's *Novum lumen chymicum* (1604), Basil Valentine's *The Triumphant Chariot of Antimony* (1604), Willem Mennens's *De aureo vellere* (1604), Alexander von Suchten's *Liber de tribus facultatibus* (1606), Oswald Croll's *Basilica chymica* (1609) and Thomas Willis's *The Search for Causes* (1616).[136] It is however important to note that many of these seemed to refer to 'water' in a metaphorical sense; what they termed 'water' was actually a primordial mercurial substance.

Van Helmont recognised the influence of medieval alchemy on his views of water. Thus, he stated that he learnt the nature of water by 'the caduceus of Mercury',[137] referring to the medieval alchemist Geber as an important inspiration.[138] Yet he must have been aware of Paracelsian precedents as well. His tactic of acknowledging the influence of medieval alchemy while omitting modern sources was undoubtedly part of the general intent of the *Ortus* to set himself apart from Paracelsus and his contemporary followers.

Geber was the author of the influential 'mercury-alone theory', according to which all metals had a common root, philosophical mercury (a watery metallic substance analogical but not identical with quicksilver). Linked to this was Geber's double-sulphureity theory of metals, according to which each individual species of metal was composed of external sulphur and philosophical mercury; the latter hid within its core an inner sulphur. As Newman has already pointed out, Van Helmont subscribed to this theory, which he saw in corpuscular terms.[139]

Van Helmont thought that the inner sulphur lay within the 'nucleus' of a mercurial particle.[140] This particle was conceived in spatial terms, in a manner that Newman traced back to Paracelsus.[141] Van Helmont believed that the inner sulphur could not be removed by any solvent. By comparison, the external sulphur, which was the carrier of 'ferments, putrefactions by heat, odours, specifical seminal flavours, for any types

136 Basil Valentine, *The Triumphant Chariot of Antimony* (London: Thomas Bruster, 1660), 159: 'The first matter therefore of all bodies is water, which by the dryness of the fire and the aire is changed into earth'; 213; Croll, 'Admonitory Preface', 58; Willem Mennens, 'De aureo vellere', in *Theatrum Chemicum*, 6 vols (Strasbourg: Lazarus Zetzner, 1622), V, 332–3; Thomas Willis, *The Search of Causes*, 11. See also Walter Pagel, *Smiling Spleen: Paracelsianism in Storm and Stress* (Basel: Karger, 1984), on Von Suchten and Fabius Violet, 14, 30.

137 Ch. 13, 'Progymnasma meteori', §8, 56: 'Fateor equidem, me non alias, naturam elementi aquae didicisse, quam sub ferula, ex caduceo Mercurii parata'.

138 On this theory and its influence see William R. Newman, 'L'influence de la *Summa Perfectionis* du Pseudo-Geber', in *Alchimie et Philosophie à la Renaissance: Actes du Colloque International de Tours (4–7 Décembre 1991)*, ed. Jean-Claude Margolin and Sylvain Matton (Paris: Vrin, 1993), 65–77.

139 Newman, 'The Corpuscular Theory', 174–5.

140 Ch. 13, 'Progymnasma meteori', §17, 57: 'Ergo interior Mercurii Nucleus, à dissolventibus non attingitur, multo minus terebratur'.

141 Newman, 'The Corpuscular Theory', 175.

of transmutation', could be removed.[142] Once this external sulphur was stripped away, only the philosophical mercury, which Van Helmont called the mercury of things (*Rerum Mercurius*) remained. This was a substance that was unchanging and resisted any attempts at division. The fundamental impassiveness of philosophical mercury was the attribute that Van Helmont extended to water.

Nevertheless, the Flemish physician's association between water and philosophical mercury is ambiguous. How does the idea of an indivisible mercury square with the theory that all things were ultimately water? What is the precise relationship between water and alchemical mercury?

The answer may be found in a different treatise, 'De tria prima chymicorum' ('On the Tria Prima of the Alchemists'), where we are told that 'the Mercury of things is nothing other than mere water, not yet sufficiently matured by the disposition of a seed'.[143] Here Van Helmont seems to reference Sendivogian theory. Like Van Helmont, Sendivogius believes that water is the first matter of all things, but goes further to identify it as philosophical mercury.[144] This implies that Sendivogius does not view the watery first matter as ordinary water; indeed, the gist of his philosophy is that the real *prima materia* cannot truly be known. Alchemists can only learn and manipulate the second matter (by which he means the seeds).[145]

Could Van Helmont also have meant that elemental water is different from ordinary water? The Flemish physician points out in 'Progymnasma meteori' that watery vapours, acceding to the middle region of the air, are there 'reduced' by the exceeding cold into the simplicity and 'primitive' purity of elementary water.[146] He further maintains that he has searched for 30 years to uncover the 'mercury' of the water. This 'mercury' can only be described partially, as it escapes his understanding. Yet, apart from these rather cryptic remarks, Van Helmont's philosophy depended on the notion that ordinary water was the prime matter of things. Otherwise, how could we make sense of his willow tree experiment or his concept of the *Alkahest*? It would hence be more accurate to say that in Van Helmont's philosophy, water took the place traditionally reserved by the alchemists for philosophical mercury. In doing so, Van Helmont borrowed the explanatory power of the alchemical principle to serve his own natural philosophy.

---

142 Ch. 13, 'Progymnasma meteori', §14, 56–57: 'In sulfure nempe, sunt fermenta, fracedines, odores, sapores specifici seminum, ad quasvis transmutationes'.

143 Ch. 56, 'Tria prima chymicorum', §66, 332: 'Rerum Mercurius autem, non est nisi mera aqua, non sat adhuc a dispositio seminis . . . maturata'.

144 Sendivogius, *De Lapide philosophorum*, 31: 'vento . . . resolvitur in Aquam, ex qua res nascuntur omnes', 85: 'Sed Aqua est omnium Elementorum dignior, quia mater est omnium rerum', 47: 'quatuor Elementa in prima operatione Naturae stillant per Archaeum Naturae, in Terrae Centrum vaporem Aquae ponderosum, qui est Metallorum Semen, & dicitur Mercurius . . . Corpus Metallorum ex Mercurio sit procreatum, quod de Mercurio Philosophorum intelligi debet'.

145 Sendivogius, *De Lapide philosophorum*, 84 : 'Materia prima, quam solus Deus scit'.

146 Ch. 13, 'Progymnasma meteori', §7, 55.

The Element of Air

Although air preoccupies Van Helmont somewhat less than water, this element has a dignity and role of its own. As seen in his interpretation of Genesis, its chief function is to separate, being appointed for this purpose by God. The air is often portrayed as being more active than the water.[147]

In his second exposition of Genesis in *Ortus medicinae* (ch. 14, 'Gas aquae'), Van Helmont makes it clear that 'air' is both 'Heaven' (*coelum*) and 'Firmament' (*firmamentum*). Air encompasses the entire sphere of the stars, the latter of which are 'lights' (*lumina*).[148] Such views of the element air display a strong departure from traditional medieval cosmology. Air is seen as the common stuff of the sublunary and superlunary world. The stars do not dwell in a mysterious *quinta essentia*, but in an element like all the others. This view was in fact Paracelsian; as Didier Kahn has pointed out, in *Philosophia de generationibus et fructibus quatuor elementorum*, Paracelsus had postulated that the *quinta essentia* does not exist, and that the four elements comprise both the sublunary and the superlunary worlds. Air, also called 'chaos', contained within itself all the other elements just like an egg-white contained all parts of the egg.[149]

Van Helmont further asserts that air's role as separator is achieved by naturally having empty spaces within it, wherein it may contain water vapours and *Gas* exhalations.[150] Indeed, he shows by a 'mechanical' operation, that in a pipe of 28 fingers, ambient air can be compressed up to a density of five fingers, this being done without a destruction of the air. He concludes that a large part of air must be made of vacuum, and that air was created for the explicit purpose of containing *Gas* and exhalations within it.[151]

Van Helmont postulates that air lacks the power of acting upon bodies. Instead, it is acted upon by the *blas* of the stars. This view is in line with his doctrine of the elements as being passive, but is not necessarily congruent with the idea of air 'actively' separating waters noted above. The explanation may be twofold: on one hand, acting on bodies could be deemed different from acting on elements; on the other hand, the 'divine' argument can be invoked: God is free to act arbitrarily and may impose an active role on an otherwise passive element.[152]

---

147 Ch. 14, 'Gas aquae', §4, 60, '... operativum illius separationis principium'.

148 Ch. 14, 'Gas aquae, §5, 60.

149 Didier Kahn, 'What can be known of Paracelsus' ideas on heaven, stars and comets from an astronomical and cosmological point of view', 16. I am grateful to Dr Kahn for providing a draft of what promises to be a ground-breaking essay on Paracelsus's cosmology.

150 Ch. 16, 'Vacuum naturae', §19–20, 70.

151 It is possible that Van Helmont may have influenced Boyle's famous air-pump experiments immortalised in Steven Shapin and Simon Schaffer's classical *Leviathan and the Air-Pump* (Princeton: Princeton University Press, 1985).

152 Because of this action of air on water, air cannot be deemed a completely inert medium as Guido Giglioni described it in 'Per una storia del termine Gas da Van Helmont a Lavoisier: Constanza e variazione del significato', *Annali della Facolta di Lettere e Filosofia dell'Università di Macerata* 25–6 (1992–1993): 440.

Air also plays an important role in the constitution of bodies. As we will see further, the spirit that acts in the seed and in the formed body, the *Archeus*, is made of very fine air. Air is essential in the production of the *spiritus vitae*, as it helps the ferment of the heart to make it.[153]

The Tria Prima*: Critique and 'Corpuscularian' Use*

The doctrine of the three Paracelsian principles, the *tria prima*, was an important component of Paracelsus's theory of matter that had been adopted by many alchemical philosophers up to Van Helmont's time.[154] The *tria prima* were usually viewed as corporeal, albeit invisible, components of matter.[155] At the same time, not all Paracelsians emphasised these three principles equally: some preferred to refer to seeds and atoms as basic material constituents.[156]

Van Helmont's works reveal that his attitude toward the *tria prima* changed from apparent adoption in youthful *Eisagoge* (1607) to apparent rejection in the *Ortus*. 'Apparent', because neither adoption nor rejection lack complexity. A closer reading shows that, despite formal allegiance in the *Eisagoge*, the Paracelsian *tria prima* are less important than the seeds.[157] In *De Spadanis fontibus* (1624), the three principles are mentioned only once as being the principles of embodiment of invisible and incorporeal seeds.[158] As time passes, and Van Helmont experiments further, doubts about the *tria prima* emerge. In Van Helmont's letters to Mersenne (1631) the principles are openly questioned for the first time. He observes that 'the principles cannot be separated by fire without a secondary artifice', and that charcoal does not transform into the three principles under the vexation of the fire.[159]

By the time the *Ortus* was written, Van Helmont had further reflected on the concept and decided to reject the theory that the *tria prima* constituted the universal principles of bodies. In the system of the *Ortus*, the three Paracelsian principles, in the sense of corporeal beginnings of bodies, were redundant. Nevertheless, the later

---

153 Ch. 25, 'Blas humanum', §34–37, 149–50.

154 The *tria prima* is prominent in the 'mature' works of Paracelsus. See particularly 'Opus Paramirum', in *Paracelsus: Essential Theoretical Writings*, 305. Many alchemical philosophers adopted his ideas to some extent, including Petrus Severinus, Joseph Du Chesne, Oswald Croll, Jean Beguin and others. For a cogent analysis of the adoption of the *tria prima* see primarily Reijer Hooykaas, 'Die Elementenlehre der Iatrochemiker', *Janus* 41 (1937): 5–18.

155 Debus, 'Fire Analysis', 128–9.

156 It is beyond the scope of this book to analyse the change in perception of the *tria prima* after Paracelsus; however, it is important to note the influence exercised by Petrus Severinus, who in his *Idea medicinae* (1571) emphasised the role of the seeds over the *tria prima*. For an analysis of Severinus's seeds, see primarily Shackelford, *A Philosophical Path*, 162–6.

157 Broeckx, 'Le premier ouvrage', 11 (1854): 125, 127, 136, 144. Hirai has argued that the seed theory is a pillar of Van Helmont's system in *Eisagoge*; Hirai, *Le concept de semence*, 441.

158 Ch. 95, 'Paradoxum alterum', §2, 548.

159 '15 Janvier 1631', 33.

Van Helmont did not completely eliminate the *tria prima* from natural explanations: he only altered their meaning and function.

Van Helmont's objections to the corporeal *tria prima* stemmed from several sources. First of all, he could not accept the tenet that all bodies could be reduced into the three principles. In Van Helmont's doctrine, the elements (water and air) were fundamentally different from other material bodies, being simple and irreduceable. This idiosyncratic understanding of elementary matter clashed with the Paracelsian *tria prima* doctrine.

Van Helmont had further objections to the three principles. While the Paracelsian *tria prima* doctrine argued that fire reduced a body to its constituents, Van Helmont thought that, on the contrary, fire always produced something new.[160] Fire had a transmutational rather than reductive power. Consequently, where many Paracelsians viewed the salt obtained by burning wood as an original, if hidden, component of wood, Van Helmont considered it a new being produced by the fire. This example points out rather forcefully Van Helmont's dissident worldview: where Paracelsians considered that the fire was making invisible things visible, he saw it as an instrument creatively transforming matter.

Consequently, for Van Helmont, the three principles produced by fire are nothing more than new products compounded through fiery action.[161] As he sums up his argument,

> Although fire does elicit the *tria prima* out of some bodies, that is not done through the separation of pre-existing things, but by a transmutation caused by the fire: they are generated as new beings, which did not exist before.[162]

The Flemish philosopher's rejection of an element's division into three principles was not only theoretical, but supported by alchemical experience. Practice in the laboratory led him to affirm that water could not be further reduced into the Paracelsian *tria prima*.[163] Moreover, salt, sulphur and mercury could not be elicited from all bodies. In some cases, the fire could only produce one or two of the *tria prima*, and in other bodies, none at all. The most resilient bodies were gold, mercury, sand, flint and stones that did not contain lime.[164]

---

160 Van Helmont's critique of Paracelsian fire analysis is well described by Debus in his 'Fire Analysis', 137–9. Debus also points out previous similar critiques by Thomas Erastus, but it is not clear that Van Helmont was influenced by him. Antonio Clericuzio draws attention to the fact that Marin Mersenne, Van Helmont's correspondent, held the view of fire as generator rather than reducer; *Elements, Principles and Corpuscles* (London: Kluwer, 2000), 49. It is possible that Mersenne influenced Van Helmont, although it might also have been vice versa.

161 Ch. 20, 'Imago fermenti', §7, 91.

162 Ch. 56, 'Tria prima chymicorum', §47, 329: 'Licet è quibusdam tria prima, pro parte per ignem eliciantur: non tamen id sit per separationem eorundem praeexistentium: sed quatenus per transmutationem ab igne factam, ibidem generentur, tanquam nova entia, fiatque, quod non erat ante'.

163 Ch. 9, 'Elementa', §14, 43.

164 Ch. 56, 'Tria prima chymicorum', §50, 329, §55–6, 330.

Van Helmont adds a third criticism to the Paracelsian theory. The *tria prima* play no role in disease, as they cannot be actually isolated from a living body.[165] Paracelsus, he believes, was mistaken in assuming that the salt in urine was one of the *tria prima*, when in fact it is only salt water that has not yet been separated into its components. In fact, the *tria prima* cannot be obtained from living things at all; only by the destruction of the living principles of the *Archeus* and the seed could the three principles be separated by fire and made into actual beings.[166] Hence diseases cannot be caused by the three principles, as the *tria prima* are chronologically posterior to the diseases. Still, as Van Helmont underlines, the fact that they are not the original principles of bodies does not make them any less useful for medicine.[167] They are in fact 'gifts from God' (*dona Dei*) to cure and heal, which to Van Helmont makes them paramount in importance.

Having criticised the *tria prima* so vehemently, we would not expect Van Helmont to use them in explanations at all. Surprisingly, this is not true: on numerous occasions, he employs the *tria prima* in his theory.[168] This may seem an inconsistency, but it is not necessarily so. In fact, having rejected the idea of the *tria prima* as corporeal principles of bodies, Van Helmont frees them for a different use: as models used to explain the behaviour of composite bodies.

A cursory look through Van Helmont's works reveals the importance he placed on unravelling the inner structure of things, a process he called 'anatomy'. In order to elucidate the nature of beings, he appealed to such principles of explanation as seeds, atoms, *Archei*, and to idiosyncratic alliances between corpuscles and incorporeal forces. The *tria prima* in particular became an important tool for explaining the behaviour of bodies.

Nowhere does the *tria prima* as a theoretical model come across more clearly than in Van Helmont's meteorological tracts 'Progymnasma meteori' and 'Gas aquae'. Here, the *tria prima* are described as constituents of water particles. Kurt Lasswitz considered that Van Helmont comes close to the formulation of a molecular theory. Lasswitz's views were repeated and re-affirmed by William Newman in his article on Van Helmont's corpuscularianism.[169]

In his meteorological tracts, Van Helmont first postulates that water presents itself in four distinct states: ice, liquid, vapour and *Gas* (see Fig. 5.1). The flexibility of water to change amongst these states without ever altering its essence implies a mechanism, and the ideal manner of explaining it is the corpuscularian theory.[170] Thus, Van Helmont contends that water is formed of 'atoms', which in turn are made up of the three principles (mercury, salt, and sulphur) in different spatial arrangements

---

165 Ch. 56, 'Tria prima chymicorum', §10–28, 326.

166 Ch. 56, 'Tria prima chymicorum', §37, 327–8.

167 Ch. 56, 'Tria prima chymicorum', §42, 328.

168 As Pagel has pointed out, Van Helmont's theory of pulse and breath makes heavy use of the *tria prima*; *Joan Baptista Van Helmont*, 90.

169 Newman, 'The Corpuscular Theory', 161–5.

170 See also the account of Newman and Principe in *Alchemy Tried in the Fire*, 64–6.

within the atom. These components give water both its resilience and diversity of character.

Thus liquid atoms of water are made up of a fluid mercury mixed with a dissolved salt, both encircling sulphur.[171] Under the application of heat, liquid water turns into vapour: this means that the salt volatilises, dragging the mercury upwards with its sulphur. The spatial structure of vapour water appears to remain the same as in liquid water. The same happens in the case of the transformation of water into ice or snow: low temperature causes the mercury and salt of water to condense and 'plate' themselves with an external coating of ice, while the sulphur becomes divided into fine particles.[172]

The upper reaches of the atmosphere, where temperatures are extremely low, cause a re-arrangement of the water atom structure. The sulphur 'extraverts', creating a shell (*cortex*) around the mercury and salt (Fig. 5.2). This forces water to take on the nature of a *Gas.*

The *tria prima* of the water cannot be isolated in any manner:

> . . . there is no hope that they can be separated from each other, because in every form of simplicity of the water, the adequate sulphur is somehow hidden, and cannot be separated from the other two; instead, they all accompany each other.[173]

Van Helmont concludes that his *tria prima* are not 'certain universal bodies that are common to all particular species' as Paracelsus would have it.[174]

Such a view of the *tria prima* has generated puzzlement and dissension amongst scholars. What is precisely the nature of this *tria prima* that Van Helmont is referring to? One is tempted to classify the *tria prima* as corporeal particles; this would be in line with Lasswitz's opinion that Van Helmont is describing a complex corpuscle similar to a molecule. In fact, the analogy is not very accurate; a more enlightening, if anachronistic, comparison would be with the inner structure of atoms.

Nevertheless, Antonio Clericuzio disputed this interpretation, arguing that Van Helmont used the *tria prima* only rhetorically, analogically, and did not mean them to be taken as physical particles at all.[175] Clercuzio's refutation is based on Van Helmont's comparison of the *tria prima* model with the astronomers' eccentrics.[176]

---

171 Ch. 14, 'Gas aquae', §8, 60.

172 Ch. 14, 'Gas aquae', §12–13, 60: 'Ideo Mercurius, & sal aquae, properant seipsos custodire ab aeris frigore. Sicque condensant, loricant, atque glacie seipsos incrustant, ut magis resistant in soliditate'.

173 Ch. 13, 'Progymnasma meteori', §20, 57: 'illudque tamen à se invicem divelli, nulla sit spes, eo quod in omnimoda aquae simplicitate, occultetur adaequatum quodammodo sulfur, quod ab aliis nequit separari, quin simul comitentur omnia'.

174 Ch. 19, 'Complexionum atque mistionum', §5, 85: 'non quidem ut corpora quaedam universalia, quae cunctis speciebus sunt communia'.

175 Clericuzio, *Elements, Principles and Corpuscles*, 57–8.

176 Ch. 14, 'Gas aquae', §8, 60: 'Haec suppono, prout Astronomi suos excentricos, ut intelligendi imbecillitati nostrae, eatur obviam'.

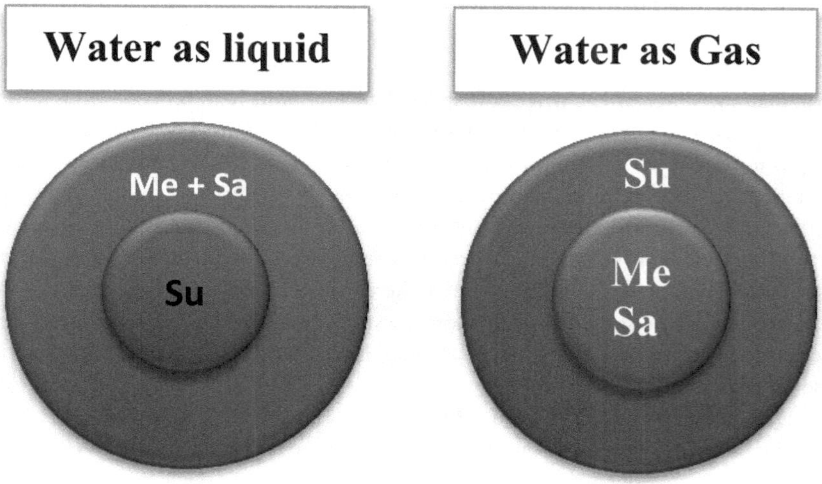

Figure 5.2 A representation of Van Helmont's structure of the water atom

This interpretation has in turn been rejected by Newman and Principe, who upheld the Lasswitz view of the reality of Helmontian particles.[177]

In fact, Van Helmont remains ambiguous in regard to the nature of his *tria prima* of the water. In 'Progymnasma meteori', he states that 'even if water . . . truly consisted of mixed salt, sulphur and mercury, it cannot suffer separation of these', while later he maintains that 'certainly there is something in water that is similar to the three principles'.[178] This equivocal language is also present in the 'Tria prima chymicorum', where Van Helmont asserts that the attribution of *tria prima* to water in anterior treatises was made by analogy; yet to this he immediately adds that 'it is then sufficient [to know] . . . that the hetereogeneal parts of the water are, in the simplest elementary body, indivisible by art, nature, and in all ages, and impossible in reality, as they consist of extreme simplicity'.[179] In this same sentence, Van Helmont calls the three principles of water 'heterogeneous parts of water' and 'impossible in reality', so it is unlikely that the debate on the subject may be clearly settled. It seems that he thought the behaviour of water betrayed the presence of something akin to the *tria prima*, but he was not certain that these parts were real.

177 Newman and Principe, *Alchemy Tried in the Fire*, 64–5, n. 83.

178 Ch. 13, 'Progymnasma meteori', §7, 55: 'Etsi aqua, polleret tribus vulgatis initiis, vereque constaret sale, sulfure, & Mercurio commistis, attamen non patitur ullum eorundem separationem'.

179 Ch. 56, 'Tria prima chymicorum', §54, 330: 'Sufficit enim, ibidem quoque monuisse, partes heterogeneas aquae esse in simplicissimo elementi corpore, arte, natura, omnibusque seculis indivisibiles, atque realiter impossibiles, constantes simplicitate extrema'.

Instead, Van Helmont encourages us to separate the *tria prima* of water in our mind. We can only conjecture, he says, that water is made up of something like sulphur, salt and mercury.[180] These statements reveal his real interest, which is in the explanatory power of the three principles. Van Helmont here proceeds in a philosophical manner, creating models that explain the behaviour of water. It is interesting to note that his contemporary Descartes used precisely the same astronomical metaphor as Van Helmont to present several possible theories of light. Commenting on this point, Osler observes that Descartes 'considered it possible to generate true observations from theoretical premises of undetermined truth-value'.[181] We are perhaps encountering something similar here: Van Helmont cannot ascertain the 'truth-value' of the statement that there are such things as the *tria prima* in water, because they cannot be empirically isolated. However, he finds them a highly useful tool of explanation.

Despite the complexity of Van Helmont's treatment of the *tria prima*, his criticism had a major influence in the eventual abandonment of the Paracelsian principles by later chemists. Today, Robert Boyle's treatise *The Sceptical Chymist* (1661) is better remembered than Van Helmont's treatises for its influential critique of the *tria prima*. In fact, in this work, Boyle mainly reiterated Van Helmont's objections that the *tria prima* could not really be found through separation by fire. Boyle famously concluded:

> For though I dare not absolutely affirme it to be impossible to Analyze these Bodies into their *Tria Prima*; yet because, neither my own Experiments, nor any competent Testimony hath hitherto either taught me how such an *Analysis* may be made, or satisfy'd me, that it hath been so, I must take the Liberty to refrain from believing it, till the Chymists prove it, or give us intelligible and practicable Processes to performe what they pretend.[182]

Spirit

The Seeds: From Incorporeal to Corporeal

The subject of spiritual nature must be treated in conjunction with Van Helmont's emphasis on the generation of bodies. This topic agitated many minds in the early modern period. The debate on generation touched not only on key philosophical concepts, including life and soul, but also on theological views, such as the role of God in embryo formation and the immortality of the soul.

Many early modern philosophers were concerned about the compatibility between Christianity and the Aristotelian view of generation. For Aristotle, matter

---

180 Ch. 56, 'Tria prima chymicorum', §54, 330: '. . . cum tamen actualiter à se invicem distrahi sit impossibile, tantum conjecturae locus est, ut quamvis ista non sint, verum sulfur, sal, & Mercurius, saltem iis quodammodo respondeant'.

181 Osler, *Divine Will*, 142.

182 Robert Boyle, *The Sceptical Chymist* (London: J. Cadwell, 1661), 176.

contained the potentiality of form; the soul of a living being emerged from the material substratum. By contrast, philosophers such as Ficino, Paracelsus, Fernel and Severinus sought to emphasise the spiritual agency over the material one. They upheld theories such as that of the world soul, the *spiritus mundi*, the elemental soul or the spiritual seeds.

Amongst these, it was particularly the seed theory that had an impact on Van Helmont's thought. The presence of a seminal theory in Van Helmont's writings has been duly emphasised by several scholars, including Hiro Hirai, Antonio Clericuzio and Jole Shackelford.[183] As they observed, Van Helmont's theory was initially influenced by Petrus Severinus. Severinus considered the seeds to be anterior to the four elements and divided them into two categories: the original ones, which remain on an ideal plane, and the 'real' ones, which are only approximations of the original.[184] The latter rested in matter until their appointed time, when they 'awoke' to action.

In essence, Severinus's seed theory was an elaboration on the Augustinian idea of the *rationes seminales*, itself drawn from Stoic sources. For Augustine, the *rationes seminales* are spiritual principles engendered by God in the world.[185] The African Father believed that these *rationes seminales* were not continuously inserted in matter, but were placed there at the beginning of time as latent causes, 'pre-programmed' to be activated at a certain point in time.[186]

The Augustinian theory of the *rationes seminales* was an important part of late Scholastic natural inquiry, perhaps because of their resemblance to Aristotelian forms.[187] The seeds were explicitly associated with a Christian interpretation of Genesis.[188] This religious connotation made them attractive to early modern philosophers such as Fernel, Paracelsus and Severinus, who were seeking to replace Aristotelian Scholasticism with a more Christian form of thought.[189]

As early as 1607, Van Helmont adopted the Severinian view of the importance of the seeds as immaterial principles of creation, although he did not embrace the dual-seed theory. In *Eisagoge* (1607) as well as *De Spadanis fontibus* (1624), the seeds are

---

183 Hirai, *Le concept de semence*, 428; Clericuzio, 'From Van Helmont to Boyle', 309–10; Shackelford, *A Philosophical Path*, 241–4. Pagel had also remarked that 'the whole of van Helmont's work can be conceived as a search for the "semina", i.e. the active principles in beings which are responsible for their specific form and function', in 'Some Religious and Philosophical Aspects of Van Helmont's Science and Medicine', *Supplements to the Bulletin of the History of Medicine* 2 (Baltimore: Johns Hopkins Press, 1944), 16.

184 Shackleford, *A Philosophical Path*, 165.

185 Hirai, *Le concept de semence*, 27.

186 Augustine, *The Literal Meaning of Genesis*, trans. John Hammond Taylor (New York, NY: Paulist Press,1982), VI, 10, 189–90; see also Hirai, *Le concept de semence*, 27.

187 For Augustine's ideas of the *rationes seminales*, see Etienne Gilson, *The Christian Philosophy of St Augustine* (New York, NY: Random House, 1960), 206–9. On the late medieval manifestations, see Steneck, *Science and Creation*, 34–5.

188 Steneck, *Science and Creation*, 34–5.

189 Jean Fernel, *On The Hidden Causes Of Things: Forms, Souls, and Occult Diseases in Renaissance Medicine* (Leiden: Brill, 2005), I, 10, 353; for Severinus's use of Fernel, see Hirai, *Le concept de semence*, 263–5.

perceived as 'invisible and incorporeal' originators of all beings in the Universe.[190] In themselves, they are free from any qualities such as colours, quantity, and taste. The seeds contain the 'essence, existence, gift, knowledge, duration and destination' of all beings.[191]

However, in the later *Opuscula medica inaudita* and *Ortus medicinae*, the seminal theory undergoes a major change, and becomes part of Van Helmont's developed philosophy of generation. This philosophy, never analysed before by scholars, posits a dual manner of coming-into-being: a pre-formed matter and a juxtaposed form.[192]

According to Van Helmont's late theory, the seed is the immediate cause of the body, but not the maker of its form.[193] This is bestowed directly by God when the embryo is prepared, or 'disposed' for it. Van Helmont clearly opposes this view to Aristotelian hylomorphism, according to which matter engendered form. Instead, the Flemish philosopher argues that an embryo must be organised and fitted with instruments in order to accommodate the light of form, which comes from a divine immaterial source.

Henceforth in this section I will focus on the seed, with the following two sections addressing the other components of the theory of generation. In the *Ortus*, Van Helmont is particularly concerned with distinguishing the seed from the Ferment, since the Ferments now take up the Augustinian role of *rationes seminales* from the seeds.[194] Thus, he asserts that the Ferment is prior to the seed, originating it and engendering it in matter.[195] By contrast, the seed is as mortal and transitory as the life that it carries within it. It is nevertheless important given that it plays the role of the efficient and immediate cause of the individual body, being both inward and essential.[196] The Ferment, on the other hand, has a more generic nature, being a power engendered in matter that can breed many individuals.

The seed is further envisaged as a corporeal container for the active seminal agent, the *Archeus*. As such, it is the carrier of hereditary information across generations: for instance, a dog's behaviour and personality are foreseen in the seed.[197] In a similar way, Van Helmont attributes specific gender behaviour, personality change, and many personal traits (inclinations) in human beings to the seed.[198]

---

190 Ch. 95, 'Paradoxum alterum', §3, 548–9.

191 Ch. 96, 'Paradoxum tertium', §1, 549: 'Natura scilicet essentia, existentia, donum, scientia, duratio, destinatio, primum in radice seminum connexa fuerunt'.

192 On preformation in early modern Europe, see Richard Arthur, 'Animal Generation and Substance in Sennert and Leibniz', in *The Problem of Animal Generation in Early Modern Philosophy*, ed. Justin E.H. Smith (Cambridge: Cambridge University Press, 2006): 148–9.

193 Ch. 22, 'Ortus formarum', §17–18, 108.

194 This was already noted by Hirai, *Le concept de semence*, 456.

195 As the *vis insita terrae* (the power inside the earth), the ferment 'impregnates' the archetypal matter, water. See ch. 5, 'Causae et initia naturalium', §29, 30.

196 Ch. 5, 'Causae et initia naturalium', §10, 28.

197 Ch. 21, 'Astra necessitant', §35, 100.

198 Ch. 21, 'Astra necessitant', §36, 101.

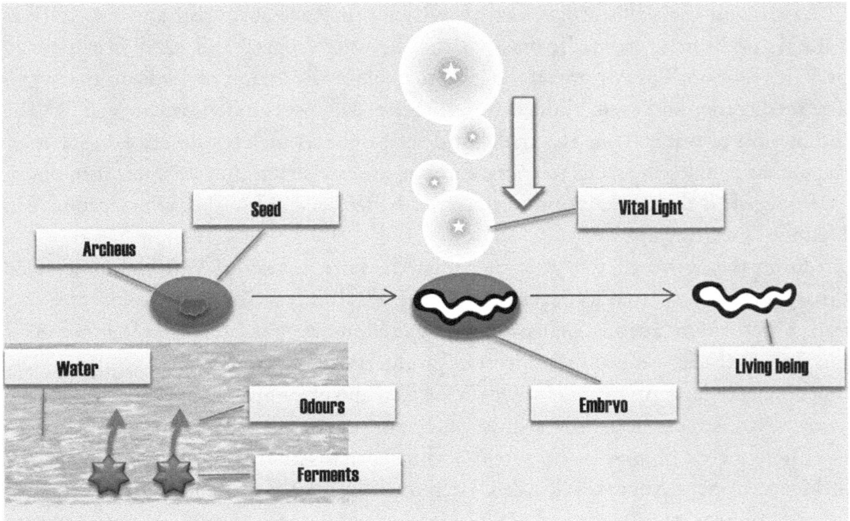

Figure 5.3 A schematic representation of Van Helmont's process of generation

To explain his perspective, the Flemish philosopher divides 'inclination' into four categories: inclination toward a certain profession, a moral or ethical inclination, a vital inclination (referring to health and length of life), and inclination to fortune. The first is given directly by God, while the rest originate from the seed. In other words, our inclination toward moral behaviour, specific diseases and good or bad luck are all dictated by heredity.

In granting the seed such a crucial function in human life, Van Helmont clearly goes beyond the Scholastic understanding of heredity, and he is strongly aware of it.[199] His motivations are drawn from Augustinian thought and its view of the seminal transmission of Original Sin. Augustine believed that seeds perpetuated the Sin and spread it through sexual intercourse.[200] Van Helmont concurred with this view. He firmly believed that God created human beings without a seed; it was only due to the corruption caused by Sin that they actually acquired one.[201] This perspective determines Van Helmont to view seeds, in the specific case of human beings, with disapprobation. All evil comes by means of the seed, including inclination toward evil, immorality, disease, and death; our only defence against it is the Grace of God. Outside humankind, however, seeds are neutral, neither good nor evil.

199 Ch. 21, 'Astra necessitant', §42–3, 102.

200 Rist, *Augustine: Ancient Thought Baptized*, 317–20, 326.

201 Ch. 91, 'Mortis introitus naturam humanam', §12–13, 514.

Although the seeds are generally invisible, they are corporeal and can reify as animal and human sperm. This association between 'sperm' and 'seed' is evidenced by Van Helmont's use of sexual imagery to explain the action of seeds on matter.[202] The seeds order and even 'abduct' water; matter is 'impregnated by the seed'.[203] The interaction of water (seen as female) and seed (seen as male) is described as a 'marriage' that enables the seeds to transmute the otherwise immutable water into physical bodies. It is this sexual connotation that Newman and Hirai have noticed in Van Helmont's writings.[204]

From the above analysis it is clear that the later version of Van Helmont's seed theory moves away from its Severinian source. His ideal of Christian philosophy now leads him to adopt a more Augustinian stance. Moreover, the later Van Helmont takes the view that the seeds are the material products of the *rationes seminales*, in a view that closely reflects Augustine in *The City of God* and elsewhere.[205] Instead, the *rationes seminales* of Augustine are something much more basic, spiritual and immutable.

The concept of seed as corporeal and its sexualised action may also reflect the influence of Sendivogius. Like Van Helmont, the Polish alchemist identified seed (*semen*) as corporeal 'congealed air'.[206] He viewed all generation as originating from the sexualised interaction between an ethereal masculine 'sperm' and a feminine earth, which is described as an empty womb inhabited only by the fiery *Archeus*.[207]

Ferments, the Incorporeal Rationes Seminales

In Van Helmont's later thought, the Ferments acquire an unparalleled importance as the universal agents of biological change. Yet, as with many of Van Helmont's ideas, the concept of the Ferments is complex, and hence liable to generate some confusion.

Van Helmont differentiates between two types of Ferments: the 'principiating' Ferment (*fermentum principians*), the primordial 'trigger' of generation, and the ordinary, seminal ferment (*fermentum seminale*). The former does not come into the composition of bodies and is immortal, while the latter is 'frail and perishing' (*caducum et interiens*).[208] This distinction is important to grasp, as it is fundamental

202 Ch. 94, 'Supplementum de Spadanis fontibus: Paradoxum primum', §11, 546; ch. 119, 'De lithiasi', ch. 1, §6–8, 651.

203 Ch. 13, 'Progymnasma meteori', §7, 55: 'Aqua semper enim ideo, tota manet ut est, vel sine ulla trium divisione, transformatur in fructus, vaditque, quo illam vocant, & abducunt semina'; ch. 5, 'Causae et initia naturalium', §32, 30: 'considero materiam plerumque semine impraegnatam'.

204 Newman, 'The Corpuscular Theory', 180–81, Hirai, *Le concept de semence*, 454.

205 Augustine, *The City of God*, in *The Works of Saint Augustine*, VI–VII, 22:14, *The Good of Marriage*, in *The Works of Saint Augustine*, IX, 20:23. See also Rist, *St Augustine: Ancient Thought Baptized*, 319, n. 4.

206 Sendivogius, *De lapide philosophorum*, 42: 'Semen enim aliud est, quam aer congelatus in aliquot corpore'.

207 Sendivogius, *De lapide philosophorum*, 18–25. It is important to note that for Sendivogius sperm denotes the ethereal air, or form, while the seed (*semen*) is the coagulated product of the interaction between sperm and earth.

208 Ch. 5, 'Causae et initia naturalium', §29, 30.

to a proper understanding of Helmontian natural philosophy. Pagel failed to distinguish between the two, with the result that his description of Ferments is less than enlightening.[209] Since for Van Helmont the 'principiating' Ferments are paramount in his Christian philosophy, I will concentrate on these here.

Van Helmont defines the 'principiating' Ferment in the following manner:

> The Ferment is a formal created being [*ens*], which is neither substance nor accident, but a neutral thing, like light, fire, the *magnale*, the forms etc., created at the beginning of the world in the place of its monarchy, so that it may prepare, excite and precede the seeds.[210]

In other words, principiating Ferments are agents that produce substantial transformation within beings. Indeed, elsewhere Van Helmont states that the Ferments 'are essential causes for the transmutations of things'.[211]

In the previous chapter I have noted that, in Van Helmont's late thought, Ferments superseded the seeds as the primordial principles of things. There are three reasons why this change might have occurred. First, as already mentioned, Van Helmont initially drew his understanding of seeds from Severinus. Yet, at some point, the Severinian account seemed unsatisfactory, and Van Helmont went back to Severinus's ultimate source, the Augustinian theory. He interpreted the African Father's account to mean that *rationes seminales* are a type of divine emanation that is physically placed in matter to actualise according to the divine plan.[212] The Ferment is hence described as 'a power implanted in specific places by the Lord the Creator, for ends ordained by Himself in the succession of days'.[213] For Van Helmont, these *rationes*, although physically present in matter, must also be divine principles (*logoi*), which cannot be removed or exhausted in any way. The Flemish philosopher describes them as divine 'gifts' (*dona*) and 'roots' (*radices*) of all things, not subject to change or death.[214] The Ferments rest at the very base of the Universe: 'they are placed by the Creator's Word, so that they may be the roots of successive seeds even to the end of the world'.[215] The language, here and elsewhere, is Severinian,[216] but the concept has changed: the Ferments are now the generic and enmattered source of beings.

---

209 Pagel's treatment of ferments is put under the heading of 'Odours'; he affirms that ferments are contained in the seeds, *Joan Baptista Van Helmont*, 71–4, see also table 2 at 73.

210 Ch. 5, 'Causae et initia naturalium', §24, 29: 'Est autem fermentum, ens creatum formale, quod neque substantia, neque accidens: sed neutrum, per modum lucis, ignis, magnalis, formarum, &c, conditum à mundi principio, in locis suae Monarchiae, ut semina praeparet, excitet, et praecedat'.

211 Ch. 27, 'Spiritus vitae', §16, 159: 'causae sunt essentiales in rerum transmutationes'.

212 Ch. 103, 'Externorum proprietas', 571.

213 Ch. 5, 'Causae et initia naturalium', §30, 30: 'Est ergo vis locis insita, à Creatore Domino, ibidem locata, ad fines in successione dierum sibi ordinatos'.

214 Ch. 5, 'Causae et initia naturalium', §27, 30.

215 Ch. 23, 'Magnum oportet', §38, 126: 'Ideoque, à Conditore verbo locata, ut sint radices seminum successivorum, in mundi usque finem'.

216 Hirai, *Le concept de semence*, 456.

The second reason for Van Helmont's change of mind might have been related to the contemporary debate on spontaneous generation, the idea that animals such as gnats, flies, and worms appeared to be generated spontaneously in matter.[217] It was an idea that Aristotle himself seemed to support,[218] but in the Renaissance, Platonists in particular saw the phenomenon as a justification of their ideas of the spiritual origin of soul. The theory was the object of lively discussion in Van Helmont's time, with Fortunio Liceti's (1577–1657) *On the Spontaneous Generation of Living Beings* (1618) making an important contribution to the debate.[219] According to Liceti, the spontaneous generation of beings was caused by the generic 'rudiments' of form lying hidden in matter. Liceti's masterly review of ideas on spontaneous generation influenced Daniel Sennert's eponymous treatment on the subject, published in 1636. Further elaborating on Liceti's insight, Sennert tied this to seed theory by positing that there is an internal engine of spontaneous generation which he calls a 'seminal principle'.[220] Van Helmont's Ferments are similar to Liceti's 'rudiments' in that they are embedded in matter, but are also tied in with seed theory, as in Sennert's case.[221]

The third reason for Van Helmont's change of mind was linked to his ongoing alchemical concerns, both theoretical and practical. The fundamentally alchemical source of Van Helmont's concept of Ferments has been duly noted by Pagel, who observed the important role given to ferments by Paracelsus, Jean d'Espagnet, Edward Jorden (1569–1633) and Anton Billich (1598–1640).[222]

In fact, the alchemical interest in Ferments went further back than Paracelsus to medieval precedents. In the pseudo-Arnaldian *Rosarium philosophorum*, first published in 1550, the ferment is compared to the soul, acting both as a life-giver and bestower of form.[223] *Clangor buccinae* similarly affirms that 'the ferment is soul', while the Pseudo-Lullian *Testamentum* and *Semita semitae* describe the ferment as a spirit that transmutes matter. [224]

Van Helmont's focus on Ferment as a cause of change is an example of how he attempted to supplant the Aristotelian-Scholastic framework with ideas originating

217 On the debates on spontaneous generation in the early modern period, see Everett Mendelsohn, 'Philosophical Biology vs Experimental Biology: Spontaneous Generation in the Seventeenth Century', in *Topics in the Philosophy of Biology*, ed. Marjorie Grene (Dordrecht: Kluwer, 1976), 37–65.

218 Aristotle wrote that 'all things are in a sense full of soul' in *Generation of Animals*, 311, 762a18–21; one of his Greek commentators, Themistius, connected the Platonic World Soul and spontaneous generation with this statement. Themistius's ideas were made popular amongst Scholastics by Averroes. See Hirai, *Medical Humanism*, 124–5.

219 On Liceti's work, see Hirai, *Medical Humanism*, 123–50.

220 Hirai, *Medical Humanism*, 166.

221 Van Helmont was probably aware of Liceti's work, but not necessarily of Sennert's.

222 Pagel, *Joan Baptista Van Helmont*, 80–82; see also Paracelsus, 'Paragranum', 227: 'No small art in nature is involved in the ripening of grain . . . For in it is the elixir and highest [degree of] fermentation, surpassing all other things in nature'.

223 *Rosarium Philosophorum*, 20.

224 *Clangor Buccinae*, in *Artis auriferae*, I, 510; 'Semita semitae', in Arthur Dee, ed., *Fasciculus chemicus* (London: J. Flesher, 1650), 113; 'Raymundi Lulli Theorica & Practica', 92.

in alchemical practice. With Scholastics, heat was a cause of change. Yet, the practice of fermentation suggested to medieval alchemists that there was something else involved in the process other than simply heat, hence the growing usage of the 'ferment' concept. Van Helmont relates his experience of seeing an alchemist

> who every year filled two thirds of a vinegar flask with water of the river Rhone: he exposed the jar to the heat of the Sun, and thus transmuted water, in itself insipid, into true vinegar, a Ferment being conceived in the flask . . . therefore, a vessel's Ferment by its odour alone changes water into vinegar.[225]

The example above suggests that, in addition to alchemical texts, it was the empirical observation of fermentation that gave Van Helmont the idea of associating it with the *rationes seminales*.

The authority of alchemy, combined with Augustine's seminal theory and the Platonic concern with spiritual agents of generation, led Van Helmont to raise the notion of Ferment to a philosophical status. Indeed, no generation or growth can take place, Van Helmont explains, without the intervention of a principiating Ferment.[226] Thus, he theorises, 'all bodies primitively and materially are made solely of water which has acquired the seed of a Ferment'.[227]

For Van Helmont, the main argument on behalf of Ferment as a principle is related to its stability: Ferments are immutable and immovable, almost resembling Aristotle's immovable mover.[228] They are forces implanted in specific locations (*vis locis insita*), which give each generated being a particular character related to its origin. For instance, if the seeds of a healthy fruit tree are moved to a different location, they produce inferior fruit.

The force of principiating Ferments, Van Helmont adds, is particularly evident in spontaneous generation, where Ferments act directly upon matter, without the means of seed. Accordingly, Van Helmont divides the principiating Ferments into two types: those that contain a flowing seminal *aura* (a future seed), and those that act without the means of a seed.[229] Alleged spontaneous generations, such as when flies are begotten of excrement, or mice are born when a foul shirt is kept in a vessel

---

225 Ch. 30, 'Sextuplex digestio alimenti humani', §59, 177: 'Memini me chymicum vidisse, qui cadum aceti ad duas tertias quotannis implebat aqua Rhodani, exponebat Solis ardoribus, & sic aquam, in se insipidam, transmutabat in verum acetum, concepto è cado fermento . . . Quapropter ut fermentum vasis, solo sui odore, aquam in acetum mutat'.

226 Ch. 58, 'Catharri deliramenta', §2, 346.

227 Ch. 20, 'Imago fermenti impregnat massam semine', §33, 95: 'Corpora cuncta primitive, atque materialiter ex aqua sola fiant; acquisito per fermentum semine'.

228 Ch. 5, 'Causae et initia naturalium', §26, 30.

229 Ch. 20, 'Imago fermenti', §8, 91. Newman, 'The Corpuscular Theory', 180, observed that this differentiation 'is trying here to explain the difference between sexual and spontaneous generation'. It should be noted, however, that Van Helmont is not talking only about sexual generation, but any type of generation by means of seed, including in his view minerals and metals.

filled with wheat, demonstrate the power of the principiating Ferment and its control over the process of generation.

For Van Helmont, it was insufficient to propose a principiating agent such as the Ferment without unpacking the mechanism by which it acts on matter. First of all, he posits that the Ferment acts by putrefaction. Putrefaction, Van Helmont points out, is not corruption, as traditional Aristotelian speculation would have it.[230] Rather, it is the process whereby a vital light (or form) is extinguished and a new one is born. For instance, the eggs of birds putrefy before constituting the chick.[231]

The source of Van Helmont's speculation is probably pseudo-Paracelsus's *De natura rerum*. The treatise focussed on the process, rather than the agent of generation, which it identified as putrefaction. The work defined putrefaction as 'the mutation and death of all things, & the destruction of the first essence of all natural things, which produces a new regeneration for us and a new birth, thousand of times better than before.[232] Importantly, *De natura rerum* also connected putrefaction with Christian thought, specifically the famous edict of Christ in John 12:24 that, unless a wheat grain putrefies, it cannot bear fruit.

Yet, in comparison with Paracelsus, who viewed generation by putrefaction in terms of a reduction to a primary matter, Van Helmont believed that the Ferment acts by division, breaking down a body into its atoms, causing it to become volatile and to transmute into another being.[233] This process, Van Helmont suggests, is similar to the alchemical maxim of *solve and coagula*: the Ferment first destroys a body then rebuilds a new one. Thus Van Helmont compares the Ferment with fire, which acts similarly, if much more weakly.

The process of fermental generation needs an instrument, and this Van Helmont identifies as the 'odour'. Pagel has already observed that that the odour is an 'active principle' of the ferment and operates by penetration.[234] Beings are disposed by the odour, arising from putrefaction, to change their nature. The odour had already been attributed important powers by the *Natura rerum*: it maintained, for instance, that the 'life' of aromatic substances was the odour, without which they are dead.[235]

Despite the tremendous power of Ferment, substantial alteration is not as simple as it sounds. As I have already noted, Van Helmont posits that matter always desires to remain in its current state, a natural instinct of preservation he calls *philautia*.[236] To defeat this desire, Van Helmont imagines the process of change as one of courtship

---

230 Van Helmont obviously has Aristotle's *On Generation and Corruption* in mind; there, corruption is the opposite process to generation.

231 Ch. 20, 'Imago fermenti', §10, 92.

232 Paracelsus, 'De natura rerum', 258–59: 'Dann die Putrefaction ist ein umbfehrung und der Todt aller dingen/und ein zerstörung deß ersten Wesens/aller natürlichen dingen/darauß uns herkommet die Widergeburt/und newe Geburt/mit taussentsacher Besserung'.

233 Ch. 20, 'Imago fermenti', §23, 25–6, 93–4; see also Newman, 'The Corpuscular Theory', 179–80.

234 Pagel, *Joan Baptista Van Helmont*, 71–4.

235 Paracelsus, 'De natura rerum', 280.

236 Ch. 23, 'Magnum oportet', §3, 121. See above, section 'Genesis and the Elements'.

and sexual attraction. Hence the odour of the Ferment essentially seduces matter by arousing its passion.[237] The odour then acts by penetrating matter and impregnating it with the seed. Once again we see the sexual metaphor in action, this time the 'male' part being represented by the odour. The odour, as the male representative of the Ferment, is so powerful that a medicine can cure by its smell alone: thus, Van Helmont maintains that colic and melancholy are healed not by ointment, but by its odour.[238] At the same time, foul odours can cause diseases such as vomiting, headaches, and apoplexy.

Hence the evanescent nature of the odour accompanies the Ferment as its active, 'masculine' instrument or its *symbolum*, as Van Helmont describes it. The Ferment and its odour are so closely linked that on some occasions Van Helmont almost makes them equivalent to each other; for instance, in the 'De lithiasi', he associates the odour with the Ferment.[239]

Archeus*: The Seminal Agent*

In Van Helmont's philosophy, the process of generation begins when the Ferment, through the instrument of the odour, 'impregnates' matter, endowing it with the seed. The seed, in turn, conceives its own active agent, which may be inherited from the parent (in sexual generation) or directly bestowed by the Ferment (in spontaneous generation). Van Helmont refers to this agent as the *Archeus*.

The term *Archeus* originated with Paracelsus, and hence represents one of the best known intellectual links between Van Helmont and the Swiss physician. The *Archeus* is found in several treatises of Paracelsus, including prominently in 'De mineralibus' as well as 'On the Matrix', 'The Lord's Supper', and *Labyrinthus medicorum errantium*. Paracelsus's *Archeus* is a spiritual principle of organisation, which intervenes and shapes passive matter.[240] The action of the *Archeus* is teleological, perfecting the body and bringing it to its entelechy.

The concept of the *Archeus* was adopted by many alchemical philosophers after Paracelsus. Martin Ruland the Younger (1569–1611) defined it as separator and organiser in the elements, the highest and most exalted of spirits that can be separated from bodies, and as the dynamic agent that separates all things from the Iliaster.[241] Basil Valentine spoke about the *Archeus* within the Element of Earth which made antimony perfect.[242] Gerhard Dorn affirmed that the *Archeus* is an inner sun, as well as a salt and natural balsam.[243]

237 Ch. 20, 'Imago fermenti', §18, 93.
238 Ch. 20, 'Imago fermenti' §17, 93.
239 'De lithiasi', ch. 1, §7, 651.
240 On Paracelsus's concept of *Archeus*, see Pagel, *Paracelsus*, 105–13.
241 Ruland, *Lexicon alchemiae*, 52–3.
242 Valentine, *A Triumphant Chariot of Antimony*, 65.
243 Gerhard Dorn, 'Speculativa philosophia', in *Theatrum Chemicum*, 6 vols (Strasbourg: Lazarus Zetzner, 1602), I, 307–8.

Not everyone employed the term *'Archeus'*, however: Severinus avoided the Paracelsian term and replaced it with *spiritus mechanicus*.[244] By doing so, Severinus wanted to draw attention to the correspondence between the Paracelsian *Archeus* and the notion of *spiritus*, which had a long tradition in Western culture.[245]

One origin can be traced back to Aristotle's description of the *pneuma*, the spirit in the seed, analogous with the stars.[246] The Stoics further elaborated on this spirit, describing it as a gentle breath or breeze (*aura*).[247] They taught that, in the body, the *pneuma* was located in the *hegemonikon*, the Stoic centre of perception and the ruler of the organs, typically situated in the heart.[248]

Another source of the notion of *spiritus* lay in the Hippocratic-Galenic notion of the formative power (*dynamis diaplastikê* or *vis formativa*), which was conceived differently from the Aristotelian form.[249] The formative power was a 'faculty' that was often described in Platonic terms as descending from the stars or originating from the world soul.[250] The formative power enjoyed strong popularity in the Renaissance and early modern periods, culminating with the notion of plastic nature amongst the Cambridge Platonists.[251]

Van Helmont's description of the *Archeus* strongly resembles the Hippocratic-Galenic formative power. However, he was unwilling to acknowledge Galen's contribution in this respect. Instead, time and again, he associates the *Archeus* with Hippocrates' *impetum faciens* (*to enormon*).[252] This is well in line with his view of Hippocrates as an inspired ancient philosopher who was later misinterpreted by Galen and his followers.

Nevertheless, Van Helmont's more direct source lay in Paracelsian speculation, as evidenced by his use of the term. Van Helmont had employed the term *'Archeus'* in

---

244 Shackelford, *A Philosophical Path*, 179–80.

245 See particularly *Spiritus: IV colloquio internazionale Roma, 7–9 gennaio 1983*, ed. Marta Fattori and Massimo Bianchi (Roma: Edizioni dell'Ateneo, 1984).

246 Aristotle, 'Generation of Animals', I, 736b30–37, 1143. On this topic see also Freudenthal, *Aristotle's Theory of Substance*, 106–48.

247 Gilles Quispel, 'Hermes Trismegistus and the Origins of Gnosticism', in *From Poimandres to Jacob Boehme: Gnosis, Hermetism and the Christian Tradition*, ed. Roelof van der Broeck and Cis van Heertum (Amsterdam: Bibliotheca Philosophica Hermetica, 2000), 150.

248 Samuel Sambursky, *Physics of the Stoics* (Westport, CN: Greenwood Press, 1959), particularly 22–4.

249 See Roger's perceptive discussion of the difference between Aristotelian form (intrinsic to matter) and the formative power, *Les sciences de la vie*, 57–62. On the subject of the Arab interpretation of the Aristotelian *pneuma*, see Gad Freudenthal, 'The Medieval Astrologization of Aristotle's Biology: Averroes on the Role of the Celestial Bodies in the Generation of Animate Beings', *Arabic Sciences and Philosophy* 12 (2002): 111–37.

250 Roger, *Les sciences de la vie*, 58. On the questioning of this medieval interpretation in Niccolo Leoniceno, see Hirai, *Medical Humanism*, 24–33.

251 On this topic, see Justin E.H. Smith, 'Leibniz and the Cambridge Platonists in the Debate over Plastic Natures', in *Leibniz and the English-Speaking World*, ed. Pauline Phemister and Stuart Brown (Dordrecht: Kluwer, 2007), 97–110; William B. Hunter Jr, 'The Seventeenth Century Doctrine of Plastic Nature', *Harvard Theological Review* 43 (1950): 197–213.

252 Ch. 71, 'De morbu Archealibus', §4, 438: 'Adeoque Archeum esse impetum faciens, apud Hippocr. extraque, vel praetor quem, nil moveri, sentiri, vel alterari in animantatis'.

his first written work *Eisagoge*.[253] This work alternates between the term *Archeus* and the Severinian term of *spiritus mechanicus*.[254] In the later *Ortus*, however, he drops the Severinian expression altogether in favour of the traditional Paracelsian term. Nevertheless, Van Helmont does not completely disavow Severinus, as he often refers to the *Archeus* as *faber*, a Latin term reflecting Severinus's Greek term *mechanikos*.

The role of the *Archeus* is closely linked with Van Helmont's theory of generation. As mentioned above, Van Helmont affirmed that form as life can only be received from God once matter is pre-disposed to receive it. To achieve this pre-formation, matter (or water) is activated first by the principiating Ferment, then by the seed that it places in its bosom. Yet the seed itself needs an inner worker, an engine (or inner efficient cause) that would arrange matter in accordance with the blueprint present within. This is the *Archeus*, the spirit Van Helmont also called *cosmocrator* and *primus motor*.[255]

Thus the *Archeus* is described as being made of a very fine air that Van Helmont calls an *aura* or *aura vitalis* surrounding an incorporeal nucleus.[256] As such, it falls into the liminal category of entities at the border between the corporeal and incorporeal. The *aura* varies in density according to species: in vegetals it is compressed into the form of a juice, while in metals it is extremely condensed. The nucleus, in turn, is the key to generation, as it contains a seminal image and the *scientia*, the stored blueprint of the future being. The idea of an 'inner *scientia*' was widespread amongst Paracelsians, including Petrus Severinus and Oswald Croll.[257]

Before generation, the *Archeus* is potential, or 'drowsy' (*sopitus*), sleeping in the seed.[258] Once the seed is activated, however, it becomes 'greedy of the splendor felt in the generator [parent]'.[259] Subsequently, the *Archeus* becomes filled with what Van Helmont calls a 'self-preserving desire' (*per philautiae desiderium*), and begins to construct the body. Van Helmont paints a colourful portrait of the spirit's activity within the seed:

> For in animated things, it walks through all the retreats of his seed, and begins to transform the matter, according to the entelechy of its own image. For here it places the heart, but there it appoints the brain, and it acts as an unmoveable indwelling president, governing its whole monarchy, according to the ends of exigence, of parts, and of destinations.[260]

---

253 See for instance in his *Eisagoge*, in Broeckx, 'Le premiere ouvrage', 122, 182, 184, 185.

254 For Severinus's use of 'mechanical' see Shackelford, *A Philosophical Path*, particularly 179–80.

255 'De febribus', ch. 9, 766, 'Spiritus ergo cosmocrator', 767, 'primus motor animantis'.

256 Ch. 72, 'Ortus imagines morbosa', §10, 443, ch. 6, 'Archeus faber', §4, 33, also 'Explicatio', no.2. The image he seems to convey is that of a spiritual centre around which gravitates the corporeal air. On the importance of image and imagination in Van Helmont's writings, see Giglioni, *Immaginazione e malattia*, 15–41.

257 See Hannaway, *The Chemists and the Word*, 26; for more on *scientia* refer to Chapter 6, section on 'The Possibility of *Scientia* of Nature'.

258 Ch. 22, 'Ortus formarum', §61, 116: 'Archeus sopitus, & coagulo seminum obdormiens'. This too is drawn from Aristotle, *Generation of Animals*, 2.1, 735a9–11.

259 Ch. 22, 'Ortus formarum', §19, 108: '... avida splendoris, sensati in generante ...'

260 Ch. 6, 'Archeus faber', §6, 33: 'In animantatis enim, perambulat sui seminis latebras omnes, & recessus, incipitque materiam transformare, juxta imagines suae entelechiam. Hic enim cor locat, ibi vero cerebrum designat, atque ubique immobilem habitatorem praesidem, ex universali sui monarchia determinat, juxta exigentiae, partium, & destinationum fines'.

The work and desire of the *Archeus* produces heat, which eventually ignites a fiery light. This light of the body is in accordance with its species: in fishes, for instance, it is a mild, cold light (*nitor*), while in animals or plants it is a strong, warm light (*splendor*).[261] Each species is further particularised by a specific quiddity through the impregnation of each light with an odour.

This light should not be mistaken for the formal light of the soul (as is explained below in the section on 'The Fiery Light'). Its role is to pre-form the body in accordance with its proper species.[262] The fiery light originates from the 'bosom of nature', has degrees and is extinguishable, while the formal light comes from God and has no degrees. After the light in the seed ignites in accordance with the species, the form-light descends upon it, penetrates it and gives it life.

Once the body is fully formed and the form is received directly from God, the *Archeus* becomes the instrument of life. Van Helmont sees it as a spiritual entity that continues to govern the body. In the case of sensitive beings (such as animals and human beings), it acts as the operative arm of the sensitive soul.[263] The scheme is Galenic-Aristotelian in spite of Van Helmont's opposition: spirits, being quasi-corporeal, must obey the soul, which is incorporeal. The soul is the true and ultimate ruler of the body as its form and life. The attention of Van Helmont is firmly concentrated on this form-life, which he understands as the entelechy of each individual.

The ruling *Archeus* is now called the 'President' (*Rector*) of the body.[264] Thus, the *Archeus* remains throughout one's life

> . . . the internal moderator and assisting architect of generation: and thus it who for an end directs everything to its purpose, causes everything for itself, and all things act for it.[265]

The presiding *Archeus* is the same as the one in the seed, as Van Helmont clearly states.[266] Yet the location of the governing spirit is not made clear in the treatise 'Archeus Faber'. The problem is that the account here may be incomplete: the treatise is just over a page long and seems woefully insufficient for the complexity of the topic. The chapter ends with the affirmation that there is further clarifying information on the *Archeus* in the *Liber de vita longa*, by which Van Helmont probably means the chapter 'Aura vitalis', where the *impetus faciens* of Hippocrates is mentioned.[267]

---

261 Ch. 22, 'Ortus formarum', §63, 117.

262 Ch. 22, 'Ortus formarum', §70–71, 117.

263 Ch. 45, 'Ignota actio regiminis', §48, 272.

264 Ch. 6, 'Archeus faber', §6, 33.

265 Ch. 63, 'Ignotus hospes morbus', §36, 393–4: 'moderator internus, vitalis, & assistens Architectus generationis: adeoque qui in finem dirigit cuncta ad scopos, sibi omnia patrat, & pro se egit universa.'

266 Ch. 27, 'Spiritus vitae', §1, 157.

267 Ch. 105, 'Aura vitalis', 576.

The ambiguity is complicated by the fact that Van Helmont posits the existence of two types of *Archei* in the grown body. One is a flowing (*'fluctuans'* or *'influus'*) spirit, which is described as being unattached to any bodily members, and is the vital spirit (*spiritus vitae*).[268] The other type of *Archeus* is described as the *Archeus insitus*, the spiritual implanted ruler of different parts of the body.

In thus distinguishing two types of *Archei*, Van Helmont was inspired not only by Paracelsus and Fernel, as scholars have pointed out,[269] but also by the Averroistic doctrine of the plurality of forms. According to Averroes, there are two kinds of forms: the *forma informans*, which unites substantially with matter, and the *forma assistens*, which is 'allied' to matter 'as the pilot is united to the ship he moves' (*sicut nauta unitur navi quam movet*).[270] This view was interpreted in various ways by Van Helmont's time, but one such interpretation was given by Daniel Sennert, who posited the existence of a 'master' form in the body with several 'servant' forms.[271] This view is clearly mirrored by Van Helmont, who calls the *Archei* of the organs 'pilots' (*naucleros*); this suggests that he was aware of the provenance of this doctrine in Averroes. Yet, as mentioned, Van Helmont's dependency on the theory of the plurality of forms is relative, as it does not extend to the *Archeus* being an actual form.

The complexity of Van Helmont's account of the *Archei* resulted in a fundamental misunderstanding in Pagel's account of the subject. Probably based on the description in 'Archeus Faber', he thinks that the governing *Archeus* is the *Archeus influus*, which is incorrect.[272] In fact, Van Helmont clearly states that the *Archeus influus* is different from the presiding *Archeus* both in 'Archeus Faber' and subsequently in 'Spiritus vitae'.[273] The presiding *Archeus* is an *Archeus insitus*, an implanted spirit.

The *Archeus influus* is the spirit of life, *spiritus vitae*. This spirit is a very fine volatile salt. It is made, Van Helmont thinks, in the left chamber of the heart. His scheme of blood circulation is based on the Galenic tradition, not on William Harvey, whose epoch-making *De motu cordis* (1628) Van Helmont ignores, whether intentionally or not. Yet it is not truly a Galenic scheme either; indeed, the Flemish philosopher takes issue with such traditional concepts as the existence of three

[268] Ch. 6, 'Archeus faber', §7, 33: 'Alter vero fluctuans, nulli assignatus membro, intuitum servat, super particulares membrorum naucleros, lucidus, at ferians nunquam'.

[269] Antonio Clericuzio, 'Spiritus Vitalis: Studio sulle teorie fisiologiche da Fernel a Boyle', *Nouvelles de la Republique des lettres* (1988): 48–52 ; Pagel, *Joan Baptista Van Helmont*, 99–100.

[270] Averroes, *De Anima*, 3c5, 138–53.

[271] Daniel Sennert, *Hypomnemata physica* (Frankfurt: Clement Schelich, 1636), 218. On this see also Des Chene, *Physiologia*, 77–9 and Hirai, *Medical Humanism*, 167.

[272] See Pagel, *Joan Baptista Van Helmont*, 98.

[273] Ch. 6, 'Archeus faber', §7, 33: 'Praeses demum ille, manet curator, rectorque internus finium, in obitum usque. Alter vero fluctuans . . .', and in ch. 27, 'Spiritus vitae', §1, 157: 'De Archeo tanquam fabro in semine, jam ante disserui, & qualiter cuncta disponat tam in generatione, quam transmutationibus ciborum per cursum vitae; Quod munus proprie spectat spiritum insitum'.

spirits (there is only one, the *Archeus-impetum faciens*) and the notions of innate heat and radical moisture.[274]

Thus, Van Helmont believes that the venous blood, found in the right chamber of the heart, penetrates the sept via triangular pores and passes to the left. There, a vital and luminous ferment 'digests' it and transforms it into arterial blood; a part of this blood is further transformed into vital spirit.[275] Van Helmont argues that the act of nourishment is always directed toward the creation of this vital spirit, as food is progressively subtilised by digestions occurring in the stomach, duodenum and liver. These digestions are similar to alchemical distillations, since the result is an increasingly volatile salt.[276] Eventually, in the last digestion of the heart, the spirit of life becomes a very fine salty *aether*. This in turn becomes enlightened by the formal light of the sensitive soul, and is spread throughout the body by means of the arterial blood.[277] The *spiritus-Archeus*, endowed with the vital light of the soul, acts as a 'balsam' for the body, preserving it from corruption.[278] The idea of the *spiritus vitae* as salt suggests Van Helmont's debt to a branch of early modern alchemical thought that associated the spirit with salt.[279]

While the *Archeus influus* circulates via the arteries throughout the body, the location of the chief implanted *Archeus* can only be identified conjecturally. Thus, Van Helmont states that diseases 'sit immediately within the nest of life, the *Archeus*, which is implanted in the orifice of the stomach',[280] and that 'the sensitive soul, and the *Archeus* himself inhabit the membrane of the stomach'.[281] The *Archeus* of the stomach is elsewhere called 'the central point', 'the principle of life', the 'vital *Archeus*', 'the Ruler of life' and 'the pilot of the Duumvirate'.[282] This evidence strongly points to a location of the presiding *Archeus* in the stomach, or perhaps more accurately in the spleen, 'the seat of the *Archeus*'. The difference between stomach and spleen is slight in Van Helmont, since he argues that together they create a close alliance, or Duumvirate.[283]

The *Archei insiti* of the organs correspond to the planets, in accordance with traditional medical astrology. The *Ortus* does not clearly identify the precise organ-planet

---

274 On this see Peter Niebyl, 'Old Age, Fever and the Lamp Metaphor', *Journal of the History of Medicine* 26 (1971): 351–68.

275 Ch. 25, 'Blas humanum', §20–23, 146–7. See also Clericuzio, 'Spiritus Vitalis', 48–52.

276 Ch. 30, 'Sextuplex digestio alimenti humani', 167–80.

277 See section below, 'The Formal or Vital Lights'.

278 Ch. 27, 'Spiritus vitae', §17, 159.

279 For instance, Clovis Hesteau de Nuysement, *Traittez du vray sel, secret des philosophes, et l'esprit général du Monde* (Paris: Jérémie Perier et Abdias Buisard, 1621). I have already mentioned Dorn's association of the *Archeus* and salt.

280 Ch. 41, 'A sede animae ad morbos', §12, 236: '. . . sedent tamen proxime intra ipsum vitae nidum, in Archeo, qui nimirum insitus est ventriculi orificio'.

281 Ch. 78, 'In Verbis, herbis, et lapidibus est magna virtus', 464: 'Nam revera, in membrana ipsa stomachi, Anima sensitiva, ipseque Archeus habitant'.

282 Ch. 40, 'Sedes animae', §6, 230, ch. 71, ch. 78, 'In verbis, herbis et lapidibus est magna virtus', 466.

283 Ch. 45, 'Ignota actio regiminis', §49, 273: 'Ex Splene namque & stomacho unicum facio connubium, & thorum unum'.

association, but the Dutch-language *Dageraad*, and following it the German version of the *Ortus*, *Aufgang der Artzney Kunst*, does.[284] The idea of the *Archei* corresponding to planets closely connects Van Helmont with Paracelsus, who envisioned the body as the microcosm, and the organs as the stars within.[285] Only the *Archeus influus* is connected with the Firmament directly, rather than the planets.

Nevertheless, as Van Helmont makes it clear, the *Archeus* does not 'obey' the planet; instead, in its healthy state it is in tune, or harmonised with it. Only in illness does the *Archeus* become dependent on the stars.[286] By this, Van Helmont seeks to demonstrate that the normal, unperturbed 'ends' of the organism do not depend on the stars, but only on the seed. The seed contains within itself the entelechy directly impressed by God.

Van Helmont argues that, in order to act within the body, each *Archeus* has an individual *blas*. This *blas* allows an *Archeus* to implement its own corporeal will: the will of the flesh already mentioned.[287] As Van Helmont conceives it, the *Archeus* is endowed with anthropomorphic traits. Thus, the *Archeus* is at times 'Architect', 'President', or 'inward worker'. Its chief role as governor of the body is to subdue the *Archei* of ingested creatures by reversing their last life into the middle life. Van Helmont posits that every being has three lives: a first life, as seed; middle life, as embryo, and last life, as fully formed creature. The last life is not useful for the nourishment of another body, so the *Archeus*, by means of digestion, must reverse the food to its middle life.[288]

The analogy between the *Archeus* and human beings is strengthened in Van Helmont's description of the former's behaviour in disease. The *Archeus* can be 'seduced', suffers 'passions', 'antipathies', 'disturbances' and 'furies'.[289] Hence, the *Archeus* can conceive a disease when it receives an unnatural idea or image. The role of images in the activity of the *Archeus*, and particularly in the process of disease, has been emphasised and explained thoroughly by Guido Giglioni in his *Immaginazione e malattia*.[290] The process of cure, consequently, is also linked with the *Archeus*: medicine must pacify, sedate, and restore it, rather than the body itself.

The Formal or Vital Lights

Van Helmont's theory of vital or formal lights is one of the most important ideas advanced by the Flemish physician and a pillar of his project of Christian

284 Joan Baptista van Helmont, *Dageraad, ofte Nieuwe Opkomst der Geneeskonst* (Rotterdam: Joannes Naeranus, 1660), 45; Von Helmont, *Aufgang der Artzney Kunst*, 41.

285 Walter Pagel, 'Medieval and Renaissance Contribution to the Knowledge and Philosophy of the Brain', in *The History of the Brain and its Functions*, ed. E.N.L. Poynter (Oxford: Oxford University Press, 1958), 108.

286 Ch. 71, 'De morbis Archealibus', §18, 440.

287 Ch.31, 'Pylorus rector', §4–8, 181; see section above, 'The Universal Force: Blas'.

288 Ch. 23, 'Magnum oportet', §42, 127.

289 Ch. 69, 'Progreditur ad morborum cognitionem', §6–8, 10, 428–9, ch. 82, 'De conceptis', §5, 482, ch. 63, 'Ignotus hospes morbus', §85, 403.

290 Giglioni, *Immaginazione e malattia*, 97–133.

philosophy.[291] He continuously reverts to his light theory throughout the *Ortus* as a mainstay of his thought. The theory is a fundamental example of Van Helmont's attempt to straddle the boundaries of theology, metaphysics and natural philosophy by offering an all-encompassing framework.

Van Helmont posits the existence of four types of light: the supreme or uncreated light, the substantial light, the vital or formal light, and the physical light. These exist in a clear hierarchy, which emanates from God and extends downwards to all creatures. From the supreme uncreated light of God comes the light of the human mind (*mens*) and that of angels, the only created substantial lights. Below these in the hierarchy are the vital lights of the sensitive souls, including those of human beings. At the bottom lies the fiery light, which is still uncorporeal but inferior. [292]

The uncreated light and substantial light are immortal, while the other two are ephemeral. The vital or formal lights represent individuals who at the end of their appointed time return to nothingness. This is the natural order of Creation, since

> Indeed all created things were made of nothing, and so they serve the nature of that principle; therefore the forms and being of things voluntarily render themselves into their former nothing.[293]

Van Helmont qualifies this rather harsh pronouncement by positing the return of the vital light to its divine source. In this sense, the light is 'preserved' and another one is produced instead of the former one.[294]

In accordance with this strict hierarchy, the superior lights can act upon the inferior lights without suffering their reaction. Van Helmont strongly rejects the Aristotelian principle of action and reaction in hierarchical relationships. He condemns as 'blasphemy' any view according to which created things can act back upon God.[295] Thus, the divine light penetrates the light of *mens*, and this, in turn, pierces the light of the sensitive soul (*anima*). This penetration allows an intermingling of rays without each losing its own characteristics.[296]

The formal lights originate from one source alone, God, or the uncreated light, that contains within itself all species and individuals. As already noted, God's active role in Creation is fundamentally that of form-giver: 'the Father of Light alone

291 Scholars have generally ignored the topic of Van Helmont's light metaphysics. Heinecke's work *Wissenschaft und Mystik bei J.B. Van Helmont*, 75–8 seems to have been the only one to briefly touch upon the importance of light to Van Helmont's philosophy. However, Heinecke exclusively referred to Van Helmont's mystical and theological use of light imagery and paid no attention to Van Helmont's employment of light in natural philosophy.

292 Ch. 22, 'Ortus formarum', §24, 109.

293 Ch. 22, 'Ortus formarum', §69, 117: 'Creata nimirum universa ex nihilo facta sunt, adeoque istius principii indolem servant, atque ideo inprimis formae & esse rerum, redit sponte in pristinum nihil'.

294 Ch. 109, 'Vita', 586.

295 Ch. 22, 'Ortus formarum', §77, 118.

296 Ch. 20, 'Ortus formarum', §75, 118, ch. 40, 'Sedes animae', §19, 232.

immediately prepares the lights of forms, and the forms of lights: who gives life and all things to all, nor is He far off from each one of us'.[297]

The concept of the vital or formal light is a reinterpretation of the traditional Aristotelian form, as Van Helmont himself admits. As already noted, Van Helmont had denied that the form is an active agent, proposing instead that it is the ultimate 'perfection' (entelechy) of a body.[298] Although Nature has a 'disposition' toward perfection, it has no power of its own to effect it. Hence form has to be given directly by God.[299]

According to Van Helmont, there are four types of forms: the essential form, the vital form, the substantial form and the formal substance.[300] Thus, the essential form is present in metals, minerals, earths and the heavens; these have 'hardly any life' and no soul. The vital form is represented by those beings that have life but do not have a soul, such as plants. The substantial form belongs to those that have life and (sensitive) soul, such as animals. Finally, the formal substance belongs only to the mind of Man and the angels, and is immortal.

This classification is important because it clarifies the relationship between form, life and soul in Van Helmont's philosophy. First, form is not the same as soul, which is an attribute that belongs only to substantial forms and the formal substance. Van Helmont did not think plants have a vegetative soul, like Aristotelians did.[301] In fact, he argued that there was no such thing as a vegetative soul, but only a growing and nourishing power. This 'vital power' that precedes the soul lies in the *Archeus*.[302] Interestingly, this denial of the existence of a vegetative soul has been associated by Mino Bergamo with a distinctly 'Rheno-Flemish' conception of the soul.[303]

The classification also suggests that form is related to life, but is not precisely the same thing. In the case of plants and animals, things are straightforward: form and life are one and the same.[304] Yet Van Helmont hesitates on whether to extend life to lower types of being, for instance to metals or to the Earth. He thinks metals and

---

297 Ch. 22, 'Ortus formarum', §65, 116: 'Quapropter solus Pater luminum, formarum lumina, & luminum formans, immediate condit, *qui vitam & omnia, dat omnibus, nec longe abest ab unoquoque nostrum*'. The last part originates from Paul, Acts 17:27: 'That they should seek the Lord, if haply they might feel after him, and find him, though he be not far from every one of us'.

298 See above, 4.2.1.4.

299 Ch. 22, 'Ortus formarum', §2, 105.

300 Ch. 22, 'Ortus formarum', §67, 116–17.

301 On this see more in Chapter 6, section on 'The Problem of the Soul: Immortality and Unity'.

302 Ch. 40, 'Sedes animae', §10, 230–31: 'Proprie quidem loquendo, & distinctim intelligendo, non est quaedam anima vegetans in plantis, aut brutis: sed est potestas quaedam vitalis, ac tanquam animae prodromus. Anima autem sensitiva, habenas illius potestatis Archealis in se sumit, vitalisque illa praecurrens potestas dispositiva in Archeo liquescit, ac deinceps se submittit animae sensitivae'. This is reaffirmed in ch. 42, 'Ius Duumviratus', 240.

303 Mino Bergamo, *L'anatomia dell'anima: De François de Sales à Fénelon* (Bologna: Il Mulino, 1991), 188.

304 This is confirmed in ch. 109 'Vita', 585: 'Nec formam ideo in vitalibus à vita distinguo, excepta utrobique mente humana'.

minerals might have a type of life by virtue of their power, but not by vital form.[305] In ch. 109 'Vita' ('Life'), he further argues that minerals have a 'mute life' that is revealed by the signs and marks of their vital force.[306] This power may be the *Archeus*, since Van Helmont talks about the '*Archeus* of minerals' in 'Ortus formarum'.[307]

In ch. 10 'Aqua', Van Helmont seems to extend some kind of life to the Earth too, since he talks about the 'vivid *Archeus* of the Earth'.[308] This is surprising, because Earth is described as a secondary element, or the offspring of the (lifeless) element of water. In his conjecture, he was probably influenced by William Gilbert's work *De magnete*, whom he actually quotes in *De magnetica vulnerum curatione*.[309]

This hesitation in regard to lower forms seems to spring from his reliance on empirical inquiry rather than philosophical distinctions. What he is certain of is that life is a formal light:

> Life is a light and formal beginning, whereby a thing acts what it is commanded to act. This light is given by the Creator of things, being infused in an instant, just as fire is struck out of a flint; it is enclosed under the identity and unity of a form, and is distinguished by genera and species.[310]

The vital light, Van Helmont further insists, is immanent to beings, and does not have degrees, like the physical light.[311] However, he argues that it has a certain colour: not red, as we might suppose, but blue.[312] The colour denotes the celestial origin of this light, but is also drawn from the practice of distillation, since Van Helmont compares it to the light of burning *aqua vitae*.[313]

The theory of vital lights sounds highly abstract, and Van Helmont is conscious of the fact that it might be objected to that this light cannot be experienced directly. In fact, Mersenne had already baulked at his theory in a 1631 letter, by pointing out that the body of animals is too opaque and gross to manifest such light. At the time,

---

305 Ch. 5, 'Causae & initia naturalium', '§17, 29: 'Mineralia ergo, & vegetabilia, si qua sorte vivente videantur, cum tantum vivant per potestatem; & non per formam viventem, in lumine animato'.

306 Ch. 109, 'Vita', 585.

307 Ch. 22, 'Ortus formarum', §86, 119, discussing the 'Archeus of minerals'.

308 Ch. 10, 'Aqua', §6, 47.

309 Ch. 113, 'De Magnetica vulnerum curatione', §19, 597; John Henry, 'Animism and Empiricism: Copernican Physics and the Origins of William Gilbert's Experimental Method', *Journal of the History of Ideas* 62 (2001): 116–19.

310 Ch. 109, 'Vita', 585: 'Inprimis ergo vita est lumen & initium formale, quo res agit, quod agere jussa est. Hoc autem lumen à Creatore rebus infusum datur unico instanti, prout à silice ignis excutitur, sub formae identitate ac unitate clauditur, perque genera & species est distinctum'.

311 Ch. 22, 'Ortus formarum', §72–3, 117.

312 Van Helmont calls it 'a sky-coloured and obscure splendour'; ch. 107, 'Fluxus ad generationem', 581: 'splendore coeruleo & obscuro'.

313 Ch. 107, 'Fluxus ad generationem', 581: 'Quae lux erat tanquam fulgor, è flamma, quam aqua vitae ardendo exhibet'.

Van Helmont replied that invisibility is not a valid critique, since the life of animals cannot be seen either, but we can perceive its effects.[314]

Later, Van Helmont may have pondered that this argument might be insufficient to convince sceptics. In *Ortus*, he changes tack; he now argues that vital light can in some conditions be directly perceived, as in the case of the tiny light that shines in the eyes of an animal or human.[315] The reason for this is that the vital lights shine under the influence of physical light. Furthermore, he argues that the vital lights can be revealed through alchemical operations: for instance, in the case of vegetables, the light can be perceived by means of the liquor *Alkahest*.[316] In this case as well, alchemy comes to the aid of his theory as an effective demonstration tool.

The doctrine of light has important applications in medicine. The light of the body can be easily penetrated by other lights which originate from the other beings it consumes as nourishment: in other words, light feeds on light.[317] Such influence can have positive and negative effects. An example of a negative effect is that of poisons being taken in by the body, which penetrate and eventually extinguish the light of the vital *Archeus*.[318] Poisons are not diseases, because they are created by God and are hence endowed with a vital light, while the latter were not.[319] By contrast, diseases, though they are real entities, do not have a formal light, but borrow it or corrupt it from human beings. Hence diseases are defined as entities created by men due to Original Sin.

One cannot overstate how important the formal light theory is to the thought of Van Helmont and his Christian philosophy. The theory allows him to postulate the theological doctrine of God's immanence in Creation in rather stark terms, as well as to provide clear principles of life and individuation in bodies.[320] Van Helmont firmly believed that his light theory was a pure and original form of 'Christian philosophy', as it originated from the Bible.[321]

Indeed, light figures prominently in the Genesis account of Creation as the first product of the Word of God, *fiat lux*, 'let there be light' (Genesis 1:3). Many Biblical commentators believed that this primordial light was not our mundane physical one but something of a different nature and significance.[322] The importance of light in the Bible was further enhanced in the New Testament, particularly in the Gospel

314 '30 Janvier 1631', 59.

315 Ch. 27, 'Spiritus vitae', 160; ch. 41, 'A sede Animae ad morbos', 237, ch. 105, 'Aura vitalis', 578.

316 Ch. 22, 'Ortus formarum', §88–9, 119.

317 Ch. 76, 'Recepta injecta', §5, 454.

318 Ch. 45, 'De ignota actio regiminis', §48, 272.

319 Ch. 68, 'In puncto vita subjectum inhaesionis morborum', 426.

320 Compare this with Descartes, whose philosophical system did not manage to provide a convincing principle of individuation; Funkenstein, *Theology*, 73.

321 Ch. 22, 'Ortus formarum', §20, 108.

322 The tradition of assigning high importance to light can be found in Basil, *Exegetic Homilies*, trans. by Sister Agnes Clare Way (Washington, DC: Catholic University of America Press, 1963), 2.7 (PG 29:43–47), Augustine, *The Literal Meaning*, 1.16–17. For typical medieval views of light, see also Steneck, *Science and Creation*, 43.

of John, where it is associated with the Word-*Logos*. Johannine writings refer to the incarnate Word as 'the life that was the light of men', and as the light of the world.[323] Van Helmont was further influenced in his theories by the identification of God as the 'Father of Lights' in James 1:17.

The prominence of light in the Old and New Testament cross-fertilised with the tradition of light derived from Platonic, Aristotelian and Neoplatonic speculation.[324] The subsequent Christian tradition of light symbolism employed Neoplatonic ideas abundantly to envision light as a manifestation of the divine.[325] For Augustine, the Trinity was a triune light, which could be perceived through the light of grace. Pseudo-Dionysius posited that there was a fundamental or 'primal' light as a spiritual manifestation of God the Father, the visible light being only an image of it.[326]

In parallel with the Christian tradition, an interest in light in the context of natural philosophy was developed within medieval Islam. An influential speculation was that of Al-Kindi (801–873), who imagined the universe as a wide network of rays emanated by every being existent in it. This remarkable idea may have encouraged Van Helmont to postulate the pervasiveness of lights in the constitution of the universe.[327] Another source for Van Helmont's theory of light as form is Avicenna, who used Aristotle's views to present the process of cognition as the Active Intellect radiating intelligible forms into the human mind in a manner similar to light.[328]

The Arab tradition of light symbolism cross-fertilised with Western Christian thought to create original syntheses. The most famous of these was the light metaphysics of Robert Grosseteste (c.1170–1253), who originated the concept that lights are the first forms, an idea that was developed by Van Helmont.[329] The Renaissance was a period when the symbolism of light reached its peak of popularity. Nicholas of Cusa, Marsilio Ficino and Pico della Mirandola used the symbolism of light in their works. Ficino in particular may have been a source for Van

---

323 John 1:4, 1:5, 1:9, 8:12, 9:5.

324 Aristotle, 'On the Soul', I, 2.7.418b13–19; Plotinus, *Enneads*, trans. A. H. Armstrong, 3 vols (London: Heinemann, 1966–67), II, 2.4.15, IV, 3.17.13; see also David C. Lindberg, 'The Genesis of Kepler's Theory of Light: Light Metaphysics from Plotinus to Kepler', *Osiris* 2 (1986): 10.

325 James McEvoy, 'The Metaphysics of Light in the Middle Ages', *Philosophical Studies* 26 (1979): 126–45.

326 Pseudo-Dionysius, 'Celestial Hierarchy', 145–6.

327 A Latin manuscript on Al Kindi's rays was found amongst the papers the Inquisition confiscated from Van Helmont in 1634.

328 Jon McGinnis, *Avicenna* (Oxford: Oxford University Press, 2010), 133–7. Note that although Avicenna's theory of the Active Intellect is formally different than that of the Given of Forms (*dator formarum*), which can also be found in his works, as McGinnis observes, 'all evidence suggests that Avicenna saw the Active Intellect and the Given of Forms as two names for a single entity', 135–6.

329 James McEvoy, *The Philosophy of Robert Grosseteste* (Oxford: Clarendon, 1982), 151, 161.

Helmont's speculation; he maintained, probably on Thomist grounds, that light is similar to soul (*anima*).[330]

The strongest contemporary influence on Van Helmont was the Paracelsian movement. As Goldammer has pointed out, the Renaissance symbolism of light was epitomised in the work of Paracelsus.[331] As mentioned in Chapter 1, the Swiss physician formulated the concept of the two lights, of Nature and of Grace, as rough analogies of the two books, of Nature and of Scripture. Paracelsus talked about the Light of Nature as present throughout the universe and immanent in beings.[332] He found it both in the soul of man, and in the fire ensuing from the internal 'star' (*astrum*) of each being. The inner light was a spirit that was equally 'life' and 'balsam' (preserver of the body).[333] This was an idea which Van Helmont clearly drew on for his own conception of light as intrinsic form and life.

Other Paracelsians might have influenced Van Helmont as well. Heinrich Khunrath affirmed that the Light of Nature is the 'Life of the World'.[334] For Khunrath, light was the essence of nature, and the common stuff of heaven, earth and water, of metals, minerals and stones. The anonymous 'Cabala chemica' maintained that 'light is form, the living state and the efficacious virtue'.[335] Johannes Ernst Burggravius, whom Van Helmont mentioned in one instance, compared life with a lamp whose light is maintained by the inner celestial balsam,[336] while Jean d'Espagnet claimed that 'the souls of creatures are the beams of that heavenly light, which do breath[e] life into them'.[337]

The Fiery Light

In his middle period (*De Spadanis fontibus*, 1624; 15 January 1631 letter to Mersenne), Van Helmont concurred with standard Paracelsian doctrine that fire was fundamentally

---

330 Kurt Goldammer, 'Lichtsymbolik in philosophischer Weltanschauung, Mystik und Theosophie vom 15. bis zum 17. Jahrhundert', *Studium Generale* 13 (1960): 670–82; Marsilio Ficino, 'De lumine', 9, 13 in *Opera omnia* (Basel: Henricpetri, 1576), I, 979, 982. Thomas Aquinas had maintained, using Aristotle and Avicenna as guidance, that the Active Intellect is a light that makes the images of the senses intelligible. See William B. Ashworth Jr, 'Light of Reason, Light of Nature: Catholic and Protestant Metaphors of Scientific Knowledge', *Science in Context* 3 (1989): 92.

331 Goldammer, 'Lichtsymbolik', 670–82.

332 Pierre Deghaye, 'La Lumière de Nature chez Paracelse', in Lucien Braun et al, *Paracelse* (Paris: Albin Michel, 1980), 55–59.

333 Philippus Theophrastus Bombast von Hohenheim Paracelsus gennant, 'De vita rerum naturalium', in *Sechster Theil Der Bücher und Schrifften*, ed. Johannes Huser (Basel: Conrad Waldkirch, 1590), 276–7.

334 Khunrath, *Vom Hylealischen . . . Chaos*, 42: '. . . Leben der Welt ist das Licht der Natur', see also Szulakowska, *Alchemy of Light*, 92.

335 'Cabala chemica', *Theatrum chemicum*, 6 vols (Strasbourg: Eberhard Zetzner, 1659), 380. This work first appeared in *Arcani artificiossimi aperta arca*, attributed to Johann Grasse and published either in 1607 or 1617.

336 Johannes Ernst Burggravius, *Biolychnium, seu Lucerna* (Franecker: Ulderick Dominick Balck, 1611), 20–21.

337 Jean d'Espagnet, *Enchyridion Physicae restitutae or The Summary of Physicks Recovered* (London: Bentley, 1651), 22.

coelum, the heavens.[338] However, later in life he disagreed with this tenet: 'Neither do we, with Paracelsus, acknowledge the fire to be a superlunary element, by the name of lights and stars, nor to have been formed from the beginning (*ab initio*)'.[339]

Despite this formal repudiation of Paracelsian doctrine, Van Helmont's belief about the nature of fire shows remarkable continuity between the middle and later phases. Thus, in the 1631 letter to Mersenne, Van Helmont argued that fire was neither a substance nor an accident; this very same idea is reiterated in *Ortus*.[340] Fire is a substantial form; in fact, it is a form of light.

In the *Ortus*, Van Helmont develops his doctrine of fire further. In the first place, he is eager to base it on the Bible, by observing that fire is not named first-born (*primigenius*) in Genesis and hence cannot claim the title of primordial element. To clarify what fire is, Van Helmont appeals to experiment. By concentrating the light of the Sun through a glass until it becomes fire, he concludes that 'all fire is essentially nothing but light', differing only in degree (*gradus*).[341]

Having experimentally determined the equality between light and fire, Van Helmont proceeds to state that light and fire are not corporeal.[342] Therefore, concludes Van Helmont, fire is 'a certain true and subsisting being, which is neither a substance, nor an accident, but a neutral creature destined by God for human use, and subordinate to their will'.[343] He further defines fire as 'an artificial death' or 'a death at the hand of the artist'.[344] In his judgment, fire acts by destroying bodies, separating their heterogeneal parts or, in the case of flammable bodies, transforming them into *Gas*.[345]

Although fire essentially consumes the seeds of a body, it cannot by itself produce a new being, as it lacks a seminal beginning.[346] Nevertheless, by its work it often promotes the creation of new beings, which are often superior to the anterior ones. In this sense, fire bestows 'a positive death',[347] which allows the alchemist to understand and experiment with nature. Hence, fire

> ... opens, teaches to solve secrets, to hasten the operations of nature, which are often slow, drowsy, and hidden. Next, it separates and expels superfluities; with the aid of an adjunct Ferment, it removes the middle life of things, which causes liveliness and increase of

---

338 Ch. 95, 'Paradoxum alterum', §5, 548; '15 Janvier 1631', 36: 'Elementum ignis nihil est aliud quam tota coelorum respublica'.

339 Ch. 9, 'Elementa', §9, 43: 'Nec item cum Paracelso, licet luminarium & stellarum nomine, ignem, supralunare elementum agnoscere, ut neque ab initio conditum'.

340 '15 Janvier 1631', 39; ch. 22, 'Ortus formarum', §25, 110.

341 Ch. 22, 'Ortus formarum', §26, 110: 'Ergo omnis prorsus ignis, est essentialiter, nil nisi lumen'.

342 Ch. 22, 'Ortus formarum', §24, 109, §35, 112; §37, 112–113. See also ch. 16, 'Vacuum naturae', §16, 69.

343 Ch. 22, 'Ortus formarum', §24, 109: 'Est ignis proinde quoddam ens verum, subsistensque, quod tamen ut non est substantia; ita neque est accidens: sed creatura neutra à Domino in hominum usus destinata, & subter nutum ejusdem data'.

344 Ch. 19, 'Complexionum atque mixtionum' §34, 89, 'mors artificialis'. This idea also originates from Paracelsus, see 'Das Buch Meteororum', in *Achter Theil Der Bücher und Schriften*, 182.

345 Ch. 22, 'Ortus formarum', §29, 111; 'Gas fuliginosum' or 'Gas sylvestre'; see section below on Gas.

346 Ch 19, 'Complexionum atque mistionum', §10, 86.

347 Ch. 22, 'Ortus formarum', §30, 111: 'positiva mors', 'mors artificialis, positiva'.

strength. It also separates the pure from the impure, the precious from the vile, the hurtful from the useful, and the crude from the ripe: indeed, it ripens crudities themselves. Then, the fire prepares the instruments of the arts, which our life needs.[348]

Here fire's value to medical alchemy is eloquently celebrated. Van Helmont posits that fire is an instrument to be employed by alchemists for artificial transmutation, since it has the property of separating substances. It is 'the Vulcan of arts, dedicated to human needs'.[349]

Fiery light bears an analogical similarity to the formal or vital light.[350] Fiery and vital light are both bright and shining, but the formal light is much more spiritual, bestowing life and quiddity on creatures. By comparison, the physical light cannot bestow life.

In this sense, Van Helmont insists on the inability of the Sun to engender new creatures. He denies the Aristotelian axiom that 'both the human being and the Sun create the human being'.[351] Based on the catoptric experiment of the connection of the beams of light, he concludes that the Sun must be 'a most fervent fire, the capital centre of created lights in Nature'.[352] Yet, despite the Sun's high place in heaven, it is unable to confer life on creatures, as its light is only physical, not spiritual. It can only 'excite' by its heat, but cannot bestow seeds. Thus, despite the analogical relationship between the light of life and that of the Sun, the vital light is clearly superior.[353]

Magnale and Vacuum

Van Helmont denied the Aristotelian principle according to which nature abhors a vacuum, arguing that in fact air easily expands or contracts itself leaving spaces (*vacua*) in between.[354] As pointed out above, he even devised an experiment to show that he is correct in his speculation. To him, the presence of empty space in the air is no surprise, as in fact the entirety of nature is close to nothing. His argument on behalf of the vacuum is drawn, again, from religious and voluntarist principles: as Nature was created by God *ex nihilo*, it is always close to nothing.[355]

---

348 Ch. 22, 'Ortus formarum', §38, 113: 'Hactenus enim aperit, docet recludere arcana, properare naturae operations, saepe alioqui desides, somnolentas & sepultas. Separat denique, expellitque superflua, submovet virtute adjuncti fermenti, vitam rerum mediam, unde alacritates, & virium augmenta. Separat quoque purum ab impuro, pretiosum à vili, nocuum ab utili, & crudum à maturo, imo maturat ipsas cruditates. Ignis dein praeparat artium organa, quibus vita nostra indiget'.

349 Ch. 22, 'Ortus formarum', §38, 113: 'Artium vulcanus, humanis necessitatibus dicatus'.

350 Ch. 112, 'Mortis occasiones', 591.

351 Aristotle, *Physics* 2.2, 194b13.

352 Ch. 114, 'In Sole tabernaculum', 620: 'Sol igitur ignis fervidissimus, luminum creatorum centrum principale in natura'.

353 Ch. 114, 'In Sole tabernaculum', 621.

354 Ch. 16, 'Vacuum naturae', §19, 70.

355 Ch. 16, 'Vacuum naturae', 71–3. For a comprehensive discussion on the debates on the vacuum in the Middle Ages and Early Modern Europe, see also Edward Grant, *Much Ado about Nothing: Theories of Space and Vacuum from the Middle Ages to the Scientific Revolution* (Cambridge: Cambridge University Press, 1981).

Van Helmont's strong claim on behalf of the existence of a vacuum in the form of air pores is softened by his belief in the universal presence of a medium he calls *magnale*. All vacuum contains this very fine matter. Particularly since Van Helmont sometimes uses *magnale* as a synonym for vacuum, he seems to support the idea of empty space only in potentiality.[356] It is likely that his beliefs may have been similar to those advocated by such philosophers as Bernardino Telesio, Francesco Patrizi and Tommaso Campanella, according to which, although vacuum was real, in practice this was completely filled by fine matter.[357]

Van Helmont further explains that *magnale* is neither a substance nor an accident.[358] The theory that there is such a thing as a 'neutral' creature is a significant departure from the standard Aristotelian view. Such an intermediary state would have been incomprehensible to Aristotle or his followers. Yet many of Van Helmont's 'entities' lie in this intermediary state, including fire, light and the forms.

The *magnale*'s role is to be the companion or assistant of the air, and to facilitate the instantaneous propagation of the *blas* of stars and of light in all directions.[359] Since *magnale* is very close to a vacuum, it acts in opposition to the corporeal air: when the air is contracted, the *magnale* is extended, and vice versa. This opposition is also related to temperature: when the air is cold, the *magnale* is warm and vice versa. This theory enables Van Helmont to ascertain that atmospheric coldness forces the *magnale* to contract, hence impeding the influence of stars.[360] The speed of propagation of light and *blas* decreases when they encounter air.

Interestingly, the *magnale* makes a rather minor appearance in *Ortus medicinae*, where it is mostly confined to chapter 15, 'Vacuum Naturae' ('A Vacuum of Nature'). Most speculations on this medium occur in the middle-period Van Helmont's correspondence with Mersenne between 1630 and 1631. His explanations there come in the wake of his brief but tantalising mention of *magnale* in *De magnetica vulnerum curatione*, where he had affirmed that *magnale* was a universal spirit that promoted action at a distance, including magnetism and attraction.[361] In the Mersenne letters, he attempts to further explain the concept: he calls the *magnale* by the Aristotelian term 'ether', and maintains that it is 'like' the 'soul of the universe' (*anima universi*).[362] Since *magnale* 'penetrates' all bodies, it is not corporeal; it lies at the boundary between matter and spirit.[363] The belief in the intermediary nature of *magnale* is further repeated in *Ortus*.[364]

---

356 Ch. 16, 'Vacuum naturae', §28, 71.

357 On their views, see Grant, *Much Ado*, 182–206. Grant points out that this idea is ultimately derived from John Philoponus (490–570), 192.

358 This idea of beings that are not substances or accidents brings Van Helmont even closer to the views of Telesio and Campanella, who advocated similar views. See Grant, *Much Ado*, 182–206.

359 Ch. 16, 'Vacuum naturae', §22, 70.

360 Ch. 16, 'Vacuum naturae', §28, 71.

361 Ch. 113, 'De magnetica vulnerum curatione', §151, 616.

362 '30 Janvier 1631', 59.

363 '21 Février 1631', 111–12.

364 Ch. 16, 'Vacuum naturae', §19–20, 70. See also *Opuscula*'s 'Explicatio', no.6.

The middle-period Van Helmont further points out that the *magnale* acts best by means of the air, where it facilitates magnetism. Yet the *magnale* can also be found in water as well, and comes into the composition of all bodies. Van Helmont states that under extreme cold, water is partially transformed into vapours, part into ice and part into its 'first *magnale*'.[365]

In his letters to Mersenne, Van Helmont also suggests that the *magnale* plays an important role in maintaining life. Thus, arguing that the presence of the medium in water allows the fishes to breathe, Van Helmont concludes that 'all things live in *magnale*' as the sustainer of life. Hence a deficit of *magnale* results in disease and death.[366]

Middle-period Van Helmont points out that the doctrine of the *magnale* belongs to 'our school'.[367] By this he probably he means the alchemical, or Hermetic school he mentions elsewhere. Indeed, Pagel observed a resemblance between Van Helmont's *magnale* and Paracelsus's *magnale*, which is also a receptacle for astral influences, although his association with the *Philosophia ad Athenienses*'s *Mysterium magnum* is less convincing.[368] Pagel is further correct that Van Helmont is influenced by Marsilio Ficino, although not by the 'astral body' but the concept of *spiritus mundi*, which was reaffirmed by Agrippa.[369] Ficino and Agrippa maintained that *spiritus mundi* was a quintessence that is neither corporeal nor incorporeal, but an intermediary between the two. This argument made its way into Paracelsian literature. Heinrich Khunrath equated Agrippa's *spiritus mundi* with the ether and *Schamaim* (*shamayim*), the primordial heaven of Genesis which we have seen piquing Van Helmont's curiosity as well.[370] The notion of quintessence-*spiritus mundi* was also supported by Joseph Du Chesne.[371]

Another less obvious source of Van Helmont's speculation on *magnale* is Hippocrates' treatise *On Breaths* (*Peri physon*). In it, Hippocrates argues that 'wind', 'aer' and bodily 'breaths' or 'flatuses' are one and the same thing, and that 'everything between earth and heaven is full of wind'.[372] He also points out in a manner that resembles Van Helmont's argument to Mersenne: 'Nay, it is clear that the sea, too, partakes of wind, for swimming creatures would not be able to live did they not partake of wind. Now how could they partake except by inhaling the air of the water?'[373]

The comparison of the letters to the *Ortus medicinae* reveals some interesting aspects regarding the progress of Van Helmont's thought. The *magnale* is present in both as an

365 '30 Janvier 1631', 60.

366 '21 Février 1631', 111, 121.

367 '21 Février 1631', 111.

368 Pagel, *Joan Baptista Van Helmont*, 94–5.

369 Heinrich Cornelius Agrippa, *De occulta philosophia* (Lyon: Beringos Fratres, 1550), liber I, chap 14, 30: 'quasi non corpus, sed quasi iam anima, sive quasi non anima, & quasi iam corpus, quo videlicet anima corpori connectatur'.

370 Khunrath, *Vom Hylealischen . . . Chaos*, 39.

371 Hiro Hirai, 'The World-Spirit and Quintessence in the Chymical Philosophy of Joseph Du Chesne', in *Chymia: Science and Nature*, 247–61.

372 Hippocrates, 'Breaths', in *Hippocrates*, trans. W.H. S. Jones, 8 vols (London: Heinemann, 1967), II, 231.

373 Hippocrates, 'Breaths', 231.

intermediary being that fills the interstices of the air.[374] Equally, the *magnale* facilitates the influences of the stars (*sidereum influxus*).[375] Yet in the Mersenne letters the *magnale* is a universal spirit that has important functions in bestowing and maintaining life. This lofty view can no longer be found in the *Ortus*. It seems that Van Helmont's enthusiasm for the concept of *magnale* faded. Although the *magnale* is still present in his speculations, it became minimised in favour of other ideas. One may speculate on the reasons for this. The later Van Helmont was keen to emphasise that the source of life was God Himself rather than the *magnale*. He also seems to have grown wary of the notions of 'soul of the world' or 'spirit of the world', which he had earlier associated with *magnale*. The concept of a 'soul of the world' was increasingly being criticised as a pagan concept, including by Mersenne; in turn, Van Helmont may have decided that the concept of *magnale* was unbefitting for a Christian philosopher due to its Platonic-Stoic connotation. Furthermore, as I have pointed out, the *Ortus* preferred more individualised spiritual concepts like *Archeus* and *Gas* over the more general *magnale*.

Gas

The concept of 'gas' has been unanimously recognised as one of the chief contributions of Van Helmont to modern science. Due to its impact and continued usage in chemistry, Van Helmont's *Gas* has been one of the most analysed ideas in his work. Pagel's *Joan Baptista Van Helmont* dedicated several pages to it; more recently, an entire monograph by Paulo Porto and an article by Guido Giglioni were dedicated to the subject.

Yet, as Giglioni has pointed out, the modern sense of gas is based mainly on Lavoisier's interpretation of Van Helmont rather than on that given by the Flemish philosopher himself.[376] Indeed, a close reading of Van Helmont's *Gas* shows that he saw it in much more complex ways than the modern term conveys. Since the term can be misleading, I have used the italicised and capitalised form '*Gas*' throughout the study to emphasise Van Helmont's particular usage.

Van Helmont invented the term *Gas*, probably drawing it from the word for 'chaos': 'by the licence of paradox, I have called that exhalation *Gas*, being not far from the secret Chaos of the Ancients'.[377] These 'Ancients', we learn in *Dageraad*, are the 'Greeks', by which Van Helmont probably means Hesiod and the pre-Socratics.[378] However, it is likely that the term is more directly inspired by Paracelsus's chaos, as Pagel and Giglioni have pointed out.[379] Another influential Paracelsian source may

374 '19 Décembre 1630', 588; '15 Janvier 1631', 34; '21 Février 1631', 111.

375 '15 Janvier 1631', 34.

376 Giglioni, 'Per una storia del termine Gas', 460–64.

377 Ch. 13, 'Progymnasma meteori', §28, 59: 'Ideo paradoxi licentia, in nominis egestate, halitum illum, Gas vocavi, non longe à Chao veterum secretum'.

378 Van Helmont, *Dageraad*, 92.

379 Giglioni, 'Per una storia del termine Gas', 434–7; Pagel, *Joan Baptista Van Helmont*, 64–5; however, Pagel believes that gas is closer to the elemental spirits of Paracelsus, see 66–70; see also Pagel, *Paracelsus*, 95, and Pagel, 'J. B. Van Helmont's Reformation of the Galenic Doctrine of Digestion—and Paracelsus', *Bulletin of the History of Medicine*, 29 (1955): 563–8.

have been Heinrich Khunrath, who talked at length about the primordial Chaos of Genesis in his *Vom Hylealischen Chaos*. Khunrath linked this chaos with the alchemical first matter of the *lapis*.[380]

Van Helmont neither defined *Gas* properly nor dedicated an explanatory treatise to the term, perhaps due to his untimely demise. Scholars have consequently had to elucidate the meaning of the notion in Helmontian thought. Most famously, Pagel argued that 'Gas is central to his naturalist philosophy and cosmosophy', defining it as 'the material or watery vector of object-specificity, the spiritual carrier of the specific life-plan of an object'.[381] This ambitious definition is undermined by the fact that Pagel bases his understanding of *Gas* almost exclusively on the treatise 'Complexionum atque mistionum elementalium figmentum', paying no attention to the previous treatise 'Gas aquae' and only briefly to the connected 'Progymnasma meteori', where the term *Gas* appears for the first time in the *Ortus*.

This oversight has been corrected by Porto and Giglioni, who have written detailed accounts of the use of the term in Van Helmont.[382] Porto observed that the concept of *Gas* is first introduced in Van Helmont's thought as part of his meteorological speculations and is related to an understanding of the cosmos and of air. Both Porto and Giglioni emphasised the role of *Gas* in Van Helmont's theory of disease.[383] Their work has done much to supersede Pagel's analysis of *Gas*; however, neither of them has actually disputed Pagel's definition of *Gas*, which, as I will show, needs correction.

In fact, the closest Van Helmont comes to a definition of *Gas* is in 'Progymnasma meteori', where he differentiates water *Gas* from water vapour, stating that:

> *Gas* is a far more subtle thing than a vapour, mist, or distilled oils, although it is much denser than air. *Gas* itself, materially speaking, is water masked with the Ferment of concrete bodies.[384]

This passage tells us that *Gas* is a corporeal entity made of water, which is subtle, invisible, and light in weight but denser than air. This vague definition is unfortunately insufficient to understand the quiddity of *Gas*. To better comprehend the concept, one must look deeper into Van Helmont's various treatises.

Gas in Van Helmont's meteorological treatises. In *Ortus medicinae, Gas* first appears in Van Helmont's meteorological treatises (ch. 13 'Progymnasma

380 Khunrath, *Vom Hylealischen . . . Chaos*, 54.

381 Pagel, *Joan Baptista Van Helmont*, 61, 63. This definition shows Pagel's tendency to philosophise on the margin of Van Helmont's work.

382 Porto, *Van Helmont e o conceito de gás*, 78–109; Giglioni, 'Per una storia del termine Gas', 437–45.

383 Giglioni, 'Per una storia del termine Gas', 442–5; Porto, *Van Helmont e o conceito de gás*, 101–9.

384 Ch. 13, 'Progymnasma meteori', §29–30, 59: 'Gas, vapore, fuligine, & stillatis oleositatibus, longe sit subtilius, quamquam multoties aere adhuc densius. Materialiter vero ipsum Gas, aquam esse, fermento concretorum larvatam adhuc'. This is a more comprehensive definition than that in the *Opuscula* (1644) 'Explicatio', where *Gas* is described as a spirit that cannot be coagulated, such as that produced by fermenting wine, or by nitric acid; this definition refers to *gas sylvestre* only.

meteori' and ch. 14 'Gas aquae'). Here it is linked with Van Helmont's 'anatomy' of water at different temperatures. As already noted, Van Helmont essentially posited four states of water: ice, water, vapour and *Gas*.[385]

Van Helmont argued that when the vapour of water rose high in the atmosphere and encountered the exceedingly cold air residing there, it did not turn into water, but became further divided until it achieved minute proportions. This state of extreme fineness and invisibility he called *Gas*. In his view, *Gas* formed a relatively stable stratum of water particles above the clouds, a stratum which Van Helmont called *Peroledes*.[386] He speculated that it was this *Gas* layer that constituted what we ordinarily call 'sky', and that it was the water of the *Gas* that bestowed it with its blue colour.

Van Helmont used his theory of the *Gas* stratum to explain meteorological changes. This layer of *Gas* was completely passive, but susceptible to activation by the *blas* of stars. This force effectively pushed the *Gas* back toward earth, where, in contact with the warmer air, it turned into clouds and rain.

Hence, in the meteorological treatises, *Gas* emerged simply as extenuated and invisible water. In Van Helmont's view, this state could only be found in the inaccessible middle region of the air. Hence it was a concept that could not be tested in the laboratory; nevertheless, it played an important explanatory role in accounting for atmospheric changes.

Gas sylvestre. In the treatises 'Complexionum atque mistionum elementalium figmentum' and 'De flatibus', Van Helmont discussed *Gas* as something else than simply minute particles of water. He affirmed that *Gas* could be elicited from many bodies which 'do not immediately depart into water', such as those that are not fixed. Under the vexation of the fire, he explained, such bodies produce a 'wild spirit' (*spiritus sylvestris*), also referred to as *Gas sylvestre*.[387] This *Gas* does not pre-exist as such within the body. It is a type of invisible exhalation, which contains the principiating Ferment that created the body, or more accurately, its odour (hence the peculiar smells of *Gases*).[388] Here, Van Helmont seems to posit that in certain cases, removing a seed by the fire can result in a 'ghost-like' corporeal body being formed.

Gas sylvestre is different from the *Gas* of water since it is an object that can be obtained and observed in the laboratory. Indeed, the concept of *Gas sylvestre* clearly reflected the knowledge acquired through alchemical practice.[389] For instance, Van Helmont relates how he obtained a wild *Gas* by combining sal ammoniac and *aqua fortis*, or by heating lead with mercury and sulphur.[390] He goes on to describe other experiments he carried out, such as roasting an apple or a piece of coal in a hermetically shut vessel, or distilling

385 See above, section on 'The Element of Water'.

386 Ch. 14, 'Gas aquae', §29, 63.

387 Ch. 19, 'Complexionum atque mistionum', §13, 86; ch. 85, 'De inspiratis', 490.

388 Ch. 85, 'De inspiratis', 490.

389 Ch. 57, 'De flatibus', §5, 336.

390 Ch. 19, 'Complexionum atque mistionum', §13–14, 86; see also the same experiment described in ch. 57, 'De flatibus', 343.

tartar of wine. Indeed, as he points out, the alchemist can draw 'a wild and pernicious *Gas* out of coals, Stygian waters and fusions of minerals'.[391] It is clearly this type of *Gas*, the *Gas sylvestre*, that we can most closely associate with our modern term 'gas'.

In ch. 57, 'De flatibus', Van Helmont connects *Gas sylvestre* with bodily flatus: 'every flatus in us is a wild *Gas*, stirred up by digestion from meats, drinks and excrements'.[392] Here, Van Helmont draws his inspiration from Hippocrates' treatise *On Breaths* (*De flatibus* or *Peri physon*).[393] As already noted, the middle Van Helmont was fascinated with Hippocrates' theory of the flatuses as 'winds in the body' and had already written a commentary on *Peri physon*. In the *Ortus*, however, Van Helmont denies that flatuses are actual winds; instead, he now argues that they are all *Gas sylvestre*. Even so, he does not completely eliminate the Hippocratic macrocosmic-microcosmic analogy he otherwise rejects, since he actually affirms the identity of bodily flatus with the outer gases obtained in alchemical experiments.[394]

Other gases. In the same treatise, 'De flatibus', Van Helmont attempted to formulate a taxonomy of *Gases*. This was very different, however, from what we might expect to see. Thus, he identifies

> a windy *Gas* (which is merely air, that is, a wind moved by the *blas* of the stars), a fat *Gas*, a dry *Gas*, which is called sublimed, a smoky or endemic *Gas*, and a wild *Gas*, or an unrestrainable one, which cannot be forced into a visible body.[395]

The 'windy *Gas*' he referred to is slightly dumbfounding, because he had insisted (in the same treatise) that *Gas* was not air or wind at all. Perhaps what he is discussing here is in fact the *Gas* of water, which is pushed down by the *blas* of the stars and causes winds. The 'fat' and 'dry' *Gases* suspiciously resemble Aristotle's dry and moist exhalations, which stood at the origin of the sulphur-mercury theory in alchemy.[396]

Finally, the 'smoky' *Gas* is described at being elicited by the fire from flammable substances such as oil. It is different from *Gas sylvestre* in that it does not burst out and can be restrained.[397] The distinction between the two seems restricted to

---

391 Ch. 85, 'De inspiratis', 490: 'Spagyrus sic quotidie, sylvestre atque perniciosum Gas, è carbonibus, aquis stygiis, & fusionibus mineralium haurit'.

392 Ch. 57, 'De flatibus', §34, 339: 'Ergo restat, quod omnis in nobis flatus sit Gas sylvestre, inter digestiones excitatum, è cibis, potibus, & excrementis'.

393 Van Helmont points this out himself by referring to Hippocrates's work from the very first lines of the treatise.

394 Ch. 57, 'De flatibus', §6–7, 336: rather shrewdly, Van Helmont picks on Paracelsus for propounding this analogy, when in fact he is referring to Hippocrates' theory in *Peri physon*.

395 Ch. 57, 'De flatibus', §4, 336: 'Gas ventosum (quod mere aer est, id est, ventus, per siderum Blas commotus), Gas pingue, Gas siccum, quod sublimatum dicitur, Gas fuliginosum, sive endimicum, & Gas sylvestre, sive incoercibile, quod in corpus cogi non potest visibile'.

396 Aristotle, 'Meteorology', in *The Complete Works of Aristotle*, I, IV; John A. Norris, 'Mineral Exhalation Theory of Metallogenesis in Pre-Modern Mineral Science', *Ambix* 53 (2006): 46–9.

397 Ch. 22, 'Ortus formarum', §29, 111.

behaviour. Their action is similar, and both can cause life-threatening diseases. Thus, observing by alchemical practice that smoky *Gases* 'defile the walls of the vessels', Van Helmont surmises that 'if they besiege the tender branches of the rough artery, they block them'.[398] He also posits that the air can become infected by principiating Ferments, breeding a wild or smoky *Gas* that causes epidemics like the plague.

In addition to these *Gases*, Van Helmont notes in 'Complexionum atque mistionum elementalium figmentum' that the *Archeus* within the body is a type of *Gas*.[399] By positing that the *Archeus* is *Gas*, he seeks to explain why gases in cellars or mines kill so quickly; he affirms that the hostile *Gas* penetrates and affects the *Archeus-spiritus vitalis* by mixing with it. Nevertheless, it is a strange affirmation: we know that the *Archeus* is composed of salty air (or *aether*). How does this square with Van Helmont's earlier emphatic rejection of *Gas* as being air? This ambiguity was not clearly and convincingly resolved.

In light of this brief analysis of the employment of the term *Gas* in Van Helmont, we can now review Pagel's definition of *Gas*. Is *Gas* 'central to his naturalist philosophy and cosmosophy'? This seems overstated. The assessment suggests that Van Helmont acted empirically in regard to *Gas*, progressively adding 'types' of *Gases* as he experimented. In the meteorological treatises he identified *Gas* as extremely subtle particles of water, while in 'Complexionum atque mistionum', 'De inspiratis' and 'De flatibus' he appears to draw on alchemical experiment to clarify what *Gas* was. Eventually, he did philosophise about *Gases* as being essentially water masked by a Ferment, and attempted a taxonomy of *Gases*, but we saw him wavering on its composition. Consequently, calling the term 'central' to Van Helmont's philosophy is surely an exaggeration. If this had been the case, surely Van Helmont would have defined *Gas* much more carefully and philosophically than he had.

We may conclude this brief analysis of the employment of the term *Gas* in Van Helmont's writings by observing that *Gas* covers a wide range of invisible but corporeal 'spirits' that are composed of matter and Ferment. They are representative of Van Helmont's tendency to posit the existence of entities that cannot be directly perceived by the senses, but can be alchemically detected. This is linked, of course, with his critique of Scholastic natural philosophy, which relied almost exclusively on unaided sense perception.[400]

Conclusions

For Van Helmont, the natural world betrays the permanent presence of God; His will is inextricably bound with that of the Universe. This is not a geometer God who

[398] Ch. 85, 'De inspiratis', 490: 'fuligines quaevis suos inquinant vasorum parietes . . . Nam si tenues asperae arteriae ramos obsederint, illos occludunt'.

[399] Ch. 19, 'Complexionum atque mistionum', §40, 89.

[400] See above the discussion of occult qualities, Chapter 2, under 'Van Helmont's Childhood and Youth'.

designs the world, impresses it with motion and then lets it unfold by itself, but a God who participates and intervenes constantly in matter.

The fundamental concept of divine intervention is articulated in two main ways: an indirect one, by means of secondary causes, and a direct one, by the injection of light-forms. In the former case, God acts by means of Nature, which is portrayed almost as a glove in His hand. All movement in Nature can eventually be traced back to God's will and intentions.

We have seen that this view approaches that of the mechanists. Nature can be seen as a *machina mundi*, being wholly controlled by God as the principle of life and intelligence in the Universe. Nevertheless, Van Helmont perceives Nature primarily as *physis*, an inner agent that impresses order and direction in the Universe. Hence Nature can be perceived as a power ordained by God to act on matter. Moreover, Van Helmont still thinks inert matter has sense and a type of yearning for form. While this may be a remnant of Aristotelianism, it may also be linked with his belief in the absolute origin of everything in God. Thus, even inactive matter shares something of God by virtue of being created by Him.

In the case of direct intervention, God becomes essentially involved in the generation of living beings. Van Helmont's theory of generation was directly linked with his doctrine of the two causes, material and efficient. Matter, or the elements, is seen as chiefly inert and needing the intervention of spiritual principles to generate formed beings. The primordial principle is the Ferment, the divine 'reason': it engenders water with seeds capable of generation. The seed is ruled by the *Archeus*, the formative principle, which prepares the body and governs it throughout its existence. The *Archeus*, however, is not capable of bestowing either soul, or life, which is a prerogative of God. As 'Father of Light', God performs what might be called a 'continuous miracle', participating directly in generation.[401] Without His vital light, any body, no matter how well formed by the inner *Archeus*, would degenerate into a dead mass, or a monster.[402] In this sense, Van Helmont's theory can be described as mildly occasionalist, without falling into the extremes that animated some Cartesians.[403]

Yet the goodwill of God manifests itself further, because there is one ultimate creation which surpasses all others. That, of course, is Man, which Van Helmont primarily viewed as 'the image of God' in accordance with Scriptural precedents. Van Helmont's natural philosophy makes no sense without the inclusion of Man in it as the vicar of God in Creation. Hence Van Helmont views Man as both the summit and the end of all Creation, which was 'laid at his feet' for enjoyment and use.

---

[401] This idea of God's permanent intervention in matter was later criticised by Leibniz, in *De usu et necessitate demonstrationum immortalitatis animae*. Leibniz firmly stood by a traducianist position according to which all souls, including human souls, were found in *semen*. On this see Arthur, 'Animal Generation and Substance', 155–7.

[402] Ch. 22, 'Ortus formarum', §65, 116.

[403] On this topic, see Steve Nadler, *Occasionalism: Causation amongst the Cartesians* (Oxford: Oxford University Press, 2011).

Chapter 6
Man

In Genesis 1:26 and 2:7, God creates Man. In the first account, Man is defined as the image and likeness of God; in the second, he is created out of 'the dust of the earth' (*limus terrae*) and animated by God breathing life into his nostrils. Van Helmont reads these passages to mean that Man is dissimilar to Nature. Man was created differently from all other things, through the 'mystery' of an anomalous creation (*mysterium hujus anomalae creationis*).[1]

Van Helmont's view of Man was strongly coloured by the dichotomy between the immortality of the human essence and his mortal dimension inflicted by Original Sin. The Sin caused a fundamental scission in the human being, whose eternal part (the *mens*) retreated deep within the body, while an ephemeral part (the quasi-animal *anima sensitiva*) took over, inflicting pain, disease and death. Felicitously, this dramatic condition can be partially alleviated through alchemical medicine, which can cure diseases and even extend life, and through self-knowledge, which allows one to discover one's true divine origin. However, the only true salvation is through Jesus Christ, who alone can regenerate the human being by means of His merciful death and the Sacraments He instituted.

The story of Man is followed by Van Helmont in different treatises throughout the *Ortus*. A key treatise is 'Imago mentis' ('The Image of the Mind'), which is the cornerstone of Van Helmont's philosophy of Man. Other important chapters are 'Mentis complementum' ('The Perfecting of the Mind') (ch. 43), 'Venatio scientiarum' ('The Hunting of the Sciences') (ch. 4), and the pieces composing the 'Tractatus de anima' ('Treatise on the Soul') (chs. 44–48). These demonstrate a comprehensive understanding of the inner structure of Man and of mystical knowledge. The 'theological' treatises, which are connected with the projected *Liber de vita longa*, augment the Helmontian doctrine of Man. Finally, Van Helmont's middle-period treatise *De magnetica vulnerum curatione* lays out his concept of the 'inner' and 'outer' man, a recurring motif that shapes his view of the relationship between the mind and the sensitive soul.

1 Ch. 22, 'Ortus formarum', §69, 117.

Van Helmont's Views on Human Nature

Man as Imago Dei

Genesis 1:26 convinced Van Helmont that 'truly, man alone was made after the image of God'[2] and that 'he is created in the living image of God, as an immortal substance, that he may know, love and worship God'.[3] In this sense, Van Helmont follows a long Christian tradition of conceiving man as *imago Dei*.[4]

This idea is a mainstay throughout Van Helmont's writings from the early to the late stages. However, his conception of what 'image of God' means changed over time. For instance, in his middle-period *De magnetica vulnerum curatione*, Van Helmont seems willing to accept that, to some extent, 'all things . . . represent that venerable Image'.[5] His views here mirror those of Pseudo-Dionysius, who had postulated that all creation emulates God and that there is a hierarchy of similarity to God amongst creatures, at whose summit stands man.[6] Van Helmont concurs with this opinion: 'the image (*imago Dei*) of God in Man is superior, dominates, and commands the images of God (*simulacris Dei*) in other creatures'.[7] Despite the differentiation between '*imago Dei*' and '*simulacrum Dei*', this quote suggests that he inclined toward a Renaissance view of Man best exemplified by Marsilio Ficino's speculation; according to him, all things imitated the divine as best they could, with Man being closest to it.[8]

By comparison, the later Van Helmont defines the concept of the 'image of God' much more anthropocentrically, applying it only to human beings. His treatment of the subject in the treatise 'Imago mentis' is also more systematic. First he seeks to clearly distinguish Man from angels, who were viewed as immortal substances by the Scholastics. Van Helmont does not deny this, but claims that angels are not the image

---

2 Ch. 4, 'Venatio scientiarum', §35, 21: 'Homo nimirum ad Dei imaginem solus factus'.

3 Ch. 4, 'Venatio scientiarum', §37, 22: 'creatus est in imaginem Dei viventem, in substantia immortali, ut Deum cognoscat, amet, & adoret'.

4 For the tradition of Man as *imago Dei* in the Western Middle Ages, see Bernard McGinn, 'The Human Person as Image of God: II, Western Christianity', in *Christian Spirituality I: Origins to the Fifth Century*, ed. Jean Leclercq, Bernard McGinn and John Meyendorff (New York: Crossroad, 1985), 312–30. The importance of the *imago Dei* motif in Van Helmont's anthropology was noted by Guido Giglioni, who called it 'the cardinal concept of Helmontian metaphysics', *Immaginazione e malattia*, 150.

5 Ch. 113, 'De magnetica vulnerum curatione', §90, 610: 'universa quoque . . . venerandum illus simulacrum referunt'.

6 Pseudo-Dionysius the Areopagite, 'The Celestial Hierarchy', in *The Complete Works*, 154–6. This hierarchical view of being had a long tradition in medieval and Renaissance thought, see Jean Leclerq, 'Influence and Noninfluence of Dionysius in the Western Middle Ages', in Pseudo-Dionysius, *The Complete Works*, trans. Colm Luibheid (New York: Paulist Press, 1987), 27–31.

7 Ch. 113, 'De magnetica vulnerum curatione', §90, 610: 'verum quia homo elegantissime, propriissime, atque proxime, ideo imago Dei in homine, praepollet, dominatur, & imperat simulacris Dei aliarum creaturarum'.

8 Marsilio Ficino, *Opera omnia*, 2 vols (Basel, 1561), I, 107. See also Paul Oskar Kristeller, *The Philosophy of Marsilio Ficino* (New York, NY: Columbia University Press, 1943), 74–92.

of God. Instead, they are a mirror of the 'uncreated light', and as such on a lower level of similitude to God than Man.[9] In his argument, Van Helmont attempts to distinguish between 'image' and 'mirror' in a way that departs from the fundamentally Pseudo-Dionysian background of his speculation. Pseudo-Dionysius had considered the notions of 'image' and 'mirror' as equivalent.[10] However, Van Helmont affirms that an 'image', as opposed to a 'mirror', is something that has its own internal and independent existence. This opinion is further clarified in his discussion of the divine light: thus, the mind as God's image is capable of 'incorporating' the divine light within, while the angel cannot. Since it is only a 'mirror', an angelic being does not contain the light, but simply reflects it on its surface.[11] It is this incapacity to internalise the light that makes angels 'deficient'. Man, by virtue of absorbing divine light within, becomes capable of life and individual existence, while the angel remains incapable of embodiment.

To shore up his theory, Van Helmont adduces a further proof of divine preference for man: 'whosoever God loves, that is nobler for that very reason: but God has loved man more than the angel'.[12] The Bible bears witness to the fact that God, out of His own sheer will, had created His own image in the shape of Adam and Eve. Nothing similar is said about the angels. Hence, man is truly 'the Temple of the Holy Spirit, wherein the thrice-glorious Trinity has made its abode'.[13]

Finally, the superiority of Man to angels is most starkly affirmed by the fact that God chose to embody Himself as a human being. Thus, out of profound love, Christ took upon himself the 'nature of a servant' in order to save mankind from sin. The fact that God the Son chose to become man shows that Man is prized by God above all other creatures.[14]

This argument on the difference between Man and angels clearly reveals the Flemish philosopher's agenda of affirming the superiority of human nature over all created things. Having clearly stated the pre-eminence of Man over the angels, it becomes easy for Van Helmont to affirm that no other creature partakes of the favour of being *imago Dei*. Thus, he states, 'no other created things participate in the dignity of [being] the image of God'.[15]

In making this panegyric to humankind, Van Helmont's stance is strongly coloured by the humanist tradition. Possible sources include the medieval preacher

---

9 Ch. 38, 'Imago mentis', §12, 214.

10 Pseudo-Dionysius had affirmed that all creatures are 'clear and spotless mirrors' of God, hence images of God; Pseudo-Dionysius, 'The Celestial Hierarchy', 154.

11 Ch. 38, 'Imago mentis', §12, 214.

12 Ch. 38, 'Imago mentis', §12, 214: 'quicquid Deus plus diligit, id eo ipso est nobilius: sed Deus dilexit plus hominem, quam Angelum'.

13 Ch. 38, 'Imago mentis', §12, 214: '. . . templum[que] Spiritus Sancti, in quo Tergloriosa Trinitas suam fecit mansionem'. The theme of man as 'Temple of God' also appears in Tauler, *The Sermons*, 228.

14 Ch. 93, 'Demonstratur thesis', §47, 530.

15 Ch. 38, 'Imago mentis', §37, 218: 'Quia dignitas imaginis Dei, non participatur ullatenus ab aliis creatis'.

Johann Tauler and the Renaissance philosopher Pico della Mirandola.[16] In addition, we must not forget that Paracelsians were also strong supporters of humanist views, which they viewed through a magical lens: similarity with the divine justified the capacity of Man to become an enlightened philosopher-magus. Indeed, we can find in Oswald Croll a similar affirmation that Man alone is the image of God, a gift neither granted to the angels nor to any other being.[17]

Man's Fall and Redemption

Van Helmont's exaltation of the human being as *imago Dei* is counterbalanced by a pessimistic view of the present state of Man as a fallen creature. In this sense, Van Helmont considers that the Image of God was embodied in the 'true Man' [*verus homo*], represented by the pre-lapsarian Adam and Eve.[18] *Verus homo* possessed an immortal mind and a body maintained by the Tree of Life. It was primarily the mind that made Man the image of God.[19]

The pristine perfection of Man was profoundly tainted by Original Sin, which to Van Helmont was a traumatic event that had separated not only Man from God, but Man from his true self. As already shown, Van Helmont believed that the sexual act was the cause of the Original Sin. This resulted in four disastrous outcomes: first, the replacement of Virgin generation by Adamic or sexual generation, second, the removal of the Tree of Life, third, the substitution of the mind with the sensitive soul at the helm of the body, and fourth, the simultaneous replacement of perfect *scientia* with false knowledge.

The first and second outcomes allowed death to enter into human nature. God did not make death; instead, by eating the Apple, Man condemned himself to suffer and die. Thus, prior to the Original Sin, death existed only in animals and plants, which suffered a natural 'extinguishment' or 'annihilation' of the vital light.[20]

The Original Sin also caused a distancing of the immortal mind from the body, which could no longer be ruled.[21] Instead, a sensitive soul similar to that of animals was introduced to govern the body. Since all *scientia* lay in the mind, the self-creation of the sensitive soul effectively 'obliterated' the perfect knowledge Adam possessed.[22] Even though this alteration put Man on a par with the animals, one should not fall into the trap of thinking, with Aristotle, that Man is an animal. Indeed, even in this

[16] Johann Tauler had argued that the soul of man resembles God more than any other creature, including the angels; Tauler, *The Sermons*, 230. In turn, Pico della Mirandola argued that 'when we have willed it, we shall not at all be below them [the angels]'; Giovanni Pico della Mirandola, 'On the Dignity of Man', in *On the Dignity of Man, On Being and the One, and Heptaplus* (Cambridge: Hackett, 1965), 7.

[17] Croll, 'Admonitory Preface', 65. At the same time, Croll had previously stated that 'the world was the first figure or image of God', 56.

[18] Ch. 93, 'Demonstratur thesis', §46, 529.

[19] Ch. 91, 'Mortis introitus naturam humanam', §10, 514.

[20] Ch. 109, 'Vita', 586.

[21] Ch. 91, 'Mortis introitus naturam humanam', §16–17, 514.

[22] Ch. 101, 'Intellectus Adamicus', 561: 'caliginoso tamen, miseroque rationis discursu obtenebratae'.

condition he still bears the vestiges of the superior status of Adam. Thus, the mind continues to imprint its seal, or image, upon the embryo, so that the human does not descend into an animal state.[23]

Van Helmont further points out that human beings have been provided with the instruments of salvation. God the Son, in His infinite love, has sacrificed Himself on the cross to provide the path for the regeneration of human souls. Hence Van Helmont does not hold a completely pessimistic view of mankind. On the contrary, he states that the Original Sin enabled us to experience something much more meaningful than Adam's existence in Paradise: the profound love of God manifested in the incarnation of Jesus Christ. Van Helmont prefers the knowledge bestowed by Christ to that provided by the pristine status of Adam: 'I would rather know those things which God has revealed through his only begotten Son, the Saviour of the World, than to have known the faculties of living creatures and herbs with a clear intellect'.[24]

In this sense, Van Helmont views human history as a transformative experience which lifts man higher than Adam to Jesus Christ, the 'second Adam'.[25] Thus, he observes,

> although throughout their entire life, as well as in death, the state of the faithful is far more miserable than the primitive state, yet it is happier than it, in the sense that there is fuller majesty in being more like the incarnated, dead and glorified Son of God, than to have lived with Adam free from diseases and eventually to be taken away without battle.[26]

Hence, regeneration, Helmont insists, can only be accomplished through the intermediation of Jesus Christ, who is 'the Father of virginity' and the 'Father of the Age to come'. As already noted, Christ transforms the tainted Adamic flesh into the Virgin flesh.[27] Van Helmont seems unconcerned about what will happen after death and has little to say about the afterlife. Having been regenerated during life, the mind will return to its rightful place at the helm of the body, glorifying it. The body will be purified and altered into the form of a Spirit.[28]

Van Helmont's thought was strongly coloured by eschatology; however, he refrained from formulating an expectation of the immediate coming of Christ. The millenarian view was a powerful current in Protestantism, and was also an important

---

23 Ch. 101, 'Intellectus Adamicus', 562.

24 Ch. 92, 'Thesis', §52, 521: 'Malim namquue scire, quae Deus per unigenitum mundi Salvatorem revelavit, quam novisse claro intellectu animantium & herbarum facultates'.

25 This idea reflects standard Pauline doctrine, see Rom 5:12–21.

26 Ch. 93, 'Demonstratur thesis', §27, 527: 'Etenim idcirco status fidelium, etsi per totam vitam, etiam in ipsa morte, sit primitivo statu longe aerumnosior; est tamen eo felicior, quanto est majestate plenius esse similiorem Filio Dei, incarnato, mortuo & glorificato: quam cum Adamo vixisse liberum à morbis, & tandem sublatum absque proelio'.

27 Van Helmont, ch. 93, 'Demonstratur thesis', §47, 530–31. See above, Chapter 4, section headed 'Christ as Saviour'.

28 Van Helmont, ch. 93, 'Demonstratur thesis', §27, 527, §56, 533.

aspect of Paracelsian speculation.[29] Yet, as a Catholic, Van Helmont could not advocate an immediate Second Coming. This is the more interesting as he was apparently exposed to the strong chiliastic Paracelsianism of the early 1600s, and not without consequence. Evidence of this can be found in his occasional mentions of the figure of Elias Artista.[30] The legend that the Second Coming would be prefigured by the imminent arrival of the greatest alchemist of all, Elias Artista, had been propagated by Paracelsus himself.[31] This idea acquired numerous adherents at the turn of the seventeenth century. However, the imminence of the Second Coming was not a doctrine the Catholic Counter-Reformation looked lightly upon. Consequently, Van Helmont made a rather subdued mention of the idea. Even so, the Flemish philosopher's allusions to Elias Artista, combined with his support of illuminationist ideas, might explain why his writings were well received by many Protestant alchemists.

Van Helmont's Theory of the Human Soul

The Problem of the Soul: Immortality and Unity

The problem of the nature of the human soul had been an essential one for philosophers seeking to reconcile natural philosophy and Christian beliefs. A fundamental tenet of Christianity posited the immortality of the human soul, while a doctrine upheld by the Council of Constantinople (AD 869) also affirmed its unity. By comparison, the towering figure of Greek philosophy, Aristotle, had approached the subject of the soul from a naturalistic perspective. Not only was he ambiguous on the subject of the immortality of the human soul, but his views also posed problems for the Christian doctrine of the unity of the soul. Thus, Aristotle postulated the existence of three increasingly complex souls in Nature: the vegetative, sensitive and rational souls.[32] Although maintaining that the rational soul existed only in Man, Aristotle argued that the other souls were contained within the human body as well.

---

29 Robin Bruce Barnes, *Prophecy and Gnosis: Apocalypticism in the Wake of the Lutheran Reformation* (Stanford, CA: Stanford University Press, 1988), 37–8; Charles Webster, *From Paracelsus to Newton Magic and the Making of the Modern Science* (Cambridge: Cambridge University Press, 1984), 21–3.

30 Ch. 60, 'Pharmacopolion ac Dispensatorium modernorum', §14, 369: 'Sed verendum ne is, qui permisit perire libros Salomonis, reservet hanc simplicium scientiam in aevum Eliae artistae'; 'De lithiasi', ch. 7, §25, 701: 'Etenim placuit Altissimo, praecursorem, ante Heliam artium, praemittere, docentem medicaminum crasin, & praeparationes: cui ut fideret mundus, data sunt signa, suam doctrinam stabilientia. Illustris nempe est illi magnorum arcanorum praeparatio, quae non nisi adeptione erat confirmanda. Dein subsecuti sunt, qui Paracelsi inventis addentes, essent repertae veritatis theoricae illustratores'.

31 On the myth of Elias Artista in Paracelsus and amongst the Paracelsians, see Antoine Faivre, 'Elie Artiste, ou le Messie des Philosophes de la Nature', *Aries* 2 (2002): 119–52.

32 On this, see Des Chene, *Life's Form*, 153–5. The fact that Galen had also placed the location of each soul in a different part of the body also created misunderstandings amongst natural philosophers. On this, see also Hasse, 'Plato Arabico-Latinus', 41–2.

For a Christian thinker seeking harmonisation with Aristotelian philosophy, the simple solution was to maintain that the rational human soul was immortal. However, this view did not solve the problem of the other two souls, the vegetative and sensitive ones. Grappling with this conundrum, Thomas Aquinas settled for the view that the rational, or purely human soul, somehow managed to synthesise and incorporate the vegetative and sensitive functions within itself. Thus, he effectively incorporated the vegetative and sensitive souls into the rational one.[33]

This standard Scholastic answer was not, however, to everyone's liking. Some theologians and Christian philosophers, particularly those of a Platonic and Neoplatonic inclination, felt that the immortal soul could not include bodily functions.

A trope related to this view was that of the 'inner' and 'outer' man, which was popular with mystical theologians and had a venerable tradition in Christian thought.[34] Developed by Philo of Alexandria (c.25 BC–c.50 AD), it was translated into Christianity chiefly by Paul. In the Apostle's letters to the Romans, he equated the inner man with the mind, or the *pneuma*, and opposed it to the outer man, who is described as carnal and inferior.[35] Based on the authority of Paul and the New Testament, the inner versus the outer man became an important theme in Augustinian thought. Augustine, influenced by both Plotinus and Paul, upheld the superior nature of the spiritual *homo interior* versus the fleshly *homo exterior*.[36] In turn, medieval thought also emphasised the idea of the human as *homo duplex*. Mystical writers particularly seized upon the Paulinian-Augustinian concept of *homo interior* to express the ideals of inner life.[37]

Yet certain interpretations of the trope of the 'outer' versus 'inner' man could jeopardise the doctrine of the unity of the soul. This unsettling possibility is acknowledged by Van Helmont himself in regard to Tauler's doctrine. In a letter to Mersenne in 1631, he points out that Tauler had affirmed that the soul was not simple, but heterogeneous, being divided into the soul (*anima*) and the bottom of the soul (*fundus*

---

33 Jason T. Eberl, 'Aquinas on the Nature of Human Beings', *The Review of Metaphysics* 58 (2004), 333–65.

34 Hans Betz argued that the concept is religious, not philosophical, but Geurt Van Kooten showed that it also appeared in Plato and Plotinus; see Hans Betz, 'The Concept of the "Inner Human Being" in the Anthropology of Paul', *New Testament Studies* 46 (2000): 316; Geurt Van Kooten, *Paul's Anthropology in Context: the Image of God, Assimilation to God, and Tripartite Anthropology in Ancient Judaism, Ancient Philosophy and Early Christianity* (Tübingen: Mohr Siebeck, 2008), 366. Van Kooten argued that Plotinus and Paul drew on common philosophical sources; *Paul's Anthropology*, 375.

35 Rom 7:22–25.

36 Augustine, *On John's Epistle*, 12, 99:4, *The City of God*, 13.24.2. For the Plotinian influence, see Denys Taylor, *The Darkness of God: Negativity in Christian Mysticism* (Cambridge: Cambridge University Press, 1998), 75–79. Augustine refers directly to Paul's inner man in *Against Faustus* 24:2.

37 Examples include Peter Damian, Hugh and Richard of St Victor, Thomas à Kempis, Meister Eckhart and Johann Tauler. See Lionel Friedman, 'Occulta Cordis', *Romance Philology*, 11:2 (1957): 105–6; Bernard McGinn, *The Harvest of Mysticism in Medieval Germany* (New York, NY: Crossroads Publishing, 2005), 123, 251–2; Hein Blommestijn, Charles Caspers, Rijcklof Hofman, *Spirituality Renewed: Studies on Significant Representatives of the Modern Devotion* (Leuven: Peeters, 2003), 35.

seu apice animae).[38] Therein lay 'the kingdom of God', where no creature could enter, least of all the devil. Further, Van Helmont explains, the Ground of the soul contemplates Christ continuously, and is betrothed to God (*Deum sponsum*). Hence this Ground does not fall under Original Sin, which affects only the *anima*.

A less controversial doctrine of the duality of the soul could be found in a particular strand of Scholasticism sympathetic to Platonic and Augustinian influences. Henry de Ghent (c.1217–1293) maintained that the soul is not the immediate form of the body, but the body has its own form; this view essentially posited the existence of two souls in the body.[39] This view was made prominent in the Late Middle Ages by William of Ockham, who explicitly stated that the human body was inhabited by two souls, an immortal, intellective one, and a mortal, sensitive one. This was not a heretical view because it maintained the unity of the immortal soul; nevertheless, Thomist Scholastics found it suspicious or erroneous.[40]

Nevertheless, the Ockhamian account of two souls in one body became increasingly popular in the environment of the late Renaissance. It was upheld by various non-Thomist intellectuals, including Francesco Piccolomini (1523–1607), Jacopo Zabarella (1533–1589), Fortunio Liceti and Pierre Gassendi.[41] Its attraction was probably based on its ability to reconcile Christian tenets with those of natural philosophy in a manner that preserved the unity and immortality of the soul.

Van Helmont's Initial Taulerian View

In *De magnetica vulnerum curatione*, middle-period Van Helmont states that 'in accordance with the consensus of mystical theologians, we divide man into external and internal, assigning to both specific mental powers'.[42] At this time, the Flemish physician clearly supported the religious trope of the inner and outer man, which benefitted from the sanction of the New Testament, Augustine, and German mystics.

---

[38] The description henceforth originates from Van Helmont's letter to Mersenne, '14 Février 1631', 98–9. Van Helmont leaves it to be understood that he sympathised with these ideas.

[39] For Henry of Ghent's views and those of his Augustinian contemporaries, see Gilson, *History of Christian Philosophy in the Middle Ages*, 451–2. Emily Michael identifies the source of this speculation to go back to the school of Scholastic pluralism, advocated particularly by Franciscans, 'Renaissance Theories of Body, Soul and Mind', in *Psyche and Soma: Physicians and Metaphysicians on the Mind-Body Problem from Antiquity to Enlightenment*, ed. John P. Wright and Paul Potter (Oxford: Clarendon Press, 2000), 160.

[40] Objections were raised particularly due to the existence of a similar, but clearly heretical theory in Averroes, according to which the intellect could be fundamentally separated from the body and united with an impersonal intellect. This view was attacked by both Albertus Magnus and Thomas Aquinas and condemned in 1270 and 1277 by the Bishop of Paris, but flourished in Italy in the late fifteenth century; see Ernest Renan, *Averroes et averroisme: essai historique* (Paris: Michel Levy, 1866), 322–435. Interestingly, as Des Chene points out in *Life's Form*, 154, n. 4, the Coimbran rejection of the view was mainly drawn from the Aristotelian account, as they maintained that two souls in a body equated with two distinct substantial forms, resulting into two distinct substances that could not interact.

[41] Michael, 'Renaissance Theories', 158–63, 165–8.

[42] Ch. 113, 'De magnetica vulnerum curatione', §74, 607: 'Consensu inprimis Theologorum mysticorum, hominem dividimus in externum & internum; utrique assignantes mentis cujusdam potestates'.

In expressing his concept of *homo interior* in *De magnetica*, middle-period Van Helmont mainly drew on Tauler's mystical theology of the Ground (*Grund*).[43] Tauler had indeed stated that 'the image [of God] lies in what is most interior, in the completely hidden deepest ground of the soul [*Grunde der Seelen*], where God is present essentially, and actively, and substantially'.[44] Without citing Tauler, Van Helmont argues that God is the innermost part of the self, referring to Luke 17:21 (read as 'The Kingdom of God is within you') in his defence.[45] He concludes that 'in the more inward ground [*fundus*] of the soul lies the Kingdom of God'.[46]

Van Helmont's understanding of Tauler at this time was further coloured by Paracelsus's and Croll's views on magic. The Flemish philosopher theorises that the inward man (*homo interior*) communicates with God, while the outward man (*homo exterior*) communicates with the physical world.[47] Next, he affirms that one of the chief powers of the *homo interior* is magic. Magic allows the inner man to consciously move far-away objects only by his thought or word. As Van Helmont explains, this capacity originates from man's similitude with God, who can create all things just by a manifestation of His will or Word. Consequently, man is an all-powerful creature who could command anything even at a distance: 'Man dominates over all other corporeal creatures, and . . . by his natural magic, he is able to tame the magical virtues of other things'.[48] This power is far stronger than the unconscious one present in animals or in the outward man.

At some point between 1631, when Van Helmont approvingly recapitulates Tauler's theory for Mersenne, and the composition of the *Ortus* in the 1640s, Van Helmont's views changed. In the *Ortus*, he expresses the concern that Tauler's doctrine denies the unity of the soul, and denounces it as unacceptable and unorthodox. Van Helmont now argues that separating the outer man or the soul (*anima*), from the inner man (*spiritus* or *fundus animae*) is equivalent to a division of the immortal soul (also called *anima*). Indeed, he claims that the soul in its entirety is the Image of God, and not only the inner man (the *spiritus*).[49]

Yet the Flemish philosopher does not completely disavow Tauler. He still cherishes Tauler's notions of the mind as *Grund* and Kingdom of God, and does not

43 This was admitted in the *Ortus*, ch. 38, 'Imago mentis', §21, 216.

44 Johann Tauler, *Sermons*, V29 (300.17–22). For a discussion of Ground in the context of late medieval German mysticism, see Bernard McGinn, *The Harvest of Mysticism*, 83–93.

45 However, Luke 17:21 is ambiguous. The common reading is 'the kingdom of God is among [*entos*] you', but *entos* could also be read as 'within' you, like Van Helmont did. See Joseph A. Fitzmeyer, *The Gospel According to Luke, X-XXIV* (Garden City, NY: Doubleday, 1985), 1160–62.

46 Ch. 113, 'De magnetica vulnerum curatione', §130, 614: '. . . in interiore nempe animae fundo, regnum Dei est, ad quod nulla creatura accessum habet'.

47 Ch. 113, 'De magnetica vulnerum curatione', §74, 607–8.

48 Ch. 113, 'De magnetica vulnerum curatione', §131, 614: '. . . hominem caeteris creaturis corporalibus predominari, & magia sua naturali domare posse aliarum rerum virtutes magicas'.

49 Ch. 38, 'Imago mentis', §21, 216: 'Certe non facile credam animae immortalis dualitatem, aut binarii alteritatem: si in sua ipsius essentia unitatem exhibere debeat'.

desist in believing that the mind is the Kingdom of God which is within.[50] Indeed, a close reading of the *Ortus* reveals that Van Helmont continues to use the Taulerian expression of the ground of the soul. For instance, he states that the senses cannot 'reach to the ground of the mind (*mentis fundus*) alone' and that the divine light shone 'inwardly, towards the ground of the soul'.[51] In 'De lithiasi', he remarks that, as he was meditating on the subject of apoplexy,

> Alas! The ground of the soul, so called by *Tauler*, manifested itself unto me. This was nothing else but the immortal mind itself, enveloped in an exterior darkness as in coats of skin, tied to it, and hosted within the abode of the sensitive soul, until the end of one's life.[52]

Thus, late Van Helmont preserves the Taulerian vision, but he now believes that the entire immortal soul is the *Grund*, relegating all the other parts to a mortal, sensitive soul. This two-soul solution is evidently Ockhamian in character. It upholds both the immortality and the unity of the soul, while allowing for the existence of a mortal soul that administers mundane affairs.

Van Helmont supports this view with his reading of the Fall of Adam as resulting in two souls in the human body: the mind and the sensitive soul. As already pointed out, the mind retreated within the body, while the sensitive soul took over the government of the body and its affairs. Van Helmont hence makes the mind almost inaccessible; it can only be reached by a mystical process of self-knowledge. Yet, as it alone contains the *imago Dei*, immortality and true *scientia*, its recovery is imperative for true knowledge.

Van Helmont's Final System: The Duumvirate

For Van Helmont, the mind (*mens*) is an 'immortal substance' (*substantia immortalis*).[53] This doctrine was embraced by most orthodox Christian thinkers of the period, but Van Helmont chooses to cite Augustine in connection with this belief. The African Father advocated the notion of the soul as an incorporeal immortal substance and affirmed that there was nothing closer to God than the mind or rational soul (*animus*).[54]

---

50 Ch. 43, 'Mentis complementum', §27, 253.

51 Ch. 43, 'Mentis complementum', §10, 252; ch. 2, 'Confessio authoris', §14, 13; see also ch. 50, 'Nexus sensitivae & mentis', §5, 285: '. . . de penitiori mentis vacuo fundo & fabrica . . .'

52 'De lithiasi', ch. 9, §54, 722: 'Proh dolor! Tum sese mihi manifestavit fundus animae, apud Taulerum sic dictus. Qui nil aliud, quam ipsa mens immortalis erat, quantis videlicet tenebris exterioribus, velut tunicis pelliceis, obvolveretur, quatenus alligata est, hospitaturque, durante termino vitae, in ipsius animae sensitive diversorio'.

53 Ch. 49, 'De nostra animae immortalitate', §13, 284.

54 Augustine, *The City of God*, 11:26; see also Roland Teske, 'Augustine's Theory of the Soul', in *The Cambridge Companion to Augustine*, ed. Eleonore Stump and Norman Kretzmann (Cambridge: Cambridge University Press, 2006), 116; Van Helmont refers to Augustine in ch. 48, 'Distinctio mentis à sensitiva anima', §5, 280.

The issue of the immortality of the soul was particularly pressing for Renaissance and early modern philosophers.[55] The challenge posed by the Averroist doctrine of the impersonal universal intellect, by Pietro Pomponazzi's treatise *De immortalitate animae* (1516), and by the naturalistic views of the so-called 'libertines' led to renewed focus on the subject by Christian philosophers.[56] Van Helmont himself felt compelled to write a short, perhaps unfinished, treatise called 'De anima nostra immortalitate' ('The Immortality of our Soul'). Nevertheless, this piece in itself is not particularly enlightening, as it is mainly focussed on attacking general atheist beliefs rather than on addressing the subject-matter at hand. Instead, it is his two-soul theory that is his proper answer to the immortality controversy. By clearly distinguishing the mortal soul (*anima sensitiva*) from the immortal one (*mens*), Van Helmont hoped to safeguard the integrity of Christian doctrine, while adding a natural-philosophical perspective in harmony with it.

The Flemish philosopher makes an extensive argument on behalf of the created nature of the mind. Thus, he follows in Augustine's footsteps in rejecting the Platonic view that considered the human soul as part of God.[57] Van Helmont's argument is simple and logical: if God is an infinite being, mind as a part of infinite would be the infinite itself, hence there would be at least two co-equal Gods, God the Creator and the human mind. This would be an absolute heresy as well as a philosophical absurdity.[58] Hence, he concludes, the human mind is not separated from the divine being, but created out of nothing as a free gift (*donum*) of God.

Van Helmont viewed the mind as being or reflecting the true *imago Dei*. He vacillates on this fine theological point: should the mind be conceived as the actual Image of God or after the Image of God? Traditional discourse held the human being to be created after the Image rather than as the Image itself; usually the Image was conceived as being God the Son or sometimes the Holy Spirit.[59] Yet some medieval theologians, particularly those belonging to the German mystical tradition, such as Meister Eckhart and Johannes Tauler, proposed that the deepest part of man was the true image of God.[60]

Van Helmont is ambivalent on this issue. Sometimes he clearly upholds the traditional doctrine, stating that Man was made after the Image of God and that the Image

---

55 Kristeller, *The Philosophy of Marsilio Ficino*, 326. Kristeller has deemed Ficino's writing as 'the most detailed refutation of Averroism after that of Aquinas', *Renaissance Concepts of Man and Other Essays* (New York: Harper Row, 1972), 33.

56 See Michael, 153–62, John S. Spink, *French Free Thought from Gassendi to Voltaire* (London: Athlone Press, 1960).

57 See Augustine, *The Literal Meaning of the Genesis*, 7.2 and 7.3 for reasons why the soul is not part of God. For further comments, see Teske, 'Augustine's Theory of the Soul', 117.

58 Ch. 38, 'Imago mentis', §10, 214.

59 Robert Javelet, *Image et resemblance au douzieme siècle*, 2 vols (Strasbourg: Université de Strasbourg, 1967), I, 72–91.

60 The equation of *imago Dei* and Man seems to pertain to a less orthodox view of the relationship between Man and God. For instance, Meister Eckhart believed that Man is *imago Dei* in the sense that the divine and human intellect is the same; McGinn, *The Harvest of Mysticism*, 147.

of God 'immediately, incomprehensibly and essentially' forms the mind.[61] Here Van Helmont identifies the Image of God as the *Logos*-Word, or the Holy Spirit. On other occasions, the boundary between the *imago Dei* and the mind becomes indistinct. Indeed, there are several instances where Van Helmont straightforwardly affirms that the mind actually is the 'glorious Image of God'.[62] This tendency seems to betray the influence of German mysticism.

Van Helmont's perspective on the mind as *imago Dei* has important consequences. He affirms that, since God is incomprehensible in Himself, the mind also shares His apophatic quality, being inexpressible and unimaginable. Thus, he develops a negative anthropology of the intellect in the style of Gregory of Nyssa, Meister Eckhart and John Scotus Erigena.[63] We can only know things about the mind by its works or by revelation, rather than directly.[64]

Furthermore, just as God is fundamentally simple and unique, so is the mind as His image. Thus, Van Helmont defines the mind as 'a unique, pure, simple, formal, homogeneous, undivided, and immortal act, which the incomprehensible Image of God immediately, incomprehensibly and essentially consists and forms'.[65]

Van Helmont's emphasis on the oneness of the mind goes even further than Augustine's, who famously posited a threefold structure of the intellect.[66] By formally rejecting the Augustinian doctrine that equated the image of God with the triad of memory, intellect and will,[67] the Flemish philosopher may have drawn his inspiration from Johann Tauler, who maintained that this doctrine was insufficient to account for the dignity of man's soul as *imago Dei*.[68] Yet Van Helmont's criticism of the Augustinian triad is more specific than Tauler's, and relates to the nature of these 'components' of the mind. In his quest for absolute unity, Van Helmont emphasises that the essence (*essentia*) of the mind is simple and unitary,

---

[61] Ch. 38, 'Imago mentis', §38, 219, ch. 101, 'Intellectus Adamicus', 562.

[62] Ch. 38, 'Imago mentis', §22, 216; see also ch. 50, 'Nexus sensitivae & mentis', §10, 286; ch. 93, 'Demonstratur thesis', §13, 524: '. . . homo autem, cum sit Dei immediata ac vera imago . . .'

[63] Giulio Maspero, *Trinity and Man: Gregory of Nyssa's Ad Ablabium* (Leiden: Brill, 2007), 113–14; Bernard McGinn, *Growth of Mysticism: Gregory the Great through the Twelfth Century* (New York, NY: Crossroads Publishing, 1994), 104–6; McGinn, *Harvest of Mysticism*, 121.

[64] Ch. 43, 'Mentis complementum', §1, 250: 'utpote quibus duobus verbis fere explicantur, quaecunque de humana Mente ex fide habemus; nihilque est prorsus, quod nos in ejus apertam cognitionem deducat. Quapropter quicquid de illa scrutamur, adhuc obvolutum est tenebris, ac plane ignotum'.

[65] Ch. 38, 'Imago mentis', §38, 219: 'Siquidem mens est unus actus, purus, simplex, formalis, homogeneus, indivisus, & immortalis, in quo, proxime Dei imago incomprehensibilis incomprehensibiliter essentialiterque consistit, mentemque format'. Here Van Helmont was probably inspired by Pseudo-Dionysius primarily, but also Thomas Aquinas, who insisted on the absolute simplicity of the divine nature. See McGinn, *The Harvest of Mysticism*, 28.

[66] Giglioni, *Immaginazione e malattia*, 172.

[67] Augustine, *On the Trinity*, I.5, 15:21:41. Augustine also talks about the triad *mens—notitia—amor* in book 9, but he eventually seems to settle on the *memoria-intelligentia-voluntas* from book 10 onwards. See also Rowan Williams, 'De Trinitate', in *Augustine Throughout the Ages: an Encyclopedia*, ed. Allan Fitzgerald and John C. Cavadini (Grand Rapids, MI: Wm B. Eerdmans, 1999),845–51.

[68] Tauler differs from Thomas Aquinas in this respect; McGinn, *The Harvest of Mysticism*, 245.

so the Augustinian triad can only be powers (*potestates*) or faculties of the mind.[69] This means that they are instruments that do not represent an intrinsic and eternal structure of the mind. Essentially, they are only useful while Man is trapped in the current fallen state.

To clarify his view, Van Helmont gives his own account of the nature of the mind in heaven. He observes that the function of memory becomes redundant, since time no longer exists, and the soul is in a tranquil state, enjoying 'naked truth' (*nudae veritatis aspectu*) without the striving of intellect.[70] Similarly, the power of will is obliterated, because in heaven one cannot will anything except what God wills.

Having thus rejected the existence of the traditional Augustinian triad within the essence of the mind, Van Helmont goes on to posit his own version of a ternary intellectual structure. According to him, the mind's substance incorporates understanding (*intellectus*), will (*voluntas*) and love (*amor*).[71] These three, he points out, should not erroneously be seen as separating the unified essence of the mind, but as being co-united with it. Van Helmont describes them as indistinct 'titles' (*tituli*) of the intellectual essence. These 'titles' always act in unison, without separation or difference: 'in its pure simplicity, the intellect understands God, is intent upon Him, and loves Him with its entire mind, with an undivided act of love, and a unique act of complacency or desire'.[72]

The understanding [*intellectus*] is that attribute of the mind that can achieve the contemplation of God not by actively, but passively, once it is penetrated by the supreme light.[73] The intellect is so fundamental to the mind that in a sense it is synonymous with it.[74]

In its turn, the 'intellective will' is a superior essence completely unlike the free will of man or the will of the flesh.[75] It is dissolved (*colliquata*) and united with intellect and love, being an image of God's own will.

The third 'title' is love, which is also described as 'perpetual desire' (*desiderium perpetuum*).[76] There is no substantial difference between love and desire in the mind. In this sense, we should understand desire to be radically different from fleshly passion, being a 'ruling essence, and a glorifying act'.[77] Moreover, in the mind, desire never ceases, even after reaching its object. Thus, Van Helmont postulates that, in heaven,

---

69 Ch. 38, 'Imago mentis', §20, 216.

70 Ch. 38, 'Imago mentis', §24, 216.

71 Ch. 38, 'Imago mentis', §31, 217.

72 Ch. 38, 'Imago mentis', §32, 217–18: 'Eo quod intellectus Deum intelligit, intendit, ac omni mente amat, indiviso amoris unicoque complacentiae, seu desiderii actu, in omnimoda simplicitate sui'.

73 Ch. 38, 'Imago mentis', §31, 217.

74 Ch. 38, 'Imago mentis', §38, 219.

75 Ch. 38, 'Imago mentis', §32, 217: 'Voluntas ergo & amor hujus loci, nequicquam commune habent, cum voluntate viri, aut carnis'.

76 Ch. 38, 'Imago mentis', §31, 217.

77 Ch. 38, 'Imago mentis', §31, 217: 'Nec demum amor ille, est passio: sed essentia dominans, & actus glorificans'.

the mind remains in a state of desire, as a feeling of full satisfaction would bring about a certain insensibility or boredom, not in keeping with his idea of paradise.

The model of Van Helmont's intellectual triad is not hard to identify: the divine Trinity itself. Just like the Trinity, Van Helmont's triad is one substance (the *mens*) and three unified hypostases (will, understanding, and love).[78] The intellect resembles God the Father, with the will being like God the Son. Love, as the Holy Spirit in the Catholic scheme, originates from both of them.[79]

Clearly, Van Helmont's triadic view of the mind is still Augustinian in spirit, as it posits a Trinitarian resemblance within the human soul. Augustine himself had proposed several triads, not attempting to be doctrinal in this respect.[80] He also described the triads as mental processes that should be seen as part of the same reality, the same substance, and the same mind. Hence Van Helmont's originality does not rest in his postulation of three inseparable titles in the mind, or even in his choice of them. Rather, it should be seen in his insistence that in order to understand the mind one must not consider it as it is now, but how it will be once the Original Sin is removed. Van Helmont challenges the reader to think of the mind from a postlapsarian perspective.

The Sensitive Soul

The nature of the mind as *imago Dei* justifies the privileged status of Man in the economy of the universe. Nevertheless, after Adam and Eve's Fall, the sensitive soul (*anima sensitiva*) assumed the role of the mind. Superseded by the sensitive soul, the mind becomes drowsy and retires into the innermost place in the body. It turns inward in order to avoid being tainted by its association with the inferior *anima sensitiva*.[81] The animal sensitive soul is an error, a projection of the body and a dark companion to the divine light of the *mens*. Yet, given the catastrophe of Original Sin, it is also necessary as a 'vicar' of the mind, performing the offices initially entrusted to the *mens*.[82]

Still, in Helmontian philosophy the mind does not completely desist from participating in human life. It does not retreat into a Platonic realm of intangible Ideas, but remains in close contact with the sensitive soul. Van Helmont imagines it as projecting an unceasing beam of vital light upon the sensitive soul, endowing it with life.[83]

Since the mind inhabits the body together with the sensitive soul, Van Helmont calls this duality a 'Duumvirate'.[84] By using this political metaphor, he expresses

---

78 For a presentation of the orthodox concept of God as one substance and three hypostases, see Lossky, *The Mystical Theology*, 41–61.

79 Ch. 43, 'Mentis complementum', §7, 251.

80 Stephen McKenna, 'Introduction', in Saint Augustine, *The Trinity* (Washington, DC: Catholic University of America, 1981), xi–xiii; Augustine, *On the Trinity*, 10.11.18.

81 Ch. 50, 'Nexus sensitivae & mentis', 285–6, §9, 286, ch. 101, 'Intellectus Adamicus', 561.

82 Ch. 22, 'Ortus formarum', §80–81, 118; also ch. 2, 'Confessio authoris', §6, 12.

83 Ch. 50, 'Nexus sensitivae & mentis', §10, 286.

84 Ch. 46, 'Duumviratus', §15, 276. The term alludes to the (failed) ancient Roman experiment of the 'triumvirate', or three equal governors of the state.

very effectively the tension that exists between the immortal and mortal soul. The double governorship of the body is a necessity brought about by Original Sin, and consequently it is particularly cumbersome for the mind. The mind, hindered by the sensitive soul, is unable to 'freely exercise its offices' and is forced to 'suffer against its will'.[85]

Van Helmont feels that he has to place the Duumvirate in a precise location in the human body. This decision reflects the medical-Galenic tradition, upheld by Paracelsus as well, according to which souls were found in specific organs.[86] Van Helmont disagrees with the Scholastic view that the soul is in the entire body.[87] Instead, he places the Duumvirate in the upper stomach, a fact which shows the supremacy of digestion and fermentation in his thought.[88]

As an inferior entity to the mind, the sensitive soul imitates the former's nature just like the mind imitates God.[89] Thus, the sensitive soul 'copies' the mind's unity of understanding, will and love by producing its own lower triad, respectively imagination (or reason), will and love (or desire).[90] The scheme is, once again, Augustinian: the African Father had also postulated that there is an inferior trinity in the 'outer man'.[91] Yet Van Helmont views this triad as a quasi-chaotic state where

> we understand things that are not desirable, we desire things we do not wish or plainly know . . . Consequently, desire overcomes will, and in turn will compels desire, so that there are mutual and contradictory commands.[92]

Such a chaotic state reflects Van Helmont's definition of the fallen state as anarchical and disordered. In this sense, Original Sin is fundamentally a disintegration of the unity of the mind.

Yet no faculty of the sensitive soul illustrates its inferior state quite as clearly as reason, by which Van Helmont understands Aristotelian logic and syllogism.[93] The Flemish physician rejects the notion that the higher mind is represented chiefly by the rational faculty. By contrast, he expels reason from the faculties of the mind and

---

85 Ch. 50, 'Nexus sensitivae & mentis', §11, 286.

86 On this tradition, see Hasse, 'Plato Arabico-Latinus', 41–2.

87 Des Chene, *Life's Form*, 191–2. Van Helmont appears to contradict Augustine in this respect.

88 For more on this see above, Chapter 4, on Nature.

89 Here again Van Helmont expresses the belief in the Pseudo-Dionysian concept of hierarchy; Pseudo-Dionysius, 'The Celestial Hierarchy', 153–5.

90 See ch. 38, 'Imago mentis', §35–6, 218. The belief that in the fallen soul understanding is the same as imagination is linked with Van Helmont's belief that we cannot comprehend anything except by images and figures.

91 Augustine, *On the Trinity*, 11.1, 5.

92 Ch. 38, 'Imago mentis', §46, 219: 'intelligimus non desiderata, desideramus quoque quae nollemus, nec plane noscimus. . . . Unde contingit, desiderium superare voluntatem, ac vicissim voluntatem cogere desiderium: adeoque esse mutua atque pugnantia imperia'.

93 Ch. 7, 'Logica inutilis', 34–5.

places it in the fallen sensitive soul. It is, as Giglioni has pointed out, a product of imagination rather than a stand-alone faculty.[94]

Henceforth, Van Helmont devotes his treatise 'Venatio scientiarum' ('The Hunting of Sciences') to refuting the commonplace idea that reason is the most important faculty of the soul.[95] In fact, it is a 'thick and dark little cloud', which beguiles men into thinking that without it they would succumb to the allurements of the senses. Reason is described as a 'pleasing flatteress and crafty seducer', a 'parasite' which enslaves and drives souls mad.[96] The irony of an irrational reason is surely intentional.

Van Helmont further points out why he thinks reason is inadequate. For one, it offers no certainty: if ten witty men are brought together, they will all differ in their position, each being deluded by his reason.[97] Reason divides rather than unites, hence belonging to the senses rather than to the unified understanding.[98] Further, it is not useful in natural inquiry, because it bears no real relationship with the object analysed. In fact, reason is wholly unsuitable to the study of Nature, as it sows discord, contention and misunderstanding. Van Helmont also takes issue with the Platonic concept of *anamnesis*, rejecting the idea that the practice of syllogism helps in the process of remembrance. He does not believe that knowledge could be obtained by memory; for Van Helmont, knowledge is not aprioristic, but something to be attained empirically.

Van Helmont is eager to differentiate between reason and understanding. The former is a cumulative process of thought, while the latter is an immediate revelation of the truth.[99] Moreover, Van Helmont observes, reason is only needed during this lifetime, not in heaven: because at that point the mind will know everything at once, without premises, syllogisms and disputations.[100] In fact, Van Helmont takes his argument further by affirming that reason is the essence of the Tree of Knowledge of Good and Evil, which has caused our downfall. He hence concludes that reason generates only a 'dark knowledge' (*caliginosa scientia*) which is not illuminated by God.[101] It pertains to the inferior sensitive soul; Van Helmont believes that animals also have reason.[102]

'Venatio scientiarum' belongs to a special strand of Western thought, which emphasised divine illumination rather than logic as source of true understanding. This was a particular characteristic of the German mystical tradition, as affirmed by Meister Eckhart (c. 1260-c.1327), Heinrich Suso (1300–1366), Tauler, Mechtilde of Magdeburg (1210–1285) and the movement of *Devotio moderna*.[103] A product of

---

94 Giglioni, *Immaginazione e malattia*, 29–30.

95 Scholars have focussed extensively on this treatise. See Pagel, *Joan Baptista Van Helmont*, 19–34; Giglioni, *Immaginazione e malattia*, 26–35; Heinecke, *Wissenschaft und Mystik bei J. B. Van Helmont*, 53–63.

96 Ch. 4, 'Venatio scientiarum', §1–14, 17–18.

97 Ch. 4, 'Venatio scientiarum', §24, 20.

98 Ch. 7, 'Logica inutilis', §7, 35.

99 Ch. 4, 'Venatio scientiarum', §31, 20.

100 Ch. 4, 'Venatio scientiarum', §10–11, 18.

101 Ch.4, 'Venatio scientiarum', §21, 19.

102 Ch.4, 'Venatio scientiarum', §34, 21.

103 The most famous product of *Devotio moderna* was Thomas à Kempis's work *Imitation of Christ*, which Van Helmont cited on several occasions.

this environment, Nicholas of Cusa introduced the idea of 'learned ignorance' into mainstream Catholic thought.[104] As Pagel has already pointed out, Van Helmont's speculation on reason is closely styled on Nicholas of Cusa's treatise, *De venatione sapientiae*.[105] The Flemish philosopher repeats many of Cusa's arguments, including those that reason is a non-entity (*non-ens*) and that logic and truth come from separate sources.

However, Van Helmont's emphatic attack on reason should not be understood as advocating irrationality. He does not deny the importance of reason and logic in social dealings, or advocate the lack of *logos* in Nature. He simply points out that knowledge obtained by Scholastic logic is incomplete and often misleading.

Self- Knowledge and *Scientia*

Mystical Knowledge of One's Mind

Since the Original Sin did not drive mind away from the body, the intellect can still be re-discovered through self-knowledge. Yet the process is not easy: Van Helmont describes it as a form of mystical experience that comes from God: 'knowledge of the self can only be expected from its Fountain and Governor'.[106] In this sense, Helmont believes that the Grace of God is the author of true knowledge.

In order to re-discover the mind, Van Helmont recommends following the venerable *gnothi seauton*, 'know thyself' edict of the Platonic-Socratic tradition. This tradition had been embraced by Christian Patristic writers as well, beginning with Augustine.[107]

For Van Helmont, true knowledge transcends any instrument of the sensitive soul, including the senses, reason or imagination.[108] To go beyond sense-perception, Van Helmont recommends ascetic practices:

> [C]ontemplations, continued prayers, vigils, fasts, and acts of mortification have the purpose that, the somnolence of the flesh being vanquished, men may obtain that nimble, active, heavenly, and free power directed toward God, and may sweetly confer with Him in His presence, He who demands to be worshipped in the Spirit, that is, in the ground [*fundus*] of the interior man.[109]

---

104 Donald Duclow, *Masters of Learned Ignorance: Eriugena, Eckhart, Cusanus* (Aldershot: Ashgate, 2006), particularly 307–25.

105 Pagel, *Joan Baptista Van Helmont*, 28–9.

106 Ch. 47, 'Tractatus de Anima', §4, 278: 'Ergo aliunde cognitio nostri sperari nequit, quam à suo fonte, & Rectore'.

107 Ch. 4, 'Venatio scientiarum', 17; Augustine, *Soliloquies*, I.3, 1.2.7; *On the Trinity*, 10.

108 Ch. 38, 'Imago mentis, §4, 213.

109 Ch. 113, 'De magnetica vulnerum curatione', §105, 611: 'spectant contemplationes, continuatae orationes, vigiliae, jejunia, & mortificationis actus, ut devicta nempe carnis somnolentia, potestatem illam agilem, coelestem, & expeditam in Deum nanciscantur, eumque coram alloquantur suaviter, qui non nisi in spiritu, id est, interioris hominis fundo, adorari efflagitat'.

According to Van Helmont, ascetic practices, such as fasting, can stir the soul in such a way that intellectual dream-visions may appear.[110] These provide insight into the mind, but are presented in the form of riddles. As Giglioni remarked, Van Helmont retains a view of the *visio intellectualis* inherited from Augustine.[111] The Flemish philosopher claims that there is a difference between the 'images of the phantasy' and the 'images of the intellect', the latter of which can penetrate to the very 'centre of the soul'. He appeals to the Biblical passage 'Night unto night showeth knowledge', to justify the view that dreams can provide insight into the mind.[112] Indeed, his own mystical experience of the self is facilitated by a sequence of dream-visions, whereby the mind reveals itself as an ineffable crystalline light.[113]

At the same time, Van Helmont emphasises that such intellectual images are neither sufficient nor revelatory enough:

> For no matter how beautiful the vision was, my mind did not obtain any perfection by it: for I knew that my mind in the dreaming vision had acted as a third person, nor was the representation worthy of my ardent wish.[114]

In other words, images maintain the dualism subject-object, and knowledge is indirect and non-unified.

Superior to this vision of the self is the exercise of something which Van Helmont calls 'the prayer of silence' (*oratio silentii*), 'wherein the delights of God are adored'.[115] This practice allows the person to distinguish between the operations of the mind and those of the sensitive soul.[116] The Christian origins of silent prayer hark back to Pseudo-Dionysius or even earlier; it can be found in the mysticism of Eckhart, Tauler or Teresa of Avila.[117] Closer to Van Helmont, the Paracelsian Oswald Croll maintained that 'God is prayed unto in the internall Spirit, not with noyse of words, but in a sacred silence'.[118]

---

110 Ch. 4, 'Venatio scientiarum', §42, 22.

111 Giglioni, *Immaginazione e malattia*, 37; Augustine, *The Literal Meaning*, 12.7.

112 Ch.4, 'Venatio scientiarum', §42, 22; the quotation originates from Psalms 19:2.

113 Ch. 38, 'Imago mentis', §13, 214–15; see also ch. 2, 'Confessio authoris', §4–8, 12.

114 Ch. 38, 'Imago mentis', §14, 215: 'Nam utut pulchra esset visio: attamen inde nequicquam perfectionis sibi mens acquivisit. Cognovi namque, quod mens in somniali visione, velut personam tertii egisset, nec esset tanto optatui digna representatio'.

115 Ch. 43, 'Mentis complementum', §15, 253: '. . . per Orationem silentii in spiritu, in quo deliciae Dei sunt adorari'. See also ch.4, 'Venatio scientiarum', §16, 19; ch. 47, 'Tractatus de Anima', §9, 279; ch. 50, 'Nexus sensitivae et mentis', 285.

116 Ch. 43, 'Mentis complementum', §6, 251. See also ch. 47, 'Tractatus de Anima', §10, 279.

117 Pseudo-Dionysius, 'The Divine Names', 50. Meister Eckhart portrays the Ground (*Grunt*) of God as 'simple silence': Pr. 48 (DW 2:240.7–421.3); see also McGinn, *The Harvest of Mysticism*, 119. For Teresa of Avila's views of the 'prayer of quiet' or 'supernatural prayer', see Harvey D. Egan, *Soundings of the Christian Mystical Tradition* (Collegeville, MI: Liturgical Press, 2010), 244–45. Van Helmont mentioned Teresa as an authority.

118 Croll, 'Admonitory Preface', 72.

This inner prayer is conceived as being wholly detached from sense and reason. Van Helmont brings Pseudo-Dionysius to bear witness to this process.[119] The mind must first disbar itself from all that keeps it attached to mortal things, thus freeing itself to look upon God alone. This activity of freeing one's self is called by Van Helmont 'unclothing' (*denudatio*). The image of removing one's garments, drawn from the Song of Songs 5:7, is given a particular twist: by clothes, Van Helmont implies the workings of reason, which must be rejected by the mind and replaced by the ardent love of God.[120]

The process of detachment needs to be accompanied by a renunciation of all thought, contemplation or images, and indeed the cessation of all mental activity:

> I desisted from my superficial wishing, seeking and investigating; I stripped myself of curiosity and appetite for knowledge, and withdrew myself in quietude or poverty of spirit, resigning myself into the most benevolent will of God, as if I were neither existing, nor moving: desiring nothing and understanding nothing.[121]

According to Van Helmont's doctrine, the mind has to render itself completely passive in order to receive the influx of the divine light.[122] Passivity had deep roots in the German mystical theology as advocated by Eckhart and Tauler and upheld by the *Theologia Germanica*.[123] Amongst the Paracelsians, Croll in particular maintained that knowledge was obtained 'by a passive reception of Divine things, not by study and paines, but by patience and submission'.[124]

Van Helmont makes it clear that illumination is received, not willed; it is 'suffered', not actively acquired.[125] In his voluntarist view, the summit of wisdom and sainthood is achieved when God is allowed to impose His own will, and Man obliterates his own. The idea of the annihilation of the will can be found in medieval mystical tradition, and also in Paracelsus.[126]

Such profound renunciation discontinues all mental activity, except for amorous desire, which lifts up the mystic to the seat of the mind.[127] Van Helmont calls this

---

119 Ch. 43, 'Mentis complementum', §13, 252.

120 Ch. 43, 'Mentis complementum', §21, 254.

121 Ch.4, 'Venatio scientiarum', §43, 22: 'destiti angustius aliquid optare, quaerere, & investigare, denudavi me omni curiositate, & appetitu sciendi, ad quietem, sive paupertatem spiritum, me recepi, resignans me in amabilissimam Dei voluntatem, in quasi non esse, non operari, in merum nil desiderare, nil intelligere'.

122 Ch. 43, 'Mentis complementum', §21, 253.

123 On Tauler's notion of passivity, see Richard Kieckhefer, 'The Notion of Passivity in the Sermons of Tauler', *Recherches de théologie ancienne et medievale* 48 (1981), 210. For the *Theologia Germanica*, see Steven E. Ozment, *Mysticism and Dissent: Religious Ideology and Social Protest in the Sixteenth Century* (London: Yale University Press, 1973), 14–45. A mention of the *Theologia Germanica* was made in Van Helmont's papers confiscated by the Inquisition.

124 Croll, 'Admonitory Preface', ch. 1, 49.

125 Ch.4, 'Venatio scientiarum', §60, 26.

126 Bergengruen, *Nachfolge Christi*, 51.

127 Ch. 43, 'Mentis complementum', §21–2, 254.

state a 'total amorous offering' (*totali oblatione amorosa*), whereby one offers his own self as sacrifice. Yet, he emphasises, not even this profound renunciation can ensure divine illumination. In the Augustinian tradition, he insists that God reserves His Grace to whom he wants.[128] Such affirmation may sound as if Van Helmont, like the Protestants, is suggesting that works are useless. However, he further states that one must strive to deserve Grace. In this sense, the bestowal of Grace can be seen as a confirmation of merit.[129]

Having reached a superior level of purification, the soul thinks of itself directly, without consideration of time, place, or circumstance.[130] The process is viewed in markedly sexual terms, resembling the mystical tradition based on the Song of Songs.[131] As the soul offers itself 'nude', the light of truth 'penetrates' it. As a result of this penetration, a divine form [*Deiforma*] is born within the mind, as in 'the foetus of a pregnant woman'. This God-form is nothing other than the purified mind, ready to receive the in-flowing grace of God.

Van Helmont's supreme expression of mystical union is metaphorically expressed by the concept of the Rabbinic *Binsica*, death by God's kiss. The idea of the *binsica* originated in a sixth-century Jewish Midrash on the Song of Songs.[132] According to this work, the soul of Moses was taken to God with a kiss. The philosopher Moses Maimonides interpreted the kiss to mean ecstatic death, while later Jewish theologians transformed the kiss 'into a metaphor for the soul's abandonment of earthly attachments and ascent toward God'.[133] The concept was transferred into Christian thought by Pico della Mirandola's *Commentary on a Canzone by Benivieni*, and adopted by other thinkers such as Francesco Giorgi and Oswald Croll.[134]

In adopting the motif of the mystical *binsica*, Van Helmont disagrees with the standard Rabbinic and Jewish medical opinion that the kiss was a fatal sickness

---

128 Ch. 60, 'Pharmacopolion ac Dispensatorium modernorum', §14, 369.

129 Ch. 101, 'Intellectus Adamicus', 562.

130 Ch.4, 'Venatio scientiarum', §48, 23: 'Ast cum anima cogitate seipsam, at aliud quidquam, tanquam seipsum, sine alteritate cogitantis, & rei cogitatae fine appendentia, extraversione, vel relatione ad durationem, locum, & circumstantias'.

131 Michael Casey, *A Thirst for God: Spiritual Desire in the Bernard of Clairvaux's Sermons on the Song of Songs* (Kalamazoo, MI, 1987), 192–5.

132 B.C. Novak, 'Giovanni Pico della Mirandola and Jochanan Alemanno', *Journal of the Warburg and Courtauld Institutes* 45 (1982): 140; see also Moshe Idel, *Kabbalah in Italy (1280–1510): A Survey* (New Haven, CT: Yale University Press, 2011), 141–6.

133 Novak, 'Giovanni Pico', 140.

134 Giovanni Pico della Mirandola, *Commentary on a Canzone of Benivieni*, trans. Sears Jayne (Frankfurt: Peter Lang, 1984), 150. It also appears in his 900 theses; Sharon Farmer and Stephen Farmer, eds., *Syncretism in the West, Pico's 900 theses (1486): the evolution of traditional, religious, and philosophical systems: with text, translation, and commentary* (Michigan: Medieval & Renaissance Texts & Studies, 1998), 489. See also Michael J.B. Allen, 'The Birth Day of Venus: Pico as Platonic Exegete in *Commento* and *Heptaplus*', in *Pico della Mirandola: New Essays*, ed. M.V. Dougherty (Cambridge: Cambridge University Press, 2008), 90. Francesco Giorgio, *De harmonia mundi totius: cantica tria* (Venice: In Aedibus Bernardini de Vitalibus, 1525), III, vi, 18; Croll, Admonitory Preface', 216.

caused by the atrophy of imagination.[135] According to Van Helmont, Pico himself wrongly believed that 'unless the operator uses a medium, he will soon die of *Binsica*, or dryness of the brain'.[136] On the contrary, *binsica* does not result in death, first, because 'the instruments of phantasy do not operate in this act, but sleep unmoved, as if they did not exist'; secondly, because the act originates from God: 'I believe that the principal act is of [divine] mercy'.[137] Hence Van Helmont defines *binsica* as a form of divine illumination that does not damage life, because God Himself is life and light. In fact, Van Helmont believes the kiss of God to be a supreme act of divine love.

Being thus close with God, the amorous soul experiences 'a liquefaction of the mind' (*liquatio mentis*), equivalent with an outpouring of light.[138] The mind essentially becomes a true mirror of God, 'reflected in the glass of its own divinity'. For all practical purposes, the intellect becomes God in that single instance.

However, Van Helmont carefully shies away from the potentially heretical affirmations of the complete loss of identity in divinity, or the assumption of divinity by the mind. He insists that the mind remains the Image of God, not God Himself: 'although the intellect in the intellectual act transforms itself into the species of the thing understood: yet it keeps its own unmixed property and essence to which it immediately returns once this act ceases'.[139] In other words, the mind remains a created being, not God Himself.

Yet there are also lasting effects, because by contact with God the mind becomes 'sprinkled with a new dew of perfection'.[140] Once the path is known, it can be reiterated; indeed, the mystic is always eager to return to this superior state. The result of several mystical experiences is a perfecting of the mind. Our intellect 'is the more ennobled, the more it suffers the light which is beyond nature'.[141]

In his theory of mystical knowledge, Van Helmont draws on the rich Pseudo-Dionysian tradition, with its emphasis on intellectual union with God. More specifically, Van Helmont is profoundly influenced by the medieval mysticism of such thinkers as Meister Eckhart and Johann Tauler. We have seen that *De magnetica vulnerum curatione* freely borrows Tauler's concepts of the inner and outer man and the Ground of the soul. In his autobiography ('Studia authoris'), Van Helmont also

---

135 Ch.4, 'Venatio scientiarum', §51, 24. For a description of the medical *binsica,* see Bartholomew Parr, *The London Medical Dictionary* (London: Ames Mitchell and White William Brown, 1819), entry 'binsica'.

136 'Tumulus Pestis', ch. 10, 'Animal phantasticum', 859: 'Opinatur tamen Picus, nisi operator utatur medio, ipsum mox oppetiturum Binsica, sive ariditate cerebri'.

137 Ch.4, 'Venatio scientiarum', §51, 24: '. . . quia organa Phantasiae in hoc non laborant, dormiunt vero immota, tanquam si non essent . . . Credo namque principalem illum actum, esse clementiae'.

138 Ch. 43, 'Mentis complementum', §21, 254; see also ch.4, 'Venatio scientiarum', §49, 24.

139 Ch.4, 'Venatio scientiarum', §52, 24: 'Sic tamen, ut licet intellectus, actu intellectuali, se transformet in speciem rei intellectae; servet tamen impermistam suam proprietatem, & essentiam, in quam iterum redit, mox atque ab illo actu cessaverit'.

140 Ch.4, 'Venatio scientiarum', §48, 23: 'novo velut perfectionis rore perfunditur'.

141 Ch.4, 'Venatio scientiarum', §63, 26: 'Quod eo plus nobilitetur, quo plus patitur à lumine, quod est supra omnem naturam'.

mentions his reading of Tauler as having increased his desire to know the supreme truth.[142] Although the late Van Helmont criticised Tauler's soul dualism and his 'Stoicism', his concepts and approach to the mystical experience remain strongly influenced by Eckhartian and Taulerian thought. The language of his mystical experience is permeated with expressions drawn from such sources. Van Helmont portrays mystical ecstasy as a 'turning inward' (*introversio*).[143] He also refers to the process of silent prayer as 'poverty of the spirit', 'naked poverty' and an emptying of one's self, a theme that recurs in German mysticism.[144]

The Possibility of Scientia *of Nature*

Like many other philosophers of his era, Van Helmont believed that Man was primordially endowed with true knowledge of Nature.[145] This conviction draws on Genesis 2:19–2:20, where undefiled Adam is asked to name all creatures. Van Helmont hence affirms that Adam, as the image of God, 'had an intimate or intuitive knowledge (*scientia*) [of things], which is called the Attainment of Nature (*Adeptum naturae*)'.[146] The first Man 'had known the essences and names of all living creatures, because he contemplated these things within, in his own divine image'.[147] He was, Van Helmont argues, the perfect natural philosopher. Consequently, the only method whereby Man can recover this profound comprehension of Nature is by restoring the image of God within the soul.

Van Helmont's notion of *scientia* has little in common with the modern term of 'science'. Helmontian *scientia* is a form of direct and intimate perception of creatures that is only given to chosen human beings. This view is clearly indebted to alchemy, where knowledge was bestowed only by God as *donum Dei*.[148] It also betrays Van Helmont's continuing indebtedness to Paracelsus, who held that the perfect *scientia* of Adam could be recovered, albeit through an arduous path.[149]

During his trial, Van Helmont was accused of barring access to knowledge to all but a chosen few. He replied that he was not disparaging all human learning, but

---

142 Ch. 3, 'Studia authoris', §7, 15.

143 Ch. 47, 'Tractatus de Anima', §3, 278.

144 Ch. 4, 'Venatio scientiarum', §43, 22.

145 See Peter Harrison, *The Fall of Man and the Foundations of Science* (Cambridge: Cambridge University Press, 2007), particularly chapter 1.

146 Ch. 101, 'Intellectus Adamicus', 561: 'Adam justa nomina indebat cunctis animalibus: horumque ideo intimam, sive intuitivam, scientiam habebat, quae Adeptum naturae dicitur'.

147 Ch. 50, 'Nexus sensitivae & mentis', §4, 285: '... Noverat quippe essentias, & nomina animantum: quia Mens, in divina sui imagine, haec intus contemplabatur dum volebat, ipsoque aspectu eadem memorabat'; see also ch. 93, 'Demonstratur thesis', 527; ch. 101, 'Intellectus Adamicus', 561; ch. 24, 'Natura contrariorum nescia', §5–6, 132.

148 On this topic, see Vladimir Karpenko, 'Alchemy as Donum Dei', *Hyle* 4 (1998), 63–80.

149 James Joseph Bono, *The Word of God and The Languages of Man: Interpreting Nature in Early Modern Science and Medicine*, 2 vols (Wisconsin: Univ of Wisconsin Press, 1995), 131–4.

that divine illumination provided a higher degree of certainty.[150] Indeed, in *De magnetica vulnerum curatione*, he had argued that there are two forms of knowledge: the inferior knowledge, which is chiefly rational inquiry, and superior or true knowledge (*scientia*), which is obtained through illumination.[151] The inferior kind gives only knowledge of surfaces, but *scientia* reveals the internal essence of things.

Such beliefs led Van Helmont to reject the view that only a 'science of appearances' can be achieved. He implicitly disagreed with his friend Gassendi on the latter's Pyrrhonian denial of *scientia* in *Exercitationes* and with Mersenne's belief that we can only know the exterior effects of things.[152] Instead, Van Helmont held that alchemy was the privileged and unique *scientia* which penetrated within beings. It alone could provide knowledge of causes and constitution of things:

> We read in our furnaces that in Nature there is no more certain kind of science, of knowing things by their radical and constitutive causes, than when we know what and how much is contained in each thing. Thus knowledge and connexion of causes are most clear when hidden things are made to become manifest, as if they talk to you. Real beings, insofar as they stand in their primordial and succeeding seminal principles, that is, in their true substantial entity, do offer knowledge.[153]

This statement is then supported by an extended quote from Pseudo-Lull's *Testamentum*, which points out that the knowledge offered by alchemy purges the intellect from errors. Van Helmont concludes Pseudo-Lull's testimony by affirming that alchemy alone is 'the mirror of true Intellect'.[154] By its method alone the truth of being can be known in its essence and the 'fantastical' presumptions of logic can be erased.

For Van Helmont, *scientia* cannot be divorced from religion. In his mind, *scientia* leads to the one true religion, that of Christianity. Van Helmont maintains that the 'art of the fire' alone is free of pagan concepts, in contrast with other forms of knowledge.[155] We have seen in Chapter 4 on God (under 'Alchemy, Paracelsianism

---

150 See Harline's useful retelling, in *Miracles at the Jesus Oak*, 213.

151 Ch. 113, 'De magnetica vulnerum curatione', §100–106, 611.

152 For Gassendi's point of view, see Richard Henry Popkin, *The History of Skepticism from Erasmus to Spinoza* (Berkeley, CA: University of California Press, 1979), 102–3; for Mersenne, 137, 141–5. Popkin does not discuss Van Helmont.

153 'De lithiasi', ch. 3, §1, 668: 'In nostris furnis legimus, non esse in natura certius sciendi genus, ad cognoscendum per causas radicales, ac constitutivas rerum; quam dum scitur quid, quantumque in re quaque, sit contentum. Ita quidem ut cognitio, & connexio causarum, non constent clarius, quam cum res ipsas ita recluseris, ut coram prodeant, ac velut tecum loquantur. Siquidem entia realia, duntaxat stantia in suis primordialibus, & succedentibus seminum principiis, adeoque in vera entitate substantiali, dant notitiam'.

154 'De lithiasi', ch. 3, §1, 668: 'Spagyria enim sola, est speculum veri Intellectus'.

155 Ch. 19, 'Complexionum atque mistionum', §40, 89: 'quotquot sine Pyrotechnia, Philosophiam aggressae, paganicis hactenus institutis fuere delusae'.

and Van Helmont's Theology') that Van Helmont advocated an important 'didactic' role for alchemy as a way to the true religion of Christianity.

The Mystical Knowledge of Nature

Van Helmont rejects any concept according to which one could successfully inquire into Nature without having previously been illuminated by God:

> Whoever is inapt to initially comprehend the motions, exercises, effects, and quiddity of the immortal mind shall also be unfit to understand the *arcana* (secrets) of nature, which are more remote from the mind than it is to itself.[156]

This Helmontian doctrine had medieval precedents: Bonaventure, for instance, believed that the internal world illuminated the external, and knowledge of the microcosm was prerequisite to that of the macrocosm, while Henry of Ghent denied natural knowledge could be obtained without divine illumination.[157] In the Renaissance, Ficino believed that the intellect contained the world of intelligible forms within, hence the macrocosm was recapitulated within the microcosm.[158]

Van Helmont points out that, contrary to common opinion, Nature is no easier to understand than the mind, being in fact just as hidden and difficult. Hence, he argues,

> [I]t is simply false that the knowledge of the mind is more difficult than the knowledge of things in themselves [*rerum nuda*]; for this reason it is postponed. In fact, all things are equally unknown to us. This is because the quiddity of all beings is their precise Truth, closed to us, but accessible to Infinity.[159]

In this same passage, Van Helmont goes beyond negative theology and negative anthropology to propose a negative cosmology. The divine essence that is found in all things is unknowable without the aid of God. Elsewhere, Van Helmont derides those that believe that knowledge of water, for instance, is easier than that of the mind, on the basis that water can be directly perceived. This is a fallacy caused by our overreliance on the senses. In reality, ordinary sight only bestows 'knowledge of observation' rather than internal knowledge of the essence. Van Helmont postulates

156 Ch. 47, 'Tractatus de Anima', §1, 277: 'Quicunque enim ineptus est initio, comprehendere mentis immortalis motus, exercitia, effectus, & quidditatem; inhabilis quoque erit, ad intelligendum naturae arcana, à mente remotiora, quam est ipsamet à seipsa'.

157 Jaroslav Pelikan, *The Growth of Medieval Theology (600–1300)* (London: University of Chicago Press, 1978), 305; Gilson, *The Christian Philosophy in the Middle Ages*, 448.

158 Kristeller, *The Philosophy of Marsilio Ficino*, 233–43.

159 Ch. 47, 'Tractatus de Anima', §4, 278: 'Ergo simpliciter falsum, quod cognitio Mentis sit rerum nuda cognitione difficilior, aut postponenda ideo. Quippe omnes sunt nobis aequaliter ignotae. Quia entium quorumcunque quidditas, est eorum veritas praecisa, clausa versus nos, & patens ad Infinitum'.

a fundamental difference between the surface and essence; indeed, 'a visible entity is just as unknown as an invisible one'.[160]

Here Van Helmont distances himself from the Renaissance doctrine of signatures, just as he does from the Paracelsian theory of microcosm–macrocosm correspondence.[161] This is because, as Van Helmont points out, Nature is not the image of Man, so we should not interpret it in accordance with our human perspective.[162] There is no clear relationship between Man and Nature except by virtue of God. In this sense, God is the mediator between the two entities, and is the only source of knowledge for human beings.

Despite this worldview, the microcosm–macrocosm analogy still creeps into Van Helmont's speculation. Perhaps nowhere is it more prominent than in positing the existence of a 'heaven' in Man's body, which generates its own *blas*. The organs are still associated with the planets and are harmonised with them. Similarly, Van Helmont posits that the flatus of the body is the same as the *Gas* in the atmosphere, an idea which he adapts from Hippocrates.

True knowledge of Nature has to be achieved by a mystical experience similar to that of the mind. To understand others, we must first perform an act of renunciation, rejecting all acquired knowledge and pushing ourselves into a state of 'despair'. This is an experience comparable to profound darkness, where no light penetrates.[163]

The process of knowledge is achieved by the 'adaptation' [*adequatio*] of the human intellect to the being investigated. This concept reflects the Thomist definition of truth as the *adequatio* of the intellect to things.[164] However, Van Helmont gives this definition a twist by also calling it in blatantly alchemical terminology a 'transmutation'.[165] Just as in the case of the primary mystical experience, the intellect must liquefy and transform itself into the essence it seeks to understand; thus it becomes its mirror or image. He underlines that such a process of transformation can only occur with the Grace of God, even if it concerns knowledge of nature.[166]

Van Helmont explains this experience further. As already mentioned, there is only one Truth, which is also Essence and Goodness: this is God or, in Van Helmont's Christocentric formulation, Christ.[167] The interpretation of God as Truth was a typically Augustinian tenet which was upheld not only by Van Helmont but also

---

160 Ch. 47, 'Tractatus de Anima', §8, 279: 'Cum tamen alioquin, in entis entitate, tam sit intellectualiter ignotum visibile, quam est invisibile'.

161 Particularly in 'De febribus', ch. 4, §19–21, 748, ch. 14, §8, 775. The Renaissance mentality, based on resemblance, has been famously described by Michel Foucault in his *Order of Things* (London: Routledge, 2002), chapter 2 'The Prose of the World', 19–50. Although Foucault might have exaggerated the universality of this theme in the Renaissance world, it was still an important part of its worldview, and was embraced by Paracelsus and many other Paracelsians, including Oswald Croll.

162 Ch. 60, 'Pharmacopolion ac Dipensatorium modernorum', §5, 367.

163 Ch. 47, 'Tractatus de Anima', §4, 278.

164 Jan A. Aertsen, 'Truth as Transcendental in Thomas Aquinas', *Topoi* 11 (1992): 163–4.

165 Ch.4, 'Venatio scientiarum', §45, 23.

166 Ch.4, 'Venatio scientiarum', §51, 24.

167 Ch.4, 'Venatio scientiarum', §54, 25. See chapter 4 God, section 'Christ as Wisdom'.

by Thomas Aquinas, the Scholastics and by 'new philosophers' such as Mersenne.[168] Thus, once the mind gains access to its own Truth through self-knowledge, it also gains access to the Truth of everything. In this sense, the act of transformation is the same as the act of looking into a mirror: the mind sees its own Truth and in this Truth the essences of all things. Or, as Van Helmont explains, 'the understanding intellect is no more different from the thing understood than a direct beam of light differs from itself, being reflected'.[169] Through the divine Truth, the Essence of the thing is united with the light of understanding.

This account seems to imply that, for Van Helmont, the laboratory can also be seen as a locus of mystical experience and practice, just like the prayer room. Knowledge of a particular natural object involves a spiritual preparation on the part of the natural philosopher. Such preparation could imply the use of ascetic practices and of concentration techniques which can stir up images of the intellect, including dreams. Eventually, for philosophers blessed by God's Grace, their intellect would transform itself into the thing understood, leading to a momentary identification with the investigated entity. This 'insight' would be equivalent to the acquisition of 'true', inner knowledge.

We have already noted above that Van Helmont believed that knowledge of Nature was subsequent to that of the mind. Does this mean that a philosopher should abandon all pursuit of natural knowledge until complete inner knowledge is attained? This may be Van Helmont's 'ideal' scenario; in practice, however, we find that he concomitantly pursued mystical self-enlightening and natural knowledge. This was more in line with the traditional motto, *Ora et labora*, beautifully depicted in Heinrich Khunrath's *Amphitheatrum Sapientiae aeternae*.[170] Moreover, mystical self-knowledge is never final: we have seen Van Helmont argue that it must be 'perfected' in time. Hence, in practical terms, the pursuit of illumination in the prayer room and the laboratory would have overlapped.

Mysticism and Empirical Investigation

Van Helmont's mystical approach towards natural knowledge should not be taken to mean that he was advocating a passive attitude of the natural philosopher, such as praying and waiting for enlightenment.[171] On the contrary, Van Helmont viewed the investigation of nature as a practical pursuit, which involved continuous labour

---

[168] For this theory in Augustine, see Etienne Gilson, *Introduction à l'étude de Saint Augustin* (Paris: Vrin, 1929), 112–25; for Mersenne and Jesuit use of this idea, Dear, *Mersenne*, 84–5, 95–6.

[169] Ch.4, 'Venatio scientiarum', §55, 25: 'Intelligens intellectus, non est diversus ab intellecto, aliter, quam sicut radius lucis directus, à seipso reflexo'.

[170] On this topic, see Peter J. Forshaw, 'Oratorium—Auditorium—Laboratorium: Early Modern Improvisations on Cabala, Music, and Alchemy', *Aries* 10:2 (2010), 169–95. The term was introduced by Benedictine monks, but appropriated by Christian alchemists.

[171] This image of the alchemist, promoted by nineteenth-century occultism, has rightfully been ridiculed by Newman and Principe, *Alchemy Tried in the Fire*, 35–8.

and questioning of received knowledge. Indeed, he scorned those who shied away from empirical investigation, since reasoning alone could never get to the truth of things.[172] He favoured the evidence of the senses, which for him chiefly meant the revelation of the unseen part of the objects by alchemical methods. However, he did not reject other types of investigation, including mechanical experiments and dissection.[173]

Still, Van Helmont did not consider empirical enquiry sufficient. Despite his contempt for university learning, the Flemish philosopher manifested a clearly pedantic attitude toward pure empirics, whom he called 'idiots'. Although the term 'idiot' is borrowed from Nicholas of Cusa, who used it favourably, Van Helmont employs it pejoratively, as denoting those practitioners who do not know the true causes of things.[174] These were primarily the alchemical artisans, but Van Helmont extended the term to false philosophers. This explains why he describes many ideas of Paracelsus as 'idiotisms': essentially, he believed that the Swiss physician was talented at healing, but had no knowledge of causes. Similarly, he dismissed many contemporary alchemists, even trained physicians, as 'beginners' (*tyrones*) who only pretended to know alchemy.

In turn, Van Helmont viewed empirical and philosophical enquiries as being two facets of the same process of knowledge. Philosophy could only be achieved by mystical experience. We are accustomed to seeing empirical scientific pursuit and mysticism as being very different, but Van Helmont lived before the modern division of science and religion. For him, the two pursuits were part of the same process; only when acting in concert could they yield universal knowledge. Both implied seeking an immutable truth outside one's everyday knowledge and preconceptions; both involved coming to terms with the unknown; both involved the revelation of a hidden world; both were embodied experiences involving first the senses and then the understanding. The continuum between the mysticism and empiricism is suggested by the Paracelsian expression 'pray, seek and knock', which ultimately originated in the New Testament. Van Helmont used this quote to express the pursuit of knowledge.[175]

The interplay between empirical analysis and religious-mystical experience can be seen in Van Helmont's description of his alleged discovery of the nature of the kidney stone (*Duelech*).[176] The story runs for several years, since he mentions that

---

172 For instance, he ironically states that 'I am so stupid that I only judge sensible objects by the senses'; 'Ita enim stupidus sum, quod de sensibilibus, non nisi sensuum judicioso', 'De febribus', ch. 1, §6, 739.

173 Although an alchemist foremost, Van Helmont did not shy away from experimenting with weather-glasses or dissecting crabs; Van Helmont, ch. 12, 'Aer', §12–13, 52–3, 'De lithiasi', ch.7, §31–2, 703.

174 'De lithiasi', ch. 2, §15, 667: 'Nam sic plurima experimenta oberrant inter idiotas, quorum tamen causas nesciunt'; see also 'Tractatus de anima', §6, 278: 'Tantum enim de ea scit rusticus, vel idiota, quantum Philosophus'.

175 'De febribus', ch. 14, §11, 776: 'Ut orando, quaerendo, & pulsando, scientiam acquirant'. The saying comes from Matthew 7:7 and Luke 11:9, though there the first verb is 'ask' ('*petite*') rather than 'pray' ('*ora*'). The 'pray, seek, knock' version appears in Theophrastus Bombastus von Hohenheim, Paracelsus gennant, 'De natura rerum', in *Sechster Theil Der Bücher und Schrifften*, ed. Johannes Huser (Basel: Conrad Waldkirch, 1590), 352.

176 This story is found in 'De lithiasi', ch. 2–3, 666–76.

he began investigating the nature of the stone when he was visiting England as a young man. He tells this anecdote: one day he was invited by some noble women to the feast of the Epiphany at Whitehall. The feast lasted for 12 hours, and during this long period he abstained from 'making water'. When he returned home, he was fearful that his urine may have been become sandy from too much coction, as Galenic medical doctrine supposed. Yet, as he pissed through a napkin, he found the liquid clear. The next day he retained the urine in a glass urinal for 12 hours, and observed that, as the urine evaporated, it left sand on the walls of the glass. This experience led him to question two key Galenic teachings: that urine sand was produced by too much coction inside the body, and that it originated from a viscous matter. This sowed the first seeds of doubt in regard to the Galenic theory on the origin of the kidney stone (calculus).

Returning to his home, he continued to think about this matter, and was suddenly 'seized by the truth', which confirmed to him that the real cause of calculus was unknown and could only be learnt from God. This moment he described as a revelation that came to him upon waking in the middle of the night. Consequently he threw away his Galenic books and decided to seek knowledge only from the 'Father of Light, Master of the unique Truth'. Van Helmont considered himself to be 'a large sea of ignorance and an abyss of many shadows, and as such to have no light'.[177]

With this humble attitude, Van Helmont proceeded to analyse urine in the hope that he would know its contents 'visually' (*optice noscere*). Importantly, this programme of experimentation was combined with frequent prayers, and perhaps other forms of devotion as well. First he kept urine until it fermented, then distilled this into a small quantity of *aqua ardens*, obtaining a stinking white salt. With this salt he tried to dissolve calculus, but to no avail. He then tried another experiment: he putrefied urine in horse-dung for 13 days, distilled the product by cohobating it four times (in accordance with Paracelsus's advice), and obtained yellow and sharp crystals. This salt too proved useless for the dissolution of calculus. In his third try, Van Helmont mixed spirit of urine with refined alcohol, and discovered that the two coagulated together in the liquid as a white lump, volatile and subtle. 'My eye especially taught me' (*Oculus inprimis me docuit*), he says, that the spirit of urine was similar to rennet, and that two spirits did not necessarily dissolve each other, but could unite to form a solid body. Van Helmont called this process a 'substantial transmutation' (*substantialis quaedam transmutatio*), since the result was a new being different from the two it originated from.[178]

This third experiment gave him courage to proceed further with his analysis of the kidney stone. He then became convinced that the answer to the calculus lay in salts. Consequently he investigated many salts, including acids and alkalis. In turn,

---

[177] 'De lithiasi', ch. 2, §13, 666: 'Me autem, mare magnum ignorantiae, & tenebrarum multiplicium abyssum, fatebar, & quasi carere omni luce'.

[178] In alchemical terms, Van Helmont calls the action of the spirits on each other as a 'marriage'.

he found that the spirit of human urine was neither acidic nor alkalinic, but a neutral, 'simple salt'.

His chemical knowledge much increased at this point, Van Helmont proceeded to 'dissect' the salt of urine further, obtaining two more fixed salts, besides the spirit of urine. One of these salts was made of sharp crystals, and Van Helmont called this a 'sea salt' (*sal marinum*). The other, which Van Helmont continued to call 'salt of urine' (*sal urinae*), was composed of cubic crystals and stank.

Moving still further, he decided to leave urine for 40 days in a horse belly to achieve further putrefaction and separation of components. He then distilled the product and left a part of it in an expensive glass vessel for 15 days. His choice of vessel was an oversight caused by him being sidetracked by family business and Pentecost celebrations. When he returned to his experiment, however, he was much aggrieved to find that the valuable glass had been stained by the liquor from the urine. Thinking about the cost of the vessel, he regretted his curiosity and, in a fit of anger, threw away the distillate. After he calmed down, however, he began wondering about the nature of the dark stain on the glass. He concluded that this was nothing other than *Duelech* adjoined to the vessel. At this moment, he realised that what he had thought to be carelessness on his part was actually the hand of God helping him out. Enthused by God's favour, he proceeded to distil the other part of the putrefied residue and obtained a spirit that could dissolve the *Duelech* and clear the glass. Van Helmont interpreted all these events as divine will answering his fervent prayers. He concluded that it was this staining spirit that caused kidney stones.

Soon, however, he began to be assailed by doubt about his findings. He realised that his results were brought about through lengthy putrefaction, but it was unlikely that urine would putrefy in the human body to such an extent. At this point he had a crisis of faith, and, he says, rejected all the knowledge he had acquired. This caused a 'torpor' to come over him. Van Helmont does not explain what this torpor is, but it seems to have been a form of illumination. The experience confirmed his belief that *scientia* could only come through intense labour, rejection of received knowledge and divine enlightenment. After further distillations, 'by prayer I attained the causes, the manner of forming and dissolving *Duelech*'. He concluded that divine Wisdom spurned those who rejected knowledge of calculus by the fire.[179]

This account reveals the unusual combination of empirical testing and religious-mystical experience that lay at the core of Van Helmont's *scientia*. It equally suggests that he viewed knowledge as a conjunction of empirical investigation and moments of insight and illumination originating directly from God. Experiment revealed to the eye what had previously been hidden, but true understanding could only be procured through divine intermediation.

179 'De lithiasi', ch. 3, §33, 675: 'Spernit enim eos Sapientia, qui materiam ex qua, dispositiones, contenta, proprietates, progressum, & significationes lotii addiscere recusarunt per ignem'.

Conclusions

For Van Helmont, a key pillar of authentic 'Christian Philosophy' is knowledge of Man as an entity situated between God and Nature. Consequently, Man defines his existence through the relationship he establishes with that which is higher: God and the Trinity, and that which is lower, created Nature.

Most important to Van Helmont is the close relationship between Man and God, as laid out in the Bible. Within the Scriptures, one may find the entire history of Man, and can understand the past, present and future condition of humanity. Yet the Bible is not the only path for the knowledge of God. Like many Paracelsians, Van Helmont firmly believed that knowledge of God is also attainable through direct experience, that is, a mystical comprehension of the self and of God as its Archetype. The mystical union is both the mirror and the kiss of God, and this divine experience bestows understanding. Thus, true knowledge can only originate from a close encounter with God, whereby the soul transforms itself into God's reflection. By becoming the image of God the mind understands itself and its divine origin.

Self-knowledge also opens up the possibility of understanding Nature. We have seen that the relationship with Nature is fundamentally expressed by the Genesis account of Adam who knew the names of all things. Thus, once man discovers himself as true Adam, he is further able to understand Nature. The illumination of the truth can, however, be achieved only through empirical investigation combined with religious and mystical practices. The knowledge of Nature is defined as an alliance of mysticism and empirical knowledge, both of which unite in the theory and practice of alchemy.

Chapter 7
Applied Philosophy: Alchemy and Medicine

Van Helmont asks himself the question that had preoccupied many Christian thinkers throughout centuries: what is the point of knowledge? We have already seen that, according to Van Helmont's orthodox Christian views, regeneration, and indeed the afterlife, is granted through the Sacraments and the intercession of Jesus Christ. Thus, knowledge does not appear to play a part in the drama of human salvation, and hence seems unnecessary.

Van Helmont's attempt at justifying the pursuit of knowledge begins with an outspoken rejection of the Stoic concept according to which the end of wisdom is the knowledge of one's own self. For him, the pursuit of knowledge for its own sake is not a truly Christian endeavour. The paradoxical result of being only concerned with knowledge is that one never reaches it:

> [T]hey who always study and enquire after truth never truly arrive at knowledge, because being inflated by letters, they have no charity: instead, their avarice and ambition hide their atheism.[1]

In Van Helmont's perspective, *scientia* means little if it is not used to help one's neighbour. In this chapter I will consider the 'applied' aspect of Helmontian philosophy. As he clearly affirms, the end of all wisdom is not wisdom itself, but the betterment of mankind. Knowledge acquires new meaning as part of a Christian duty to help others. Charity justifies knowledge and allows the philosopher to increase the divine Grace bestowed upon him. In this sense, Van Helmont saw the natural consummation of all enquiry in practical outcomes; this pragmatism is an essential part of Helmontian philosophy and one of the strongest legacies of his thought.

The highest form of charity that Van Helmont envisions is medicine. However, in order to be effective, medicine must be rooted in alchemy, which offers the keys to expelling diseases and to extending life. It is through the 'attainment of great secrets

---

[1] Ch. 38, 'Imago mentis', §4, 213: '. . . qui studendo inquirunt semper veritatem; nunquam tamen ad ejus scientiam perveniant. Quia litera inflati, charitatem non habent: sed avaritia & ambitio in illis occultat Atheismum'. The source of this idea is Paul, 1 Cor 8:1: 'Knowledge puffeth up, but charity edifieth'.

of nature' that the physician can reverse some of the consequences of the Fall of Man by curing diseases and extending life to that of the Patriarchs.[2]

Christian Charity and Medicine

Love toward others or charity (*caritas*) is the true responsibility of the illuminated soul. Van Helmont clearly states that 'I call charity the ultimate purpose of wisdom, and the reward of the entire course of our life; it alone will accompany us, when other things had deserted us'.[3] Thus, the acquisition of knowledge justifies itself through the free gifts an enlightened person might bestow on fellow human beings.

In his development of the charity idea, Van Helmont reverts back to Paul and Augustine, for whom *caritas* (or *agape*) ranks first amongst Christian virtues.[4] God Himself is absolute charity, Van Helmont argues. The supreme model of *caritas* is Jesus Christ, whose love for other human beings led him to die on the cross.[5]

Charity, for Van Helmont, includes the Christian duty of following one's calling, as well as of doing one's best for one's neighbours.[6] Charity is best achieved through continuous efforts and struggle against adversity. By 'fighting' the vagaries of fate, human beings can hope to achieve a 'Crown', or an 'increase of Grace'.[7]

Charity can be expressed in many ways, but for Van Helmont the supreme form lies in the practice of medicine. He exalts the virtues of the physician, who was created by God to cure the illnesses of fellow human beings: 'for the Highest created the physician, as well as medicine out of the earth'.[8] This belief is drawn on the Deutero-canonical book of Ecclesiasticus (Jesus Sirach) 38:1, a text that many Paracelsians were fond of quoting.[9] Of all people, the physician is called to imitate God the healer by spreading his gifts of medicine to others.[10] Van Helmont believes that physicians should not accept payment from the poor, because they receive their reward from God Himself.[11]

---

2 Ch.4, 'Venatio scientiarum', §1, 17.

3 Ch. 38, 'Imago mentis', §1, 213: 'Ego, finem ultimatum Sapientiae totiusque vitae nostrae cursus bravium, voco charitatem, quae nos comitatur, postquam caetera nos deseruerint'.

4 Servais Pinckaers, *The Sources of Christian Ethics*, 3rd edition (Washington, DC: Catholic University of America, 1995), 27; Mary T. Clark, *Augustine* (Washington, DC: Continuum International Publishing Group, 2005), 44.

5 'Tumulus Pestis', ch. 1, 835, ch. 109, 'Vita', 586.

6 Van Helmont, ch. 101, 'Intellectus Adamicus', 562: 'Quotusquisque vero adspiraverit ad altiorem charitatis gradum, studeat, quantum ex talentu suo poterit, per actus amorosos excurrere ad abstracta per fidem credita . . .'

7 'Tumulus Pestis', ch. 1, 832.

8 'Tumulus Pestis', ch. 1, 830: 'Etenim de terra Medicum, ut Medicinam, creavit Altissimus'.

9 Debus, *The Chemical Philosophy*, 52; *The French Paracelsians*, 11. For Van Helmont's reiteration of this Paracelsian motif, see also *The Chemical Philosophy*, 357–9. The Latin Vulgate reads: 'Honora medicum propter necessitate etenim illum creavit Altissimus'.

10 Ch. 1, 'Promissa authoris', II, §6–8, 10.

11 'Tumulus Pestis', ch. 1, 832, 834.

Furthermore, the Flemish philosopher firmly believes in the 'election' of physicians by God, and the very mark of this predestination is their inclination toward healing: 'For those physicians whom the Almighty has created, are not flute-players: but out of charity's compassion cure the poor'.[12] Hence, Van Helmont does not shy away from affirming that physicians are destined 'from their mother's womb' to practise their craft.[13] As noted above, he firmly believed that the inclination toward a certain craft or occupation was bestowed directly by God, rather than by heredity.

It is clear that for Van Helmont, the true physician is more than simply a healer: he is a Christian mystic and a natural philosopher as well. These dimensions are so thoroughly intertwined in Van Helmont's view that no clear separation can be made between them. It is in the figure of the mystical thinker who is a healer and an enquirer into the secrets of nature that Van Helmont's worldview of Christian philosophy emerges. The ideal, again, comes from the Bible, more precisely the story of the craftsmen Bezalel and Aholiab, who were enlightened by God: 'The Father of Light therefore must be entreated, that He may give us knowledge, such as he once did to Bezaleel and Aholiab, for the glory of his own Name and His true Charity'.[14]

Despite his own example, Van Helmont did not believe that medicine should be practised by solitary physicians. Instead, he formulates a vision of a united congregation of doctors who act in 'blessed harmony' (*sub beato unisono*) and support each other in finding cures for their patients.[15] Their society must be pervaded by charity and religious feeling. For Van Helmont, the true physician is the one who is able to annihilate his own will for the sake of God's; he does not follow his own ends but those established by divinity. Here again we encounter the fundamental Helmontian voluntarism which emphasises the all-knowing divine will.

Undoubtedly, such exaltation of the virtues of the physician originates mainly from Paracelsian doctrine. Paracelsus viewed the physician in profoundly Christian terms. He believed that a healer must be possessed by love for one's neighbour and must fear God at all times. In accordance with the example of Christ, the true physician must sell what he has and go out to heal the sick.[16] Paracelsus does not shy away from calling the physician 'the God of the microcosm' whose duty is to imitate God, the great physician.[17] Although Van Helmont avoids such stark expressions, it is clear that he shares the same worldview with the Swiss physician.

---

12 Ch. 66, 'Respondet Author', §14, 422: 'Non enim auloedi sunt medici, quos creavit omnipotens: sed in compassione charitatis medentur peculiariter pauperibus, eoque symbolo agnoscuntur'.

13 Ch. 1, 'Promissa authoris', II, §7, 10: 'ejusmodi autem Medici sunt, in ventre matris praeparati'.

14 Ch. 60, 'Pharmacopolion ac Dispensatorium modernorum', §14, 369: 'Exorandus itaque est Pater luminum, ut dignetur nobis dare cognitionem, qualem olim dedit Bezeliel, & Ooliab, in nominis sui gloriam, & nudam proximorum charitatem'. For Bezalel and Aholiab, see Exod. 31:2–8.

15 Ch. 82, 'De conceptis', 487.

16 Theophrastus von Hohenheim Called Paracelsus, 'Seven Defensiones', in *Four Treatises of Theophrastus von Hohenheim Called Paracelsus*, trans. C. Lilian Temkin, ed. Henry E. Sigerist (London: Johns Hopkins University Press, 1941), 29–33.

17 Heinrich Schipperges, 'Paracelsus and His Followers', in *Modern Esoteric Spirituality*, ed. Antoine Faivre and Jacob Needleman (New York: Crossroad, 1992), 156.

Alchemy and Adeptship

The previous chapters have shown that Van Helmont's philosophy was strongly coloured by an alchemical worldview. There is no doubt in Van Helmont's mind that alchemy (or *spagyria*, the art of separation) is the supreme science, the apex of natural philosophy. It alone gives knowledge of causes and the structure of all things. Indeed, alchemy illuminates philosophy: 'the consideration of the fire, by physiology, unlocks the gates of nature and illuminates the whole of philosophy, excluding all despisers of *pyrotechnia*'.[18] In his view, no philosopher can reach the roots of Nature without the art of the fire.

Van Helmont calls alchemy the 'adept' science, the 'labour of Wisdom' (*labor Sophiae*) and 'the mirror of true Intellect'.[19] As he points out, 'that science, which teaches to look into closed bodies through their analysis, and to extract their hidden virtues is not the servant of pharmaceutics, but is the most powerful interpreter of natural history'.[20] Since alchemy is the apex of all knowledge, he believes that a reform of education would imply studying it only after mastering the inferior forms of knowledge. According to his plan, the first three years of university study, plus the other four years of advanced study should be dedicated to mathematics, geography and natural history. Only after those seven years should a person approach the true study of nature, which is alchemy. He would then learn the true principles of nature both theoretically and practically, in the laboratory.[21]

Van Helmont's interest in alchemy was coloured by his belief in an ancient wisdom, a medical-alchemical *prisca sapientia* that was compatible with Christianity and the Bible.[22] It was a deeper form of knowledge that he worked hard to uncover throughout his life and which he thought he had achieved in his old age.

Young Van Helmont believed that he belonged to a select class of alchemical philosophers, first mentioned in his unpublished *Eisagoge*.[23] Here, Hermes Trismegistus is portrayed as the originator of all true medicine and the leader of the Hermetic school.[24] This belief in the 'Hermetic school' continued through Van Helmont's mature years. In *Supplementum de Spadanis fontibus*, he refers to the concordance between the School of Hermes and Pythagorean philosophy.[25] In *De magnetica*

---

18 Ch. 22, 'Ortus formarum', §68, 117: 'ignis consideratio, per Physiologiam, referet naturae januam, totamque illuminet Philosophiam, omnesque excluserit Pyrotechniae despectores'.

19 'De lithiasi', ch. 3, §1, 668, 'Spagyria enim sola, est speculum veri Intellectus'.

20 'De febribus', ch. 15, §8, 778: 'Illa ergo scientia, quae docet inspicere corpora occlusa, per sui analysin, ac virtutes illorum absconditas eruere, famula Pharmaceutices non est (ut sonant ignorantium convicia) sed potissima historiae naturalis tecmarsis est'.

21 Ch.8, 'Physica Aristotelis et Galeni ignara', §9–10, 40–41.

22 For the medical side of the *prisca theologia*, see also Hiro Hirai, '"Prisca Theologia" and Neoplatonic Reading of Hippocrates in Fernel, Cardano and Gemma', in *Cornelius Gemma: Cosmology, Medicine and Natural Philosophy in Renaissance Louvain* (Rome: Serra, 2008), 91–5.

23 Broeckx, 'Le premier ouvrage', I, 360.

24 Broeckx, 'Le premier ouvrage', I, 358.

25 Ch. 100, 'Supplementorum Paradoxum numero criticum', §53–5, 560.

vulnerum curatione and 'Ad Judicem', Van Helmont often speaks in the plural, suggesting he is backed in his affirmations by the entire Hermetic School.[26] In fact, he even sets out to describe what he calls the fundamental 'Theorem of the Hermetic School' (*Scholae Hermeticae Theoremata*), which states that no axiom of physics will be admitted unless proven by the fire.[27] This pattern of self-representation is present in his letters to Mersenne, where Van Helmont talks in the plural about 'our principles' or 'our school'.[28]

However, the later Van Helmont seems to have distanced himself from the idea of belonging to a school. Although the praise of alchemy continues in 'De lithiasi', where Van Helmont cites the *Tabula smaragdina* of Hermes Trismegistus, he does not claim to belong to any alchemical community. This fact can undoubtedly be connected with the claim of originality that the later Van Helmont made: at this point he was no longer willing to be considered as one of the many alchemists of his age, but as a truly original natural philosopher.

Instead of the concept of a 'Hermetic School', the later Van Helmont emphasised the idea of the 'Adepts'. Van Helmont's concept of 'Adept' originates from Paracelsus's *Astronomia magna*, in which the Swiss physician mentioned the Adept Philosophy (*philosophia adepta*), Medicine (*medicina adepta*) and Mathematics (*mathematica adepta*).[29] These are presented as three of the nine parts of celestial astronomy. By the term *philosophia adepta*, Paracelsus may have meant nothing more than a perfected knowledge of philosophy, yet this has to be understood in the context of his idiosyncratic view that superior knowledge originated from the stars.

The followers of Paracelsus adopted and expanded on the concept of the 'adept philosophy'. In 1570 or 1571, Petrus Severinus published *Epistola scripta Theophrasto Paracelso*, which purported to summarise the method of 'Adept Philosophy'.[30] Severinus also mentioned 'Adept Philosophy' in his magnum opus *Idea medicinae* (1571), where he referred once to it as the source of true natural knowledge.[31] Later on, Croll talked at length of the 'Adept Philosophers' and 'Adept Philosophy' in his *Basilica chymica*.[32]

It was through the *Astronomia magna* and Paracelsian treatises like those of Severinus and Croll that Van Helmont encountered the concept of the 'Adepts'. However, his understanding of the term varied over time. The prologue to the *Eisagoge* suggests that Hermes Trismegistus could be seen not only as the founder of the

26 Broeckx, 'J.B. Van Helmont: Ad Judicem', 85, 108.

27 Broeckx, 'J.B. Van Helmont: Ad Judicem', 105–6.

28 'Jean Baptiste Van Helmont, à Bruxelles, à Mersenne à Paris, 30 Janvier 1631', III, 63; 'Jean Baptiste Van Helmont, à Bruxelles, à Mersenne, à Paris, 21 Février 1631', III, 111.

29 Paracelsus, 'Astronomia magna sive Tota philosophia sagax', X, 205–17.

30 Petrus Severinus, *Epistola scripta Theophrasto Paracelso* (Basel: Henricpetri, 1570/1), [4] (the *Epistola* does not contain page numbers). It is not clear whether this short work was published before or at the same time as *Idea medicinae*; Shackelford argues that it was most likely before; *A Philosophical Path,* 55 n.10.

31 Severinus, *Idea medicinae philosophicae*, 80.

32 Croll, 'The Admonitory Preface', 91, 96.

Hermetic school, but also as the chief of the 'Adept' philosophers. This does not mean that all alchemists can be described as 'Adepts'; in fact, Van Helmont's dream-vision suggests that only 'chosen' natural philosophers achieved divine knowledge and sanctity. It is implied that becoming an 'Adept' is a long-term ideal to be pursued, rather than a simple adhesion to a certain doctrine or practice like that of alchemy. This perspective indeed justifies Van Helmont's modesty when Mersenne addresses him with the title 'Adept': 'Please do not give me the title of Adept, because I do not deserve it'.[33] The ideal of becoming an 'Adept' is further described in the later Van Helmont's autobiography from the *Ortus*.[34]

One may rightfully wonder if the later Van Helmont thought he had finally attained the title of 'Adept'. It seems that he did, since he claims to know the secrets of *Arcana* such as Paracelsus's Elixir of Propriety, Butler's *Drif* and the 'Tree of Life', all three of which are described as part of the Adepts' knowledge.[35] Even further, Van Helmont admits on another occasion that 'I am esteemed an *Adeptus*, or attainer of *Arcana*'.[36] Although he never assumes the title directly, the later Van Helmont seems content to leave his readers to believe that he had eventually become an 'Adept'. One may even speculate that the entire *Ortus medicinae* is Van Helmont's attempt to justify his election to this higher 'Adept' status.

Alchemical Medicine

It is in the 'School of Hermes' and the art of the Adepts that Van Helmont finds the practice, the philosophy and the true charitable worldview that he was searching for. This art is not simply a rediscovery of old knowledge; it is a living tradition that allows the discovery of new medicines.[37] Thus Van Helmont criticises those who are content with the discoveries of the ancients as being slothful and lethargic. Physicians should not restrict themselves to Hippocrates or Galen, not only because those lived in a 'happier time' when fewer diseases plagued humanity, but also because the ancients did not know everything. In fact, Galen set medical knowledge on the wrong path. Van Helmont criticises the followers of Galen as promoting methods that are half-hearted, ineffective, and in some cases downright harmful. The worst of their sins is the rejection of alchemy, which stands at the root of medicine.

According to Van Helmont's belief, alchemical practice allows the creation of powerful and accessible medicine that can be administered to the poor. Indeed, in his view, all diseases are curable if the right medicine is discovered.[38] Van Helmont

---

[33] 'Jean Baptiste Van Helmont, à Bruxelles, à Mersenne, à Paris, 14 Février 1631', 108.

[34] Ch. 1, 'Promissa authoris', III, §7–8, 11.

[35] Ch. 79, 'Butler', 474.

[36] Ch. 88, 'Praefatio', 501: '. . . aestimorque tamen arcanorum aliquot Adeptus'.

[37] 'De lithiasi', ch. 7, §7, 692. Here he criticises those that think the 'powers of divine Wisdom are exhausted'.

[38] 'De lithiasi', ch. 7, §4, 694: 'Profecto, nullus in specie morbi est incurabilis'.

believed that remedies could be found in all natural kingdoms. He paid particular attention to plant medicine, going so far as to affirm that the cures of all diseases lay in the vegetable family.[39] Nor did he reject medicine drawn from animal sources, since 'all things have been created for the use of ungrateful man'.[40] For instance, he believed that 'the first matter of a frog' (*prima materia ranae*) could cure cancers.[41]

In line with other Paracelsians, Van Helmont particularly praised the use of metals and minerals such as vitriol, lead and mercury in medical preparations. He believed that, by proper chemical treatment, these could be transformed into outstanding medicines. Naturally, Van Helmont was aware of the attacks of Galenist physicians on alchemical medicine as being damaging to health. Yet he condemned their views as ignorance. Alchemical medicines were delivered in small doses not because they were poisonous, but because they were highly potent.[42] The poison inherent in such minerals as antimony and mercury was eliminated by alchemical means and turned into medical virtue.

Van Helmont took pride in his alchemical knowledge, and often polemicised with other alchemists whom he deemed inferior or misguided. For instance, he wrote a long disquisition on vitriol, whereby he rejected the authority of the famed alchemist Isaac Hollandus on the topic.[43] Thus, Van Helmont criticised artificial vitriol made by rusting copper for being medically weak. Instead, he maintained that one must seek the 'Chalcanthum', an acid mineral salt that could be found near copper veins; this salt, called 'vitriol' by alchemists, is medically powerful.[44]

The Flemish philosopher also had a strong interest in salts. He confessed to having spent a great deal of time analysing them: he concluded that salts could be of three kinds: acids, alkalis and the essential sulphurs in vegetables.[45] His interest lay primarily in volatile alkalis. As we have seen, he maintained that the spirit of life was a volatile salt, capable of being ignited by the formal light of the soul. He also theorised that all combustible bodies contained a volatile salt. Van Helmont further argued that volatile salts and alkali salts had the same constitution, and that volatile salts could be easily turned into alkalis by the intervention of fire.[46] However, the analogous transformation of alkalis into volatile salts was much more difficult. Yet by means of this change, ordinary fixed salts could be transformed into powerful medicines that penetrated bodies and cleansed them of illness.

---

39 Ch. 79, 'Butler', 473: 'in herbacea familia remedium latere pro omni morbos'.

40 'De lithiasi', ch. 7, §20, 699 : 'Itaque cum universa in ingrati hominis usum sint creata'. He prescribed 'powder of crabs' and the 'stone of bezoar'.

41 Ch. 22 'Ortus formarum', §57, 115.

42 'De febribus', ch. 15, §5, 778.

43 'De lithiasi', ch. 8, §21–2, 708. On Isaac Hollandus, see Annelies van Gijsen, 'Isaac Hollandus Revisited', in *Chymia: Science and Nature in Medieval and Early Modern Europe*, ed. Miguel López Pérez, Didier Kahn and Mar Rey Bueno (Newcastle upon Tyne: Cambridge Publishers, 2010), 310–30.

44 'De lithiasi', ch. 8, §10–23, 706–8.

45 'De lithiasi', ch.3, 669–70.

46 Ch. 25, 'Blas humanum', §39, 151.

The reason volatile salts had such a great medical effect was because they acted directly on the spirit of life. To illustrate their action, Van Helmont offered the simple example of the spirit of wine. He noted that due to its volatile salts, the spirit of wine was immediately absorbed by arterial blood and led to the heart and the head.[47] Here it was transformed into the spirit of life, thus strengthening the vital powers of the organism.

Volatile salts could be extracted from many sources, including ordinary plants.[48] For instance, he maintained that marjoram, rosemary and sage could yield volatile salts that cured fevers if drunk in wine on an empty stomach.[49] Even *Arcana* could be made from volatile salts: for instance, he described the *Ludus Paracelsi* as a 'volatile salt' that could cure kidney stones, and claimed that the volatile salt of tartar could act as a substitute of the *Alkahest*.[50] At least in one occasion, the *Alkahest* itself was described as a type of volatile salt.[51]

Van Helmont emphasised that the action of such medicines was not only highly effective, but also mild to the body. He condemned the Galenic use of harsh purgatives that caused vomiting and frequent stools. He believed that these were harmful to the body because they weakened the *Archeus*. As Giglioni has pointed out, the Flemish philosopher believed that illnesses fundamentally arose from an altered imagination, which corrupted the *Archei* of the body.[52] By consequence, efficient medicine acted by appeasing the passions of the *Archei*.[53] This process of pacification was achieved by mild and comforting medicines which acted by light or even 'invisible' perspiration. Van Helmont explains the process as follows: 'As soon as it is suddenly touched with the *caduceus* [of Hermes], the furious *Archeus* falls asleep, or being corrected, abstains from his prior mischief'.[54] This image suggests the efficacy of the appeasing alchemical medicine (symbolised by the Hermetic wand), whereby the *Archeus* is calmed and restored.

The achievement of powerful medicines was the result of a long-term process that involved perfecting one's skills and understanding. In accordance with his belief in 'Adeptship', Van Helmont viewed alchemists as belonging to three chief categories according to their skill: beginners (*tyrones*), advanced practitioners, and Adepts. This perspective reflected Van Helmont's understanding of alchemy as an art, *technē*; this

47 Ch. 27, 'Spiritus vitae', §12–13, 158–9.

48 'De febribus', ch. 15, §7, 778.

49 'De febribus', ch. 14, §12, 776.

50 'De lithiasi', ch. 7, §22, 700, 'De febribus', ch. 15, §26, 780.

51 Ch. 61, 'Potestas medicaminum', §24, 380.

52 Giglioni, *Immaginazione e malattia*, 97–133.

53 Giglioni, *Immaginazione e malattia*, 135–42.

54 Ch. 79, 'Butler', 472: 'scilicet Archeus in furorem actus, tam repente vel caduceo attractus obdormiat, vel correctus, abstineat ab incepto sui nocumento'.

threefold division probably emulated that of the medieval guild system of apprentice, journeyman and master.[55]

In accordance with this division, Van Helmont held to a hierarchical view of alchemical secrecy.[56] He was committed to the idea of a framework of competence whereby beginners would be allowed to advance toward the highest ideal of Adeptship. As a self-fashioned Adept, he thought it was his duty to act in a mentorship role, and believed that the disclosures of the *Ortus* could encourage and guide newcomers in the art of alchemy. Thus, he offered general advice on the importance of learning to remove poison from sulphurs,[57] or 'philosophical sentences' that include the statement that it is much easier to artificially construct gold rather than to destroy the natural one.[58] In a rare gesture, he provided a recipe for an 'unusual distillation' of vitriol.[59]

Yet these pronouncements are much less useful than the 'chemical textbooks' of the period, such as the *Tyrocinium chymicum* of Jean Béguin. These contained the basic principles of Paracelsian alchemy together with compendia of recipes. Van Helmont is frequently critical of these manuals, dismissing their writers as 'beginners' or empirics.[60] But he offers little in return. His limited disclosure is linked to the belief that an alchemist's advancement should be done through his own efforts, by experimenting and toiling with the art of the fire. This is a *scientia* that cannot really be learned; it has to be practised.[61] As he points out, 'it is not sufficient to pore over books, but one must buy coals and vessels and devote many a watchful night' to gain knowledge.[62]

Van Helmont's attitude toward advanced practitioners is similarly ambivalent. At least theoretically, he was willing to provide some direction to those mature alchemists who had sufficient experience and understanding to investigate nature.[63] Yet the usefulness of these 'disclosures' is debatable: for instance, a *Ludus Paracelsi* recipe he provides to those knowledgeable in the 'art of the fire' reveals little about the composition of the *Ludus* itself, except that it is extracted from a 'salt of urine of the earth' (*ex*

---

55 On the subject of apprenticeships in medieval guilds, see for instance S.R. Epstein, 'Craft Guilds, Apprenticeship, and Technological Change in Preindustrial Europe', *The Journal of Economic History* 58:3 (1998): 684–713.

56 On the complex relationship between openness and secrecy in the Renaissance, alchemy and Paracelsianism, see Pamela Long, *Openness, Secrecy, Authorship: Technical Arts and the Culture of Knowledge from Antiquity to the Renaissance* (London: Johns Hopkins University, 2001), especially 143–74.

57 Ch. 78, 'In verbis', 460.

58 'De lithiasi', ch. 8, §10, 705: 'Quod sit longe facilius, aurum ex non auro construere, quam aurum naturale destruere'.

59 'De lithiasi', ch. 8, §19–22, 707.

60 E.g. ch. 66, 'Respondet author', §1, 419, '. . . Scriptores, quam Basilicas & Tyrocinia chymica ingenti gloriolae pruritus, adhuc ipsimet Tyrones ediderunt'; 'De febribus', ch. 15, §13, 779: 'Adeo que sciant Galenistae, quod scriptores tyrocinii Chymici adhuc Tyrones sint', 'De lithiasi', ch. 8, §9, 705: 'Tyrocinii chymici Scriptores'.

61 Ch. 118, 'Arbor vitae', 635.

62 'De febribus', ch. 14, §26, 780, 'non sufficit libros terere: sed insuper oportet carbones atque vasa emere, & vigilatas ex ordine noctes impendere'.

63 Ch. 118, 'Arbor vitae', 635.

sale urinae liquorum terrae).[64] Often his 'recipes' are coded, abstruse or incomplete: for instance, the *Aroph Paracelsi* is described as being drawn from the tincture of gold, heated in dung, mixed with rye-bread and extracted with *spiritus vini*.[65]

Finally, Van Helmont believed that the highest alchemical secrets could not be shared in any form. Supreme *Arcana*, such as the *Alkahest*, the *Arbor vitae* or the Philosophers' Stone, commanded silence. He accepted that 'other things . . . shall be buried with me; for the world is not capable of [ready for?] them'.[66] The reason for this supreme secrecy is religious. As Van Helmont puts it, 'it is not lawful to profane the *Arcana* of God, who dispenses to whom He will'.[67] The highest secrets belong to God alone. Van Helmont appeals to the New Testament for confirmation that the greatest secrets should not be divulged, lest 'I shall cast roses before swine'.[68]

In his hierarchy of alchemical knowledge, at the bottom of the ladder Van Helmont places remedies from simples (*simplicia*), which can be prepared by decoction or distillation.[69] Even so, Van Helmont believed sympathetical remedies to be more potent. Amongst these was the famous weapon salve, whose defence got Van Helmont into trouble with the Inquisition. The sympathetical cures, are given by divine love for the comfort of the poor. They are not Satanic remedies or superstitions as its detractors claimed.[70]

An important place is occupied by the *Arcana* or the alchemical secrets of Paracelsus. Van Helmont considered that Paracelsus had been the vindicator and healer of almost all diseases.[71] According to the treatise 'Arcana Paracelsi', the greatest medicines included the tincture of Lily (*tinctura lili*), *Mercurius vitae*, the tincture of antimony, *Mercurius diaphoreticus*, horizontal gold, the Element of the Fire of Copper (*elementum ignis à cupro*), *lac margaritarum*, the *corallatus*, and the greatest of all, the liquor *Alkahest*.[72] In other treatises, however, he also mentions other Paracelsian cures such as the *Aroph Paracelsi*, or the *Ludus Paracelsi*.[73] Van Helmont praises these medicines for curing such life-threatening or debilitating diseases as leprosy, asthma, paralysis, gout and cancer. He calls them 'universal medicines', although this term here should be taken in the traditional medieval sense to mean powerful and effective

---

64 'De lithiasi', ch. 7, §22, 700.

65 'De lithiasi', ch. 7, §14, 697.

66 Ch. 118, 'Arbor vitae', 635: 'Caetera autem quae de Cedro sunt, mecum sepelientur. Nam mundus capax non est'.

67 'De febribus', ch. 14, §11, 776: '. . . nec liceat profanare Arcana Dei, qui horum disponsator manere voluit'.

68 Ch. 79, 'Butler', 474: 'ne rosas ante porcos prostravero'. The original saying is in Matthew 7:6, but refers to 'pearls' not 'roses'. By using 'roses' instead of 'pearls' Van Helmont was actually referring to the Flemish paraphrase of the original Biblical text; the proverb had been superbly illustrated by Pieter Breugel the Elder in his *Netherlandish Proverbs* painting of 1559.

69 Ch. 118, 'Arbor vitae', 635.

70 Ch. 79, 'Butler' 473: 'amoris divini emanantia pentacula . . . in miserorum pauperumque solamen'.

71 Ch. 116, 'Arcana Paracelsi', 626: 'Atque hactenus fuit morborum fere omnium vindex, & sanator'.

72 Ch. 116, 'Arcana Paracelsi', 628.

73 'De lithiasi', ch. 7, §14, 697.

remedies.[74] The action of the *Arcana*, he explains, is to cleanse or refine specific organs (*Archei*) of the body.[75] For instance, the *Mercurius diaphoreticus*, the 'universal medicine for fevers',[76] eliminates the source of disease without sweat; it cures fever with a single dose, and cancers, dropsy and asthma in one month.

Most, if not all of these *Arcana* seem to involve the alchemical treatment of metals or minerals. In 'Arcana Paracelsi', Van Helmont gives little indication as to their composition, but he had been more forthcoming in 'De febribus', where he described the manner of preparing *Mercurius diaphoreticus*. The recipe included copper vitriol, *aqua regia* and *aqua vitae*, as well as 'powder of Johannes de Vigo' (a mercury-based medicine).[77] The fixed and sweet powder of mercury thus obtained seems to be similar, if not the same, to that employed to prepare the *Arcanum corallinum* (probably the same as the *corallatus* in 'Arcana Paracelsi'). This was extracted by means of the universal solvent, the liquor *Alkahest*, out of mercury.

Another important *Arcanum* is the 'Element of the Fire of Copper' (*elementum ignis à cupro*), also called 'Element of the Fire of Venus' (*elementum ignis Veneris*), and assimilated to the Sulphur of Philosophers (*sulfur Philosophorum*). This is not rectified spirit of vitriol, but a 'volatile sulphur of copper, in the shape of a green oil, sweeter than honey and clearly separated from its mercurial body'.[78] The separation renders a white metallic body from the mercury of copper and a volatile sulphur that can be transformed into oil. Unfortunately, even this preparation seems to need the *Alkahest* as solvent.[79] Indeed most, if not all, of the *Arcana* were based on the knowledge of the *Alkahest* as prerequisite.

Alkahest

Of Van Helmont's legacies, perhaps none was as significant and influential as his 'liquor *Alkahest*', a universal solvent that was supposed to reduce all things to their first being (*primum ens* or *prima materia liquida*). Numerous natural philosophers, including Robert Boyle, Otto Tachenius and Georg Stahl laboured to achieve this

74 Ch. 116, 'Arcana Paracelsi', 626: 'Fateor lubens, me ex ejus scriptis profecisse multum: illumque potuisse per remedia ad unitatis symbolum ascendentia, sanare lepram, Asthma, Tabem, Paralysin, Epilepsiam, Calculum, Hydropem, Podagram, Cancrum, atque ejusmodi vulgo incurabiles morbos'.

75 Ch. 116, 'Arcana Paracelsi', 627.

76 'De febribus', ch. 14, §8–9, 775. I'm making the assumption that the 'precipitatus diaphoreticus Paracelsi' of 'De febribus' and the 'Mercurius diaphoreticus' in 'Arcana Paracelsi' are the same thing.

77 'De febribus', ch. 14, §9, 775–6. The 'powder of Johannes de Vigo' appears to denote a mercurial preparation, since in 'De lithiasi', ch. 7, §5, 705, he describes a 'sulphur of mercury' obtained from the powder with the aid of mineral sulphurs.

78 'De lithiasi', ch. 8, §5, 705: 'ignis ille est sulfur cupri volatile, in forma olei viridis, melle dulcius, & à corpore mercuriali sui cupri plane separatum'.

79 At least that seems to be suggested by Van Helmont, 'De lithiasi', ch. 8, §21, 708 by the claim that the fire of Venus is done in a 'far more occult manner' (*via longe occultiore*).

mythical solvent. Naturally, it has long been one of the favourite topics of Helmontian scholarship.[80]

Van Helmont's term *Alkahest* has been a constant source of bafflement from the early modern period until today. Modern scholarship has established the origin of the term, but not necessarily its real sources of inspiration. The term originates in a minor work of Paracelsus, 'De viribus membrorum', where it referred to a medicine for fortifying the liver, not to a solvent.[81] Paracelsus claimed that even if the liver were destroyed, the *Alkahest* would restore it: hence, he recommended every true physician to spend time learning its preparation. Nevertheless, Paracelsus did not give a lot of information about it. He described it as being dissolved from coagulation and then coagulated again into a new transmuted form.

Early Paracelsian followers inherited the term, but did not pay much attention to it. In spite of this, they made an important step that would prove consequential to Van Helmont's reinterpretation of the *Alkahest*. In his *Onomastica* (1574), Michael Toxites defined the *alkahest* as a liver medicine made of 'mercury'. This did not appear in 'De viribus membrorum', but was surmised by Toxites by association with another book, 'De gradibus', where Paracelsus mentions an anonymous mercurial liquor as remedy for the liver. This unnamed liquor Toxites took to be the *Alkahest*.[82] This interpretation was passed into Paracelsian circles by other dictionaries, such as that of Roch le Baillif, Gerhard Dorn and Martin Ruland. By Van Helmont's time, it was natural to believe that the medicine *alkahest* had a mercurial origin. Nevertheless, there was still nothing to suggest that this was a solvent, let alone a universal one.

In fact, the association between the universal solvent and the *Alkahest* seems to have been made rather late in Van Helmont's career. When the universal solvent makes its first appearance in the Flemish alchemist's middle-period writings, it is not called *Alkahest*. This name only occurs in Van Helmont's late work, and first in *De febribus* (1642).

Indeed, we have no reason to believe that Van Helmont even entertained the name of *Alkahest* when he first mentioned his belief in the existence of a universal solvent in a letter dated 15 January 1631 to Mersenne. Here, responding to a question from Mersenne on the Paracelsian *tria prima*, Van Helmont cast doubt on their primacy. Thus, he claimed that the three principles can be divided by means of a universal solvent, 'which dissolves, changes, separates and reduces all bodies . . . This solvent Paracelsus, in the "Book on Renovation and Restauration", called dissolved salt (*sal solutum*) or *sal circulatum*, which is the first being of salts (*primum ens salis*)'.[83]

[80] The main contributions on the subject are Reti, 'Van Helmont, Boyle and the Alkahest', 3–19; Joly, 'L'alkahest', 305–44, Newman, 'Corpuscular Theories', and Porto, '"*Summus atque felicissimus salium*"'.

[81] Philippus Theophrastus Bombast von Hohenheim, Paracelsus gennant, 'De viribus memborum', in *Dritter Theil Der Bücher und Schrifften*, 8–9. This was first pointed out by Joly, 'L'alkahest', 314.

[82] Joly, 'L'alkahest', 315.

[83] '15 Janvier 1631', 33: 'Est itaque solvens quoddam universale, quod omnia corpora resolvit, commutat, separat et reducit . . . Hoc solvens Paracelsus vocat, Lib. *De Renovatione et Restauratione*, *sal solutum* sive *sal circulatum*, et est *primum ens salis*'.

This letter is intriguing, because it suggests that Van Helmont actually 'found' the universal solvent not in Paracelsus's 'De viribus membrorum' but in the treatise 'De renovatione et restauratione'. Here Paracelsus maintains: 'This doctrine [of renovation] cannot be conveyed better than by the first entity [*primum ens*], in which there is a singular nature for operating upon the body and transmuting its essence'.[84] What does the *primum ens* mean? In *Archidoxis*, the first entity is described as a watery essence of physical bodies.[85] In 'De renovatione et restauratione', Paracelsus further compares this *primum ens* with mercury and proceeds to describe many types of *primum ens*, such as that of trees, gems, salts, metals, etc.[86] To obtain this *primum ens*, Paracelsus maintains in 'De renovatione' that the body's form must be removed in the same way as it is done in the preparation of *arcana* and *quinta essentia*.[87] The solvent that can achieve this dissolution is either *sal circulatum* or *sal solutum*. There is no equivalence between the *sal circulatum* (solvent) and the *primum ens* (essence), as Van Helmont postulated.

Nevertheless, the Flemish philosopher may have been aware of the spurious *Clavis archidoxorum*, first published in 1624 and presented as the secret tenth book of the *Archidoxis*, which had been suppressed from common knowledge until then.[88] This work clearly equates *sal circulatum* with the *primum ens salis*. Even if Van Helmont did not know this work, he could have reached the same conclusion by observing that, in 'De renovatione et restauratione', Paracelsus described the *primum ens* of salt as a special type of first being that had the dual power of turning all metals into their *primum ens* and of coagulating them into an exalted form.[89] Hence *primum ens salis* could be interpreted as an alchemical solvent similar to the traditional philosophical Mercury.

Van Helmont's analysis of Paracelsus's 'De renovatione' does not, of course, mean that his idea of a universal solvent came only from there. In fact, there seem to have been other sources of inspiration. This possibility is raised by the alternative name Van Helmont gives to the universal solvent, *ignis aqua* (fire water). This appears prominently in a dream-vision he recounted in the *Ortus* treatise 'Potestas medicaminum' ('The Power of Medicines'). Here Van Helmont described a dream he had in his youth about a bottle of liquor called *ignis-aqua*. This bottle was briefly shown to him before being taken away for 30 years. If this testimony has any truth to it, the revelation of the *ignis aqua* would have occurred around 1613–1614 (assuming

84 Philippus Theophrastus Bombast von Hohenheim, Paracelsus gennant, 'Liber de renovatione et restauratione', in *Sechster Theil Der Bücher und Schrifften*, 108.

85 See also Porto, '"*Summus atque felicissimus salium*"', 13.

86 Paracelsus's own source seems to have been Pseudo-Lull's *Testamentum*, where the *ens primum* is associated with philosophical mercury and is called 'the material terminus in nature'. 'Raymundi Lulli Theorica et Practica', 60.

87 Paracelsus, 'Liber de renovatione', 112. The reference is to the procedures described in *Archidoxis*.

88 *Clavis, Oder, Das Zehende Buch der Archidoxen D. Philippi Theophrasti Paracelsi, Bombast von Hohenheim* (Magdeburg: Johann Francken, 1624) was published by Johannes Staricius, who is also supposed to be its author.

89 Paracelsus, 'Liber de renovatione', 109.

a date of 1643/1644 for the 'Potestas'). We know that this *ignis aqua* is the universal solvent because in 'De lithiasi', Van Helmont used *ignis aqua* as a synonym for the *Alkahest*.[90] Moreover, this correlates with a further letter to Mersenne on 30 January 1631 where he referred to the *sal circulatum* as 'liquid fire' (*liquore ignis*).[91]

At first glance, the term *ignis aqua* seems to indicate the Rupescissan tradition of the *spiritus vini*, the supreme quintessence or heaven (*coelum*), often called *aqua ardens*.[92] Yet Van Helmont describes *ignis aqua* as 'one word' which is 'simple, singular, indeclinable, immutable and immortal', giving it clear spiritual connotations.[93] I would speculate that he may be referring here to the Hebrew word *shamayim* (heaven), which, as previously noted, was often translated as fire-water.[94] If this conjecture is true, Van Helmont may have been influenced, at least partially, by alchemical speculation on the term *shamayim*. Heinrich Khunrath, for instance, equated *shamayim* with the primordial chaos and the Spirit of God hovering over the waters, and thought this manifested itself in the laboratory as a solvent called by such names as 'our Vitriol', 'universal fire of nature' (*Igne Naturae Catholicae*), 'chaos' or 'salt of wisdom' (*Sal Sapientiae*).[95]

Other contemporary sources for Van Helmont's solvent may have been Croll and Sendivogius. Croll talked about a 'universal mercurial menstruum of the philosophers',[96] while Sendivogius also described a solvent called the 'menstruum of the world'.[97] It is not clear if these solvents should be deemed universal, because both Croll and Sendivogius were interested mainly in the dissolution of metals and minerals. Yet the term *menstruum* was chiefly based on Pseudo-Lull, who described this as a solvent by which all things can be converted into their first nature.[98]

Sources aside, it seems that the universal solvent obsessed Van Helmont in the latter part of his life. Despite his letter to Mersenne, the Flemish physician did not claim to possess the solvent at this stage. The key to it was still missing—a key which he believed to have eventually found in Paracelsus's 'De viribus membrorum'. Indeed,

---

90 'De lithiasi', ch. 4, 677, under headline 6, 'Ignis aqua, ens homogeneus, sibi par non habens'.

91 '30 Janvier 1631', 66–7.

92 On Johannes de Rupescissa's notion of quintessence, see Leah DeVun, *Prophecy, Alchemy, and the End of Time: John of Rupescissa in the Late Middle Ages* (New York: Columbia University Press, 2009), 64–79.

93 Ch. 61, 'Potestas medicaminum', §3, 377: 'Tum dein alter spiritus, superioris ordinis, dedit mihi lagenam, in qua erat unius verbi, Ignis-aqua. Nomen prorsus simplex, singulare, indeclinabile, inseparabile, immutabile & immortale'.

94 See Chapter 5 above, section headed 'Genesis and the Elements'. His opposition to this traditional etymology may have occurred at a later date than his 'dream'.

95 Heinrich Khunrath, *Confessio de chao physicorum* (Strasbourg: Johann Albert Dolhopff, 1609), 6–11, 21, 24.

96 Croll, 'Admonitory Preface', 204.

97 Sendivogius, *De lapide philosophorum*, 72.

98 'Raymundi Lulli Theorica & Practica', 143. On *menstruum* in Pseudo-Lull, see Pereira, 'L'alchimista come medico perfetto', 103–8.

we may conjecture that it was this 'key' that led him to rename the universal solvent by the odd name '*Alkahest*'.

If so, we may legitimately wonder what made Van Helmont assimilate the *sal circulatum* of 'De renovatione et restauratione' and the *Alkahest* of 'De viribus membrorum'. A possible answer is that the equivalence was given by Van Helmont's decision that the solvent was a mercury-based product. This fact is stated in 'De febribus': 'From vulgar, commercial mercury you should abstract the liquor *Alkahest* . . . '.[99] Here the *materia prima* of the *Alkahest* is readily—perhaps too readily—given. Van Helmont never provided such a clear clue in his post-1642 work.

In 'De febribus', Van Helmont gave us more than the source for the *Alkahest*. He offered us a glimpse into his alchemical thought-processes, as well as an indication of the medieval sources of his theory. In his search for the universal solvent, Van Helmont seemed to have connected Paracelsus's *primum ens* with Pseudo-Lullian alchemy. This was in fact an easy connection to make, given that the notion of the *primum ens* originated with Pseudo-Lull himself.[100]

In accordance with this insight, Van Helmont turned to Pseudo-Lull's work for guidance. In the latter's *Testamentum*, he thought he uncovered what he had been looking for. In this important treatise, Pseudo-Lull affirms that bodies are susceptible to dissolution, and when a body 'is well dissolved, the spiritus is congealed; and when the spiritus is congealed, the body is dissolved'.[101] Hence, concludes Pseudo-Lull, 'our final secret is the congelation of mercury'.[102]

It appears that it is this congelation of mercury that gave Van Helmont the supposed key to the *materia prima* of the universal solvent. Indeed, in 'De febribus', Van Helmont directly compares Paracelsus's *Alkahest* to Pseudo-Lull's process:

> For, says Raymundus [Lull], in the presence of my friends and the King, I coagulated quicksilver, and no one outside the King knew how this was done. In which coagulation that is singular. That liquor *Alkahest*, remaining the same in number, weight and activity, is as powerful in its thousandth action as it was in the first.[103]

To Van Helmont, this concordance between Pseudo-Lull and Paracelsus must have come across as an important discovery. The language seemed similar, as both Pseudo-Lull and Paracelsus discussed a process of dissolution and recoagulation in their

99 'De febribus', ch. 14, §10, 776: 'A Mercurio vulgo venali, abstrahe liquorem Alkahest, cujus meminit 2. de viribus membrorum c. de hepate. Quod fit unius horae quadrante'.

100 'Raymundi Lulli Theorica & Practica', 60.

101 'Raymundi Lulli Theorica & Practica', 32.

102 'Raymundi Lulli Theorica & Practica', 48: 'nostrum finale secretum est, nostri arg.vivi congelatio'.

103 'De febribus', ch. 14, §10, 776: 'Nam, inquit Raymundus, astantibus amicis & praesente Rege, coagulavi argentum vivum, & nemo praeter Regem, scivit modum. In quo coagulatione istud est singulare. Quod liquor Alkahest, idem in numero, pondere & activitate tantum valet millesima actione, quantum prima'. Van Helmont is referring to Pseudo-Lull's account in *Testamentum*; 'Raymundi Lulli Theorica & Practica', 124–5.

works. Van Helmont considered that both talked about the same thing, the universal solvent, as originating from mercury. In fact, Pseudo-Lull's quicksilver may have been a *Decknamen* for philosophical mercury.

Van Helmont's incursion into medieval alchemy did not, apparently, end with Pseudo-Lull. Having established the *materia prima* of the *Alkahest* as mercury, he looked for further guidance on how to prepare it. This he found in the Latin Geber, the originator of the 'mercury alone theory' of alchemy. As Newman has pointed out, Van Helmont drew on Geberian ideas to describe the *Alkahest*. The 'discovery' of the *Alkahest*'s source in mercury, nevertheless, seems to have originated primarily from Paracelsus and Pseudo-Lull.

The importance of the ultimate acquisition of the Liquor *Alkahest* for Van Helmont cannot be understated. It is clear that Van Helmont regarded the acquisition of this Liquor *Alkahest* as one of the greatest secrets he had achieved. Indeed, he describes the attainment of the *Alkahest* as the most laborious part of alchemy.[104] He also adds that the knowledge was only granted to some.[105]

In the most comprehensive analysis of the Helmontian *Alkahest* to date, Paulo Porto rightfully emphasised the *Alkahest*'s medical uses. Its value as a medicine is that of a purgative like the fire, which clarifies organs 'not unlike [the way that] stibium purifies gold'.[106] Yet Porto's mild polemical intention of showing that the *Alkahest* was not 'a mere chemical used in the manipulation of matter'[107] leads him to distinguish perhaps too sharply between its medical and chemical uses ('the Helmontian alkahest belonged not only to a chymical system, but was an important feature of... a whole new medical theory').[108] In fact, 'medical', 'alchemical'/'chemical' or 'natural philosophical' uses did not necessarily represent distinct categories in Van Helmont's mind, since they were all part of his comprehensive 'applied philosophy' system.

In this sense, it is important to position the *Alkahest* in Van Helmont's greater framework of Christian philosophy. We must recognise that it was a cornerstone of his system in both natural-philosophical and applied (medical) ways. The *Alkahest* as a 'universal agent' had innumerable uses. It could prove that all bodies were formed of water, just as it could be employed to extract the medical *primum ens* of beings. Not only did it reveal the essence of all bodies, but it also allowed the harnessing of the primeval power resident in this inner core. Consequently, it could also prove the fundamental principles of Van Helmont's natural philosophy. The *Alkahest* vindicated his view that all wisdom had to have a practical purpose, extinguishing the traditional distinction between science and art. It was this simultaneity of uses and meanings that stood at the core of his vision of Christian philosophy.

---

104 'De lithiasi', ch. 7, §5, 700 : '... non sit operosior pars in tota Chymeia'.

105 Ch. 118, 'Arbor vitae', 635: 'Requiritur Alkahest. Quod non datur putantibus: sed solis scientibus, & quidem duplatis'.

106 Ch. 61, 'Potestas medicaminum', §44, 384.

107 Porto, '"*Summus atque felicissimus salium*"', 3.

108 Porto, '"*Summus atque felicissimus salium*"', 29.

Drif or *Lapillus*, the Universal Medicine

Despite his praise for Paracelsus's *Arcana*, Van Helmont's ambitions reached even higher than those, to the ideal of the one supreme universal medicine. This idea originated with medieval alchemical views advocated by Pseudo-Lull, Rupescissa and Arnau of Villanova.[109] In early modern alchemy, important theoreticians of this idea were Severinus and Croll, who made this idea particularly popular.[110]

Van Helmont maintained that the idea of a unique universal medicine was first conveyed to him by the mysterious Irish alchemist Butler.[111] The latter teased him that 'unless you arrive at the unique remedy that cures every disease, you shall remain a beginner, no matter how old you are'.[112] Despite Butler's provocative words, Van Helmont was not immediately swayed by the idea, seeking both empirical and religious proofs of the existence of a unique universal medicine. He hence proceeded to carefully document the results he observed in regard to Butler's medicine, the 'little Stone' (*Lapillus*). Van Helmont recorded the positive effect that the *Lapillus* had on several persons, including himself. The *Lapillus* was efficient, he found, either as an ointment or by simply touching it with the tip of the tongue.[113]

Van Helmont found another 'proof' for the universal medicine: the power of poisons. He noted that the venom of a viper or the saliva of an enraged dog killed instantly in any quantity.[114] If poisons could do such damage in such small amounts and by contact alone, Van Helmont reasoned, there must be antidotes: powerful

---

109 On the concept of universal medicine in the Middle Ages, see Chiara Crisciani and Michela Pereira, 'Black Death and Golden Remedies: Some Remarks on Alchemy and the Plague', in *The Regulation of Evil: Social and Cultural Attitudes to Epidemics in the Later Middle Ages*, ed. Agostino Paravicini Bagliani and Francisco Santi (Florence: Sismel, 1998), 7–41; Chiara Crisciani, 'Il farmaco d'oro, alcuni testi tra i secoli XIV e XV', in *Alchimia e medicina nel medioevo*, ed. Chiara Crisciani e Agostino Paravicini Bagliani (Florence: Sismel, 2003), 217–47; Michela Pereira, 'Un tesoro inestimabile: elixir e 'prolongatio vitae' nell'alchimia del 300' in *Micrologus* 1 (1993), 161–87; Michela Pereira, 'L'alchimista come medico perfetto nel 'Testamentum' pseudo-Lulliano', in *Alchimia e medicina nel medioevo*, ed. Chiara Crisciani and Agostino Paravicini Bagliani (Florence: Sismel, 2003), 77–109; Antoine Calvet, 'À la recherche de la médicine universelle: Questions sur l'élixir et la thériaque au 14e siècle', in *Alchimia e medicina nel medioevo*, ed. Chiara Crisciani e Agostino Paravicini Bagliani (Florence: Sismel, 2003), 177–87. The topic of the universal medicine became an important Paracelsian theme at the turn of the 1600s, with the works of Joachim Tanckius, Johann Dienheim and Oswald Croll. See Thorndike, *A History of Magic and Experimental Science*, VII, 161.

110 See above, Chapter 3. Croll makes a passionate defence of universal medicine in chapter 6 of his 'Admonitory Preface', 179–226.

111 Ch. 79, 'Butler', 468. However, it is hard to believe that Van Helmont was not aware of the current of thought on behalf of the universal medicine, of which we have seen Croll to be a foremost proponent.

112 Ch. 79, 'Butler', 468: 'Charissime, nisi eo devenias, quod unico remedio queas curare quemlibet morbum, manebis in tyrocinio, utcunque senex evaseris'. This remark highlights the competition amongst the alchemists in Van Helmont's age.

113 Ch. 79, 'Butler', 472.

114 Ch. 79, 'Butler', 474.

remedies that could restore health rather than kill. The remedies had to match the diseases, since the universe was made by a benevolent God.[115]

Van Helmont sought support for the concept of universal medicine in the Bible. This he found in Ecclesiasticus (Jesus Sirach) 38. In 38:4, Sirach writes 'Altissimus creavit de terra medicinam', which the Douay-Rheims Catholic Bible renders as 'the most High hath created medicines out of the earth'. Yet Van Helmont maintains that 'medicinam' is singular, hence what is meant by the text is that one single medicine is enough to cure all diseases.[116] He finds further support for his contention in Ecclesiasticus 38:8. The Latin Vulgate reads 'unguentarius faciet pigmenta suavitatis, et unctiones conficiet sanitatis, et non consummabuntur opera ejus', translated in the Douay-Rheims Catholic Bible as 'of these the apothecary shall make sweet confections, and shall make up ointments of health, and of his works there shall be no end'. Van Helmont interprets 'pigments' (*pigmenta*) to mean medicines whose power is infinite, hence universal.[117] Thus, he later states that the *Lapillus* of Butler is referred to in the Sirach verse.[118] Another verse he discovers in the Bible in support of universal medicine originates from another Deuterocanonical book, the Book of the Wisdom of Solomon, where it is stated that 'For he created all things that they might be: and he made the nations of the earth for health: and there is no poison of destruction in them, nor kingdom of hell upon the earth'.[119]

Further reflection on the nature of God strengthened his conviction that the universal medicine existed. Van Helmont argued that, as God was fundamentally 'divine goodness', He must be the true and direct author of medicine. In essence, medicine had to reflect both the power and the will of God. Since God was both One and omnipotent, Van Helmont inferred that the most divine medicine had to be small in quantity and swift in action.[120] Hence, there must be one supreme remedy

---

115 Ch. 79, 'Butler', 471. On the theme of God as divine goodness see above, Chapter 4.

116 Ch. 66, 'Respondet author', §1, 419: 'Ergo unum ex illis Arcanis est sufficiens pro omni, & quolibetcunque morbo. Atque ideo non dicit Textus, quod Omnipotens creaverit de terra medicinas; sed medicinam: numero singulari. Quam alioqui medicinam, jam praeparatam, pro Arte medendi non creavit de terra'. It is true that he mentions this in connection with three of the *Arcana* of Paracelsus: the liquor *Alkahest*, *tinctura lili* and *Mercurius diaphoreticus*, rather than Butler's *Lapillus*, but here he makes a theoretical defence of the idea of universal medicine in general.

117 Ch. 79, 'Butler', 472: 'Ast loquor hoc loco de pigmento, cujus opera non consummabuntur; nec erit morbus, aut venenum illi resistens, sive ut habet Textus'.

118 Ch. 79, 'Butler', 472–3: 'an non poterit lapillus, oleum suo contactu imbuere, ut queat illitum subter se sanare morbum latentem? Siquidem nil aliud his denotatur verbis: *Unguentarius conficiat pigmenta suavitatis, nec consummabuntur opera ejus*'. See also ch. 79, 'Butler', 473: 'Butlerus scilicet vi propemodum infinita aliquot myriades quotannis sanare posset, persuasit mihi Textus, quod opera unguentarii illius non consummabuntur'.

119 Douay-Rheims Bible, Book of Wisdom 1:14; The Latin Vulgate reads 'Creavit enim, ut essent Omnia: et sanabiles fecit nations orbis terrarium, et non est in illis medicamentum exterminii, nec inferorum regnum in terra'.

120 Ch. 79, 'Butler', 472. He is adamant that in the universal medicine the quality is more important than quantity, see also 474.

that can cure all diseases, and this medicine should be so powerful that 'one *lapillus* is sufficient for many tens of thousands'.[121]

Having thus established the possibility of a universal medicine, Van Helmont considers how it might be obtained. At this point, he reverts to Christian alchemical imagery to maintain that the body of this medicine must first be destroyed and then resurrected in a glorified form: 'that remedy first requires . . . that it be a body once resuscitated, and once destroyed, and afterwards, as it were by resurrection, in a certain sense glorified'.[122]

The death of the body, he argues, must be violent rather than natural, in order to preserve the virtues within.[123] This alchemical notion was popular in the Paracelsian thought of the period, being expressed in the notion of the 'mummy'.[124] Alchemical philosophers believed that the spirit of a being suffering violent death could be harnessed for medical purposes.

Once freed from its corporeal bonds, the body of the future medicine becomes volatile and must be sublimed twice or three times.[125] The artist must then unite this volatile spirit with disposed matter; the result is a 'glorified' or perfected body. The process increases its power a thousandfold; here Van Helmont seems to refer to the process of multiplication of the philosophers' stone.[126] This power he compares with that of a 'principiating' Ferment, which is both constant and able to extend its virtues to surrounding bodies by the means of its active instrument, the odour.[127]

Having established the manner of obtaining the medicine, Van Helmont proceeds to identify its constitutive matter. As it suits a *philosophus per ignem*, his criterion of selection is the fire: he attempts to find a resilient body that cannot be easily annihilated by its force. Consequently, he dismisses the vegetal and animal kingdoms

121 Ch. 79, 'Butler', 471: 'unicus lapillus multis sufficiat myriadibus'.

122 Ch. 79, 'Butler', 472: 'istud remedium inprimis requirere, ut sit corpus semel resuscitatum, semelque destructum, ac deinde quasi à resurrectio quodammodo glorificatum'.

123 Ch. 79, 'Butler' 474–5: 'Requirit Drif, ut sit è corpore naturali quidem, metallicae benignitatis participe, sed quod antea, prius sit per mortem obsequens, apertumque. Non quidem cum extinctio suarum virium, velut sponte defuncti cadaveris instar: sed retentis suarum dotium beneficiis, sit ab Artifice reseratum, repagulis liberum, ac velut resuscitatum'.

124 On the concept of the mummy amongst early modern physicians, see Poma, *Magie et guérison*, 88–119. In Van Helmont's time, the mummy was seen as a spiritual and aethereal substance which contained the inner properties of the living being, and which survived for a period after that being's death; Poma, 110.

125 Ch. 79, 'Butler', 475: 'Adeoque debuit resurrexisse velut à morte omnino volatile, atque spirituale, sive bis terve sublimari'.

126 Ch. 79, 'Butler', 472: 'Ast in remedio resurgente à morte, benefica sui simplicis bonitas adaugetur in millecuplum'.

127 Ch. 79, 'Butler', 475. On the importance of principiating Ferments see above, Chapter 5.

as he considers them too frail.[128] He believes that herbs 'could not ascend to a universal and supreme government over every disease'.[129]

Somewhat predictably, Van Helmont concludes that the secret of the *Lapillus* must lie in the metallic kingdom. He argues that the power of minerals and metals originate from an original metallic faculty (*vis metallica*) which in its supreme form can be found in gold, silver, and mercury.[130] Still, he points out, powerful medicines can also be extracted from the other four metals, on condition that the poisonous sulphur within them is rendered 'familiar', or friendly, to the human being.

In traditional alchemical fashion, Van Helmont does not share the entire recipe of the universal medicine. Yet he gives some tantalising details. *Drif*, he explains, involves the distillation of sea salt in a special manner, and the separation of mercury from the sulphur in the *primum ens* of copper (*primum ens veneris*).[131] His wording is elusive, but he suggests that the *Drif* should involve the extraction of the *ens veneris* from vitriol faeces.

In envisioning a universal medicine, alternatively called Butler's *Lapillus* or *Drif*, Van Helmont was drawing heavily on the practice, language and vision of medieval alchemy. His views here are coloured by traditional descriptions of the philosophers' stone. Intriguingly, he prefers the religious alchemical language that associated the *lapis* with the glorified body and even with Christ. Although he does not directly say so, the image of the tortured and killed metallic body which is then glorified reminds one of the traditional *lapis*-Christ association.[132]

The Philosophers' Stone

The commitment to a universal medicine conceived in traditional alchemical terms raises the question of Van Helmont's views concerning the philosophers' stone itself. In fact, his writings show that not only did he believe in the *lapis philosophorum*, but he also manifested a keen interest in making it.

Despite his support of Pseudo-Lullian alchemy, Van Helmont did not advocate its dual view of the philosophers' stone as a gold-making substance and a universal medicine.[133] In the treatise 'Arcana Paracelsi', he affirmed that he could not be easily persuaded that the philosophers' stone could be a medicine for the human body, since

128 Ch. 79, 'Butler', 472: 'Consideravi denique, hoc remedium non esse de monarchia vegetabili . . . [u]tpote quae instar animalium, facile sub artista moriuntur; vix tamen à morte resurgunt, cum sub ignis tentamine vel prorsus pereant, viresque priores omittant: vel si quasi resurgere videantur, nova tamen sunt potius entia, à priorum ac parentum orbita omnino seclusa'.

129 Ch. 79, 'Butler', 473: '. . . nequaquam autem ascendere ad universalem augustamque in omnem morbum gubernationem'.

130 Ch. 79, 'Butler', 473.

131 Ch. 79, 'Butler', 475.

132 See above a brief discussion of this, Chapter 3.

133 On this subject, see Pereira, 'Teorie dell'Elixir', 135–46.

it was indissolubly homogeneous and unchanged by the 'torture of fire'.[134] Hence, this would make it unable to 'vitally' unite with the human body. Consequently, Van Helmont rejected the philosophers' stone as a medicine for the prolongation of life. The language suggests some uncertainty: clearly, Van Helmont had investigated the medical properties of the *lapis* and was unable to understand how it could act as a remedy for the human body.

Certainly, his views of the universal medicine of Butler betray a debt to the philosophers' stone. Van Helmont described the universal medicine as a 'small stone' and used overt medieval alchemical terms to describe its preparation. Nevertheless, the relationship of the *Drif* to the philosophers' stone seems one of analogy, not identity.

Whether or not Van Helmont saw *Drif* as the *lapis philosophorum*, there is proof in a letter to Mersenne that he was interested in making the philosophers' stone. In 1631, the French philosopher assailed him with an entire battery of questions regarding the *lapis philosophorum*: whether he knows what the matter of the stone is, what is the philosophical sublimation, what is permanent oil, what is the fire that vivifies metals etc.[135] This incidentally suggests that Van Helmont had acquired something of an alchemical reputation even during his lifetime, even before the *Opuscula* and the *Ortus* were published.

To Mersenne's questions, Van Helmont responded simply and cryptically that all the above were one and the same thing: philosophical mercury.[136] He further added that, in accordance with the alchemical philosophers, the first matter of the stone was not vulgar gold or any other metal, but a 'medium between metals and mercury'. Here he seems to be referring to the theory of the *primum ens*, which he drew from Paracelsus.[137] This *primum ens* is probably identical to the *Bur* he mentions as being the liquid of metals.[138]

Van Helmont immediately accompanies this explanation with a qualified negation that he has ever made the philosophers' stone. His rhetoric is subtle. First of all, he contends that it is inappropriate for anyone to discuss something he had not made, so the matter should immediately be laid to rest. Moreover, he adds, had he made it, he would talk about it even less.[139] Clearly, Van Helmont is uncomfortable talking

---

134 Ch 116, 'Arcana Paracelsi', 628: 'Imo neque unquam facile inducar ut credam, lapidem philosophorum nobis vitaliter uniri posse, propter sui summe immutabilem substantiam, incredibiliter fixam contra ignis torturas, indissolubiliter homogeneam. Id est, propter separationis, destructionis, & digestionis omnimodam impossibilitatem. Tantum abest, quod ad vitam longam faciat'.

135 '21 Février 1631', 118. Theoretically these questions come from Mersenne's alchemical aide, but it is more likely that they belong to Mersenne himself.

136 '21 Février 1631', 118.

137 See above, Chapter 7, section on '*Alkahest*'.

138 For *Bur* or *Gur* in theories of metallogenesis see John A. Norris, 'Early Theories of Aqueous Mineral Genesis in the Sixteenth Century', *Ambix* 54 (2007): 69–86.

139 '21 Février 1631', 119: 'Et disputer de la pierre me semble malseant pour un chacun qui ne l'a faicte . . . C'est pourquoy je ne dispute de la pierre volontiers, veu que tandis que je ne l'ay, je dispute par opinion, et lorsque je l'aurois, je n'en voudrois parler pourtout'.

about the philosophers' stone. To support his reluctance, he quotes from one of his favourite medieval alchemists, Pseudo-Lull, according to whom those that have not achieved the philosophers' stone are called fools and talk likewise, but those who have, are called sages and do not talk at all.[140] Consequently, he adds, he hopes one day to make the philosophers' stone himself to escape the title of fool. This rather mysterious reply suggests, at least at face value, that Van Helmont makes no claims to have prepared the philosophers' stone by 1631. Yet the letter also shows that he was preoccupied with the subject.

In the treatise 'Arbor vitae' ('The Tree of Life'), written more than 10 years later, Van Helmont makes a further surprising statement: he had made a projection of the philosophers' stone in front of a large audience.[141] Having made such an exciting claim, Van Helmont immediately distances himself from any suggestion that he had prepared the stone. In fact, he points out, an unnamed alchemical Adept had given him a gold-making powder. One may be tempted to think this was the same Butler Van Helmont talked about in the eponymous treatise, except the Flemish philosopher maintains that this one was only 'a friend of one evening's acquaintance'.[142] This story sounds absurd today, but the early modern reader would have recognised the account as fitting within a lore of alchemical Adepts leaving proofs of their virtuosity behind.[143]

In 'Vita aeterna', Van Helmont details his test of the gold-making powder.[144] He had received a quarter of a grain of powder (a grain being described as the six-hundreth part of an ounce). The powder had the colour of saffron, looked like 'powdered glass', and was very heavy. He projected it onto eight ounces of mercury heated in a crucible; the mercury immediately congealed into a yellow lump. Having extracted the lump, he found eight ounces minus eleven grains of the purest gold. Clearly, his point in both 'Arbor vitae' and 'Vita aeterna' is to re-affirm his belief in the existence of the stone and convince the reader of its authenticity, rather than to argue that he has made it himself.

What strikes the reader when perusing Van Helmont's accounts of the philosophers' stone is their opaque, elusive and reserved style. He is being deliberately diffident about the subject, as the Mersenne correspondence amply proves. It is clear that Van Helmont is interested in the stone, but is averse to discussing details. This does not stop him from upholding its existence and even using it as an analogical symbol of the Eucharist.[145] Given his tendency to demand silence on the subject, one cannot

---

[140] '21 Février 1631', 119–20; the term Van Helmont uses is the French 'sot'; one wonders if there is a link with Cusanus's term of idiot.

[141] Ch. 118, 'Arbor vitae', 630.

[142] Ch. 118, 'Arbor vitae', 630: 'Istud autem aurum dedit mihi vir peregrinus, unius vesperi amicus'.

[143] On some popular histories of transmutation in the period see also E.J. Holmyard, *Alchemy* (Harmondsworth: Penguin, 1957), 128–34; and Lawrence M. Principe, *The Aspiring Adept: Robert Boyle and his Alchemical Quest* (Princeton, NJ: Princeton University Press, 1998), 93–113.

[144] Ch. 111, 'Vita aeterna', 590.

[145] See above, Chapter 4, section on 'Alchemy, Paracelsianism and Van Helmont's Theology'.

determine if Van Helmont believed to have made the philosophers' stone, thus supposedly escaping the Lullian title of 'fool'.

The *Arbor Vitae* and the Alchemical Prolongation of Life

Van Helmont viewed the *Alkahest* as a powerful instrument that could lead him to the supreme goal of all medical pursuits, which was the prolongation of life. For the Flemish philosopher, the supreme medicine is the *Arbor Vitae*, the Tree of Life. We may be tempted to argue that the universal medicine of the *Drif* extends life by taking away diseases, hence *Drif* is already an *Arbor Vitae*. In fact, Van Helmont made a clear distinction between curing illness and prolonging life. By postulating that death is fundamentally different from disease, he argued that healing a disease does not stave off the decline of vital powers that result in death.[146] In this distinction between life extension and removal of disease, Van Helmont follows the mainstream Galenic-Scholastic tradition.[147]

The Flemish philosopher maintained that, despite their potency, the *Arcana* of Paracelsus could not fundamentally extend life.[148] Even Paracelsus's celebrated *elixir proprietatis*, which renewed the hair, nails and teeth of the elderly, did not actually qualify as a medicine for the prolongation of life.[149] Van Helmont criticised this remedy as inducing only a superficial change in the human being, rather than renewing his vital powers. Hence he concluded that Paracelsus did not have a solid understanding of the prolongation of life, despite writing a treatise on the subject.

Van Helmont links the concept of the prolongation of life with Biblical speculation on the Tree of Life.[150] According to Van Helmont, Adam would have lived indefinitely with the help of the Tree, which re-vitalised his body.[151] Yet, once he had sinned and became cut off from the Tree of Life, he was destined to die. At first, Adam and his direct descendants, the Patriarchs, lived for hundreds of years due to the effects of the Tree of Life, but in time due to the loss of vigour, injuries and diseases, the lifespan of subsequent generations was greatly reduced.[152]

In the Christian tradition, the Biblical Tree of Life was often interpreted allegorically, symbolising the cross of Jesus Christ, the Eucharist or Christ himself.[153]

---

146 Ch. 112, 'Mortis occasiones', 591: 'Enimvero tametsi nullo infestarent infirmitates, non tamen ob id mors desineret in dies, viam ad suum introitum sternere. Nam etsi sanitas vitam, ut basin respiciat: non tamen vita sanitatem includit'.

147 Niebyl, 'Old Age', 351–68.

148 Ch. 112, 'Mortis occasiones', 591.

149 Ch. 116, 'Arcana Paracelsi', 626.

150 On this topic, see my article 'Reproducing the Tree of Life'.

151 Ch. 109, 'Vita', 586.

152 Van Helmont rejects Paracelsus's speculation that Adam prolonged his life by art, as he believed the *scientia* was lost at the moment Adam ate the apple.

153 See Paul Morris and Deborah Sawyer, *A Walk in the Garden: Biblical, Iconographical and Literary Images of Eden* (Sheffield: Sheffield Academic Press, 1992), 134, 136, 176–8.

Yet, literal medical interpretations appeared in the High Middle Ages. Thus, the Franciscan friar and alchemist Roger Bacon conceived the Tree of Life as an alchemical medicine which could bestow long life.[154] It also figured as a remedy in the early modern alchemical works of Giovanni Bracesco's *Dialogue on the Tree of Life* (1561) and Pierre-Jean Fabre's *Alchymista Christianus* (1632).[155]

Following this tradition, Van Helmont viewed the Tree of Life as a real medicine that could be extracted through alchemical means. It was the most desirable of alchemical remedies, since it could partially reverse the effects of the Fall of Man. He believed that acquiring the *Arbor vitae* was destined only for chosen ones: 'the attainment of the Tree of Life is most difficult, of much labour, and revealed unto few'.[156] He obviously believed himself to be one of those rare Adepts who attained that knowledge. The *Arbor vitae* was undoubtedly the noblest of enterprises, allowing him to claim himself a master and teacher of the true Christian philosophy.[157]

Van Helmont kept as best as he could within the bounds of orthodoxy in his opinions on the prolongation of life. He maintained that the medical Tree of Life does not, for instance, free man from passion, pain or death, as the Tree in Paradise did.[158] The reason is not the potency of the medicine, but the fact that man has acquired Original Sin, which the Tree of Life cannot erase. No medicine can mitigate our fallen nature except for Christ.[159]

To discover the secret of the Tree of Life, Van Helmont explored the mineral, animal and vegetable kingdom. His alchemical investigations into these made him conclude that the *Arbor vitae* must originate from a tree that is perennial and odoriferous.[160] However, the species of the tree was only revealed to him in a dream: the Tree of Life was the Cedar of Lebanon, mentioned several times in the Scriptures. Van Helmont believed it was only the Cedar of Lebanon that survived Noah's Flood, and it was in fact the very wood the Ark was made out of.[161] Van Helmont points out

---

154 Roger Bacon believed, like Van Helmont, that the medical *Arbor vitae* could extend life but not bestow immortality. On Bacon's ideas of prolongation of life see Agostino Paravicini Bagliani, 'Ruggero Bacone e l'alchimia di lunga vita: Riflessioni sui testi', in *Il Potere Del Papa: Corporeità, autorappresentazione, simboli* (Florence: Sismel, 2009), 75–96, as well as Faye Getz, 'Roger Bacon and Medicine: The Paradox of the Forbidden Fruit and the Secrets of Long Life', in *Roger Bacon and the Sciences: Commemorative Essays*, ed. Jeremiah Hackett (Leiden: Brill, 1997), 337–64.

155 Giovanni Bracesco, 'Dialogo Lignum Vitae', in *Vera Alchemiae Doctrina*, ed. Guglielmo Gratarolo (Basel: Petrus Perna, 1561), 3–47.

156 Ch. 118, 'Arbor vitae', 631: 'Est quippe acquisitio arboris vitae difficillima, multum operosa, & paucis revelata'.

157 There are several passages where Van Helmont maintains that the knowledge of long life was rare even amongst adepts (in his view, Paracelsus did not know it). See for instance, ch. 115, 'Infantis nutritio ad vitam longam', 622, 'Sed doctrina vitae longae valde est heteroclita: scioque illam inter divinitus electos filios artis usque in occultatis permansuram'; ch. 118, 'Arbor vitae', 633: 'Sed ejus praeparatio, omnium illorum, quae subter laborem Sophiae cadunt, est longe difficilissima'.

158 Ch. 109, 'Vita', 587.

159 Ch. 118, 'Arbor vitae', 632.

160 Ch. 118, 'Arbor vitae', 631.

161 Ch. 118, 'Arbor vitae', 633.

that the ancient Hebrews revered the Cedar, and even built the doors of the Temple of Solomon from the tree.

Van Helmont was aware that other alchemists had unsuccessfully tried to extract medicine out of the Cedar. This failure was due to using the wrong method. First of all, Van Helmont argued that not all parts of the Cedar could be used, but only the trunk; the rest he deemed too frail for use.[162] Secondly, he dismissed ordinary distillation as a method of extracting the Tree of Life. Instead, the liquor *Alkahest* had to be used.[163] Van Helmont complemented this with a description of how the process must be undertaken; needless to say, the recipe was useless without knowledge of the *Alkahest*. It was hence unsurprising that the next generation of alchemists were obsessed first and foremost with finding the mysterious *Alkahest*, as numerous treatises and correspondence on this subject attest.

The acquisition of the *Arbor vitae* represented the most powerful and meaningful result of a lifetime dedicated to a practice of mystical knowledge and charity. Due to his commitment to Christian philosophy, Van Helmont believed that the prolongation of life was the highest form of practical knowledge, and that this had to be duly imparted, to the extent permitted, to other alchemists. He was, of course, keenly aware that he would be attacked as a quack; to this he simply replied that 'God knows that I write things which I know to be true'.[164]

162 Ch. 118, 'Arbor vitae', 633.
163 Ch. 118, 'Arbor vitae', 633.
164 Ch. 118, 'Arbor vitae', 634: 'Novit Deus, quod scribam quae scio vera'.

General Conclusions

Van Helmont's Universal Vision

In the treatise 'Mentis complementum' ('The Perfecting of the Mind'), Van Helmont referred to himself as 'but a bell, calling the faithful to the Temple, which remains at the top of the tower, outside'.[1] This pointed imagery suggests that he viewed his life as a mission to reform Christian thought by introducing a new perspective on the natural world and Man's position in it. Thus, Van Helmont firmly believed he had been called to create a 'Christian philosophy' that would combine Christian religion and natural philosophy in a harmonious whole.

Van Helmont's 'Christian philosophy' argues that God is at the core of all that is, including Nature and Man. In his view, God in His essence is absolutely transcendent and unknowable. He could have chosen to remain as such forever, but it 'pleased' Him to express Himself by creating the world. In this sense, His 'divine goodness' and omnipotent will are the primordial aspects of manifest divinity. Van Helmont did not concur with those interpretations that saw Nature as the 'image' of God, but it is clear that for him the Universe was permeated by the divine *Logos*, expressing the divine attributes of Goodness, Truth and Unity.

Van Helmont's views of Nature begin with the Christian doctrine that the Universe was the result of the mind and sheer will of God. The primordial Creation happened in the beginning of time and can only be known by the means of the revealed Word of the Bible, which tells us how the world and Man came to be. Yet Scripture is no easy reading: there are mysteries within it that require clarification. Like many Paracelsians, Van Helmont believed that the Book of Genesis hid secrets that could open a new era for natural philosophy. It was only through its proper interpretation that a Christian philosopher could unveil the mystery of the origin of the Universe and of matter.

The greatest secret of Genesis, Van Helmont believed, was found in the mysterious word *shamayim*, which he interpreted to mean 'waters'. In this sense, he followed a strong alchemical tradition that considered bodies to be the offspring of water. The Book of Genesis was in essential harmony with the alchemical experience of reducing solid matter into a liquid or volatile state.

By positing the fundamental constituent of all bodies to be water, Van Helmont was essentially maintaining that all things, even the most solid, can be dissolved

[1] Ch. 43, 'Mentis complementum', §25, 254: 'Profecto vereor ne sim campana, fideles ad Templum convocans, quae ipsamet manet in celsitudine turris, foris'.

and transformed. The alchemical maxim *solve and coagula* stood at the basis of his philosophy. It is true that he did not share the naïve belief that transmutation was always a good thing and, on the authority of Hippocrates, criticised those alchemists who thought that Man should intervene in all courses of Nature. Van Helmont particularly mistrusted the universal application of fire in alchemy, since he essentially viewed fire as the 'death' of creatures. Fire with the help of Ferment could create new beings, but these were not necessarily better than the previous ones. Instead, Van Helmont was much more favourable to the idea of the reduction of bodies by means of solvents, particularly the *Alkahest*. He believed that reducing matter into its *primum ens* would allow the creation of powerful medicines, including the panacea of long life. The difference between fire and the *Alkahest* lay in the ability of the latter to preserve the seeds of the bodies, and hence recover their 'pure', incorrupt state. The process is similar to Geberian alchemical theory, according to which mercury had to be deprived of its sulphur in order to become 'immortal'.

Van Helmont drew yet more detail in regard to the make-up of the world from the Book of Genesis. The elements, water and air, were naturally passive: in the absence of spirit, they remained *tohu va-bohu*, empty and void. Van Helmont viewed them as the framework, or stage, on which the 'tragedy' of existence took place. They could not be altered. This belief in the fundamental immutability of the elements led Van Helmont to postulate the principle of the conservation of matter, whereby the quantity of air and water always remained the same.

In order for anything to exist, God sowed 'seminal reasons' (*rationes seminales*) in matter by means of the divine Word. These *rationes* were the 'principiating' Ferments, spiritual powers implanted in specific places to initiate generation. They did so by putrefying matter, 'penetrating' it with its odour and planting a corporeal seed in its bosom. This primordial process was proven by spontaneous generation, which Van Helmont saw as the archetype of ordinary generation.

Within the seed lay the formative power, the Paracelsian *Archeus*. The *Archeus* lay at the boundary between matter and spirit, being a liminal figure made up of very fine matter and an incorporeal nucleus. It had the capacity to transform potentiality into actuality by following a pre-existing divine plan, the archetype or entelechy. Hence, the function of the *Archeus* was to translate this 'seal'—which Van Helmont described as an idea or an image—into a real, visible form. Consequently, the *Archeus*'s main role was to pre-form the individual according to its proper species. Only when the being is fully formed, could it be infused with life coming from God Himself.

Van Helmont envisioned life as light. In doing so, he creatively employed a rich tradition of light symbolism that was present in Christian and philosophical thought, and above all Paracelsus's concept of the 'Light of Nature', or the immanent light within beings. Van Helmont's formal lights played several roles: they were guarantors of intelligibility and mediators of the communication that took place between all living things.

If forms were 'lights', they had to originate from some ultimate source, just as the rays of the Sun originate from its body. It is here that Van Helmont made one of his chief philosophical claims: the formal lights were given by God directly. As we have seen, the doctrine of *dator formarum*, the direct provision of form by God, was not a mainstream concept. Many philosophers tended to think that the direct intervention of God in matter was unacceptable, resembling a *deus ex machina* proposal. In fact, many sought to distance God from Creation: the overarching image of divinity was one of an omnipotent maker who had created such a perfect machine that He no longer needed to intervene in it. This went in line with the Protestant rejection of miracles, and it was precisely a perpetual miracle that seemed to be implied by Van Helmont's views of generation.

In his philosophy, Van Helmont argued that God was not only the source of the Universe as a whole, but of the individual in particular. This worldview went hand in hand with the concept of an absolute necessity in Nature: no creature, except for human beings, had free will. We must not forget that for Van Helmont, all beings 'depend originally, totally, absolutely, and intimately on God'.[2] Thus, the whole Universe was pre-programmed and governed by God with his 'hand' or his 'finger', as Van Helmont presented it. The stars themselves bore upon themselves the signs of this 'universal predestination', which the chosen prophets of God could interpret.

Of course, Van Helmont's view of God as involved in the minutest details of life could be criticised as promoting a despotic perception of divinity. This would, however, be an inaccurate image of Van Helmont's understanding of God. Van Helmont was an alchemical philosopher, and for him God was not an exterior force. Instead, God was the divine essence of things, their Goodness, Truth, and *Scientia*. This view was not pantheistic, since Van Helmont believed along orthodox lines that God is in all things, but He is not all things. His essence, as the Flemish philosopher underlined, remained inaccessible and unknowable.

Van Helmont also held a Christocentric view of Creation as a way of expressing the immanence of God in the world. His Christocentrism had deep roots in alchemical philosophy, which sometimes presented the philosophers' stone as an analogy of Christ. Van Helmont's Christocentrism draws on the assumptions of medieval Christian alchemy, which emphasised the hidden presence of Christ in Creation, the possibility of transmutation, the intrusion of the supernatural into the ordinary, and the confirmation of the special status of human beings as creators of divine things.

For Van Helmont, Christ was not only Pantocrator, the ruler of the Universe, He was also 'true man'. In this fundamentally Pauline perspective, Christ was brought close to human nature as 'second Adam', or the true face of man before the Fall. The intimate relationship between Christ and human beings is also implied in Van Helmont's view that sinless Men were 'sons of God'. It is not that Van Helmont wished to diminish Christ as something less than divine. Instead, he wished to lift the stature of Man as something akin to God: it is in such a way that one must interpret his insistence on Man as *imago Dei*. This expression is more than a figure of

---

[2] Ch. 89, 'De tempore', §26, 504: '. . . originaliter, totaliter, absolute scilicet & intime à Deo'.

speech for him: it meant that we were originally created to be the mirror image of divinity, God's counterpart in the created world.

The close kinship between God and Man was emphasised by Van Helmont in his most controversial doctrine: that of the Original Sin as the sexual act. Despite its eccentricity, it is essential to understand it as an extension of his strong humanist views. By postulating that God's original plan was to generate human beings through a direct intervention into the womb of Eve, Van Helmont brought mankind even closer to divinity. The only real difference between Man and God resided in the human's creaturely nature and his material side, obtained from Eve. Otherwise, being generated by the Holy Spirit, Man would have been virginal, immortal, and in close contact with his only 'Father', God. This status also implied that he would have had a direct and unmediated understanding (*scientia*) of Nature, which would have given him dominion over all creatures.

Given such a lofty view of the original status of mankind, the Original Sin appears nothing short of a disaster. Viewing the knowledge of 'good and evil' as essentially sexual comprehension, Van Helmont concluded that Adam's 'rape' of Eve effectively resulted in the obliteration of the divine generation of Man. The Original Sin was a tragedy that caused Man to lose his status of 'son of God', his universal knowledge and his immortality. Van Helmont's vision is peculiar in that he emphasised the organic consequences of the Original Sin: the replacement of a spiritual generation with a sexual one, and of the mind with a newly produced sensitive soul.

The retreat of the mind into the recesses of the body and the assumption of governorship by the *anima sensitiva* make the present man a debilitated, fallen creature. The sensitive (animal) soul does not know how to rule a human body, and hence it subjects it to suffering, disease, and eventually death. The present status of the human being is all the more pitiful given the greatness of his initial destiny. It is the more dramatic as men do not even realise how deeply their present status is altered by Original Sin. The worst blindness, Van Helmont believes, is to mistake reason for a higher faculty, and to demand knowledge from it. Reason is the product of the Tree of Knowledge of Good and Evil: it is an animal trait that belongs to the *anima sensitiva* and not to the mind.

This bleak state of affairs can, felicitously, be remedied. God, whose love toward mankind never wavered, sent His only Son, Jesus Christ, to die on the Cross for mankind's sins. Christ instituted the essential Sacraments of Baptism and Eucharist so that human beings may be regenerated into His virginal body. Van Helmont, like Paracelsus, viewed these Sacraments in organic terms: they act at a subtle level, restoring the lost purity of the soul and body. This transformation is, unsurprisingly, compared with the transmutation of base metals into the philosophers' stone. Both, in essence, are miracles that are effected by divine will in the sublunary world.

Although Man is essentially saved by faith, good works and the Sacraments, Van Helmont formulates a vision for knowledge in his system. Knowledge can approximate the lost Adamic status of man and also bring an 'increase of Grace' if used for charity. By anchoring knowledge in the act of charity, Van Helmont gives a meaning to both mystical experience and the practice of alchemy. However, to be able to produce powerful alchemical medicine is no trivial matter: nothing less than a mystical enlightening must occur. Indeed, Van Helmont posits a spiritual ladder that a novice must climb to achieve the highest status of Adept, the possessor of universal knowledge.

No one can achieve wisdom except through self-knowledge, in Van Helmont's Christianised version of the Greek edict *gnothi seauton*. Self-knowledge begins with silent prayer, ascetic practices and vigils; eventually, intellectual visions appear to guide the soul to illumination. Yet the highest status is imageless: one must reject all representations in order to reach a state of profound darkness, whereby everything, including the human will, is renounced. When everything is thrown away except desire, God may bestow upon man the bounty of His light and the supreme gift of the kiss (*binsica*). For a moment, the soul of man is then transformed into the mirror of God. This mystical state is also one of pure knowledge, when man meets God and draws wisdom from divine revelation.

The knowledge of the self as *imago Dei* is hence the fundamental premise and archetype of other types of knowledge. The wise man, Van Helmont thinks, cannot be content with directly experiencing God for his own sake: he must go further and use that knowledge to serve other people. The greatest mystics, in his eyes, are the physicians, who not only acquire wisdom by direct illumination, but also put it in practice for the sake of less fortunate human beings.

In Van Helmont's peculiar worldview, the true Adept is the one who has mastered the highest secrets of alchemy and medicine. For this purpose, laboratory practice is insufficient: it must be accompanied by mystical techniques of a similar kind to those needed for self-knowledge. The assumption in this view is that the essence of all things is divine: hence reaching it requires contact with divinity. The process is described as a transformation of the intellect into the thing understood, or the attainment of the Truth within things. The assumption, of course, is the intelligibility of the natural world by the human mind based on the universality of the Word.

This process, is, of course, not easy, and from Van Helmont's own testimony it appears that the attainment of true *scientia* is an arduous process that requires years and years of prayer and laboratory practice. It is clear that practical work must go hand in hand with a pious attitude and the recognition of the limits of the unenlightened human mind. Only God bestows knowledge, and He gives it to whom He will. Illumination is the product of God's Grace, and not of one's own merits.

In the *Ortus*, Van Helmont claims that he achieved some of the highest forms of knowledge that may be given to mankind. Amongst these are the universal medicine, the medicine of long life, and the philosophers' stone. He maintains to have obtained two of these three, with the issue of the stone being left unclear. However, true to his beliefs in charity, he concentrates his efforts on alchemical medicine, including the miraculous *Drif* and the *Arbor vitae*. Unfortunately, such secrets cannot be shared with other human beings. This is because these are 'the secrets of God' not of man, and for Him to impart alone.

This summary has, hopefully, shown that Van Helmont's views were unitary and coherent. It has also conveyed the close relationship between his Christian views and those drawn from medical alchemy. Indeed, they meet at a fundamental level in concepts that transcend the boundaries between religion and natural philosophy. The idea of 'universal concepts' or 'principles' are those that manifest 'simultaneity', or the ultimate unity and harmony of the world.

One such 'universal' Helmontian principle is transmutation, which stands not only for the transformation of poisons into medicine or of metals into gold. Indeed, there are many types of transmutations, such as the change of mortal bodies into Christ by means of the Sacraments, the action of the fire on bodies, as well as the means by which Ferments act on passive water to create completely new and different beings. Furthermore, the concept of transmutation also applies to epistemology, that is, in the process of acquiring *scientia* and achieving the mystical experience of the mind.

Another important 'universal concept' is light, which is not only the appearance of the Taboric Christ and the experience of the divine in the soul, but also the fire the alchemist uses to create improved bodies in the laboratory, the Light of Nature that is intrinsic within things as their life, and even the principle whereby alchemy illuminates philosophy.

Two additional themes pervading Van Helmont's work are the triadic scheme and virginity. The threefold pattern is found foremost in the Trinity, but also, on the lower scale, in the three 'titles' of the mind, the three components of the sensitive soul and even in the unseparable *tria prima* of water.[3] Virginity is not only the spiritual power emerging from Christ; it is the glorified human body of eschatology, the unspotted body of the Virgin Mary, the potent state of the philosophical mercury, the true state of the elements and the purity of women who dedicate their lives to Christ.

Van Helmont also gives overwhelming importance to subtle, invisible entities. Contrary to modern views, he believes that the less material a being is the more powerful it is. Hence his attention is concentrated on invisible entities like the *magnale*, *Archeus*, the soul or *Gas*. He is also interested in those alchemical processes that imply the subtilisation of matter, such as distillation or sublimation. By these methods, he believes that extraordinarily powerful medicines can be achieved. Most of the *Arcana* he discusses are achieved by volatilising matter, particularly salts; only volatile beings have the power to 'penetrate' the diseased organs and cleanse them.

Besides volatility, Van Helmont is also interested in the liquid state. For him, liquidity is not only the state of the Universe at the beginning of Creation: it is also the purifying water of Baptism, the first being of all things, the reduction of every being into their first *ens* by means of the *Alkahest*, the watery *ignis gehennae* of the

[3] In doing this Van Helmont is clearly following Augustine's exhortation of uncovering the vestiges of the Trinity in the human mind, the 'outer man' and in the natural world. See Augustine, *On the Trinity*, 15.2, 4.

Alkahest itself, the loss of the will in God as well as the state of the mind when it becomes ready to receive God.

Van Helmont's 'Christian philosophy' can hence be seen as a coherent worldview, impaired only by its mildly sometimes confusing presentation in the *Ortus medicinae*. I have already observed that reasons for this could be linked with a relatively sudden death and to Van Helmont's own style, which he sought to make as remotely Scholastic as possible. Perhaps other obstacles he might have encountered included the inherent difficulty in articulating mystical experiences and the problem of expressing a worldview that originated from a long, but generally secretive, tradition of alchemical philosophy. Van Helmont's perspective is deeply tied to his Paracelsian background, despite his attempts to free himself from it. In fact, as I have endeavoured to show, many of Van Helmont's ideas were a reiteration of Paracelsian and alchemical views. Where he went further was in attempting to articulate a coherent Christian natural philosophy that could act as a new synthesis capable of replacing Scholasticism.

The Influence of Van Helmont's Philosophy: Some Future Directions

In 1707, a new edition of Van Helmont's *Opera omnia* appeared in Frankfurt with an introduction by Michael Bernhard Valentin. In it, the editor associated Van Helmont with Francis Bacon and René Descartes as one of the luminaries of the previous century.[4] In modern eyes, this statement may seem a gross exaggeration of Van Helmont's intellectual importance, but recent scholarship has already shown that many philosophers of the so-called 'Scientific Revolution' read Van Helmont's work attentively. Thus, Gottfried Leibniz and Hermann Boerhaave considered the *Alkahest* as a possibility.[5] Young John Locke compared favourably Van Helmont's theory of disease to that of the Cartesians, while Isaac Newton read the *Ortus medicinae*, annotating it carefully.[6] Thomas Willis, Franciscus de la Boë Sylvius and Georg Ernst Stahl formulated an extensive 'fermentational programme' based on Helmontian notions.[7]

Although there has been no comprehensive study of 'Helmontianism' to date, scholars have made important inroads into the subject. Charles Webster, Allen

[4] Michael Bernhard Valentin, 'Introductio', in Johannes Baptista Van Helmont, *Opera Omnia* (Frankfurt: Hieronymi Christiani Paulli, 1707), [1].

[5] Joly, 'L'alkahest', 312–13.

[6] J. R. Milton, 'Locke, Medicine and the Mechanical Philosophy', *British Journal for the History of Philosophy* 9 (2001): 221–43; Anstey, 'John Locke and Helmontian medicine', 93–117; Piyo Rattansi, 'Newton's Alchemical Studies', in *Science, Medicine and Society in the Renaissance: Essays to Honor Walter Pagel*, 2 vols, ed. Allen Debus (New York: Science History Publications, 1982), II, 175–80.

[7] Ku-Ming (Kevin) Chang, 'Fermentation, Phlogiston and Matter Theory: Chemistry and Natural Philosophy in Georg Ernst Stahl's "Zymotechnia Fundamentalis"', *Early Science and Medicine* 7 (2002): 31–64.

Debus, Antonio Clericuzio, William Newman and Lawrence Principe, Ole Peter Grell, Anna Marie Roos, amongst others, have revealed Van Helmont's popularity in England in the second half of the seventeenth century.[8] The influence of Van Helmont in France has also begun to be analysed.[9] Sietske Fransen's dissertation has recently reviewed the reception of Van Helmont through the lens of various editions and translations of the *Ortus medicinae*.[10]

Nevertheless, much work remains to be done in order to gauge Van Helmont's impact not only on seventeenth- but also on eighteenth-century culture. I hope that, by firmly anchoring Van Helmont in the intellectual climate of his age, the present book might lay a strong foundation for research on Helmontianism. In this concluding part, I will sketch out some further directions in which research on Van Helmont's influence may be carried out.

One line of enquiry that has been little addressed in scholarship is the impact of Helmontian rhetoric on late seventeenth-century discourse. As I mentioned, Van Helmont made stark claims of originality for his theories. This self-fashioning was grasped by the following generation; Robert Boyle, for instance called Van Helmont 'bold and ingenious' and a 'paradoxical writer'.[11]

More importantly, Boyle not only observed, but also internalised Helmontian rhetoric of novelty himself. He called his own conjectures 'paradoxes'[12] and presented himself as an original thinker. In this category we can place Boyle's claim that 'I had the good Fortune to Learn the Operations from illiterate Persons'.[13] This argument has been thoroughly rejected by William Newman and Lawrence Principe, who have shown that Boyle had learned most of his chemistry from George Starkey and some from Benjamin Worsley, neither of whom was 'illiterate'.[14] Yet Boyle's claim was closely fashioned after Van Helmont, who similarly argued that he had learned the art of the fire from an 'idiot'. Moreover, Boyle's obvious disdain for the 'vulgar chymists' mirrored Van Helmont's contempt for purely empirical practitioners and false alchemical philosophers.[15]

Ironically, Boyle's internalisation of Van Helmont's 'paradoxical' rhetoric determined the former to ultimately distance himself from the Flemish philosopher. For instance, while the *Sceptical Chymist* appropriates much of Van Helmont's rejection of the *tria prima*,[16] Boyle's mentions of Van Helmont make him to be a great

8 See the references in the Preface.

9 Boantza, 'Alkahest and Fire', 75–92.

10 Sietske Fransen, 'Jan Baptista van Helmont and his Editors and Translators in the Seventeenth Century' (London, Warburg Institute dissertation, 2014).

11 Boyle, *Sceptical Chymist*, 118–23.

12 Boyle, *Sceptical Chymist*, 347, 430.

13 Boyle, 'A Preface Introductory', [14–15].

14 Newman and Principe, *Alchemy Tried in the Fire*, 25–7.

15 See above, Chapter 6, section on 'Mysticism and Empirical Investigation'.

16 Boyle, 'A Preface Introductory', [8]: 'From whose writings it is not Therefore like he should receive any intentionall assistance, except from some Passages of the Bold and Ingenious Helmont'.

experimenter,[17] but not a similarly gifted philosopher. For, as Boyle puts in the introduction, of Van Helmont's philosophical arguments 'not only some seem very Extravagant, but even the Rest are not wont to be as considerable as his Experiments'.[18] Boyle's case of using Van Helmont's rhetoric against himself could signal a peculiarity to Helmontian reception, as it encouraged intellectual originality at the expense of acknowledging one's true source of speculation.

Another important research topic could be the influence of Van Helmont's lesser-known ideas. While notions like *Gas*, the *Alkahest* or fermentation have captivated the attention of scholars, the same cannot be said for other concepts like *blas*, the formal lights, Van Helmont's belief in an unconscious will, his theory of generation, his views on the *Arcana*, the universal medicine and the Tree of Life. Since these concepts are often less distinct than the *Alkahest* or *Gas*, research must take into account their use under modified names. In this sense, this book has attempted to clarify the essence of some Helmontian concepts so that they can be more readily identified when they occur in disguise.

Another line of enquiry would be to investigate Van Helmont's influence on the concept of 'Adeptship'. Lawrence Principe has pointed out the importance of this idea in the work of Robert Boyle;[19] however, Boyle most likely borrowed the term and the idea from Van Helmont. As I mentioned, the notion of the 'Adept' was first popularised by Paracelsus, but it was Van Helmont who introduced it into a wider use.[20] Importantly, the term did not exclusively mean *chrysopoeia*; Van Helmont described it as knowledge of the higher secrets of alchemy, including the medical *Arcana* and the medicine of long life. In this sense, it would be worth further investigating the meaning and transformations of this concept, which scholars still use today in reference to alchemists.

A distinct area that could be explored in the future is the impact of Van Helmont's Christian philosophy on further attempts to merge natural philosophy and religion. For instance, the influence of Van Helmont on Robert Boyle's own vision of a pious philosophy merits further assessment.[21] The reception of Van Helmont's alliance of religion and philosophy could occasionally take a distinctly theosophical form,

---

[17] Take, for instance, Boyle's remarks on Van Helmont as 'an Author more considerable for his Experiments than many Learned men are pleas'd to think him', 112, or 'much better vers'd in the Chymical Anatomizing of Bodies then either Sennertus or I', 178.

[18] Boyle, 'A Preface Introductory', [9].

[19] Principe, *The Aspiring Adept*.

[20] See Chapter 7 above, section on 'Alchemy and Adeptship'. See also my upcoming article 'The Rise of the Concept of the "Adept" in Early Modern Alchemy: From Theophrastus Paracelsus to Oswald Croll', in *The Old in the New: Historical Transformations in Western Esotericism from the Renaissance to the Present*, ed. Georgiana D. Hedesan and Tim Rudbøg (in development).

[21] Ducheyne has already pointed out the appeal of Van Helmont's combination of religion and philosophy in Boyle's notion of experience, 'A Preliminary Study', 130–31. For Robert Boyle's take on religious philosophy see particularly Jan W. Wojcik, *Robert Boyle and the Limits of Reason* (Cambridge: Cambridge Univ. Press, 1997) and Michael Ben-Chaim, 'Empowering Lay Belief: Robert Boyle and the Moral Economy of Experiment', *Science in Context* 15 (2002): 51–77.

as in the work of Joachim Poleman, who praised both Jacob Boehme and J.B. Van Helmont as 'two bright-shining torches for this present age' which might enable people to 'open their eyes and see by this double Light of His mercy set up, and so acknowledge or own God their most blessed Creator, walk in his light, and bring forth the due fruit of Holinesse and Obedience'.[22] This particular, 'theosophical' reception would merit further analysis.

It has long been supposed that Van Helmont was popular in Protestant, and particularly Puritan, circles.[23] Yet Catholic physicians such as Pierre-Jean Fabre or Franz Oswald Grembs were equally interested in his philosophy. In fact, it seems that Helmontianism was eminently capable of transcending religious boundaries. For instance, Van Helmont's work was embraced even by members of the Eastern Orthodox clergy. The bishop Meletios of Athens (1660–1714), who studied medicine and philosophy in Venice and Padua, appears to have been a keen teacher of Helmontian ideas at the end of the seventeenth century. One of his pupils, the Prince of Moldavia and later Counsellor of Tsar Peter the Great, Dimitrie Cantemir (1673–1723), eagerly absorbed Helmontian ideas. Cantemir's learned treatise *Sacro-sanctae scientiae indepingibilis imago* (1700) reveals a thorough grounding in Van Helmont's philosophy.[24] The work was a statement of Cantemir's beliefs that natural philosophy and Christianity should be seen as two aspects of the same discourse. In promoting Helmontian notions, Cantemir very clearly distanced himself from the mainstream neo-Aristotelian school that was fashionable at the Patriarchal Academy of Constantinople during this period. This advocated the double-truth doctrine of Averroes, which promoted the separation of religion and natural philosophy.[25] In this sense, it would be worth considering how Van Helmont's authority was used to support the concept of Christian science at the turn of the eighteenth century.

The case of Cantemir's *Sacro-sanctae scientiae* points to another issue that should be further investigated: the duration of Van Helmont's influence. Indeed, interest in Helmontian ideas faded in England at the close of the seventeenth century, and it is easy to dismiss Cantemir's work as a belated homage to an already obsolete Van Helmont. Yet England's case should not be taken as representative for the continent as well. In France, for instance, Helmontian ideas seem to have been passed on within the Académie Royale des Sciences from Wilhelm Homberg to Antoine Lavoisier.[26] We have seen Boerhaave in the Netherlands expressing a positive opinion

[22] Joachim Poleman, *Novum lumen medicum wherein the excellent and most necessary doctrine of the highly-gifted philosopher Helmont concerning the great mystery of the philosophers sulphur* (London: J.C., 1662), 116.

[23] Charles Webster, *The Great Instauration*, 273–288; J.A. Mendelsohn, 'Alchemy and Politics in England (1649–1665)', *Past and Present* 135 (1992), 30–78.

[24] This work has been recently published in a dual Latin-Italian edition as Dimitrie Cantemir, *L'immagine irrafigurabile della Scienza Sacro-Sancta*, ed. Vlad Alexandrescu (Milan: Mondadori, 2012).

[25] Cléobule Tsourkas, *Les debuts de l'enseignement philosophique et de la libre pensée dans les Balkans: La vie et l'oeuvre de Theophyle Corydalee (1570–1646)* (Thessaloniki: Institute for Balkan Studies, 1967), 211–16.

[26] Newman and Principe, *Alchemy Tried in the Fire*, 296–311.

of Van Helmont. In Germany, the Flemish philosopher's ideas were still read by physicians such as Johann Conrad Barchusen (1710) and Gottlieb Stolle (1731) and even by young Wolfgang von Goethe. Further research could enlighten us on the fortunes of Helmontian philosophy in the eighteenth century.

We may conclude by legitimately wondering why Van Helmont's work proved so popular in some circles. Beyond the strength of Van Helmont's chemical experimentation, we can recognise something profoundly enticing about his attempt at unifying knowledge. Van Helmont's Christian philosophy offered new avenues and justifications for the exploration of Nature; it was an ideology that seemed to be in harmony with Christian views. In Van Helmont's mindset, Man was free to reach as high as he could: there were no injunctions against knowledge, as long as it was accompanied by a Christian attitude of devotion and charity. The Flemish philosopher also seemed to indicate the path to the fulfilment of dreams that had haunted mankind forever: the universal medicine for diseases, the extension of life and renewal of youth. Seen in such light, it is unsurprising that so many late seventeenth- and early eighteenth-century philosophers would have been drawn to the Helmontian vision.

Appendix
List of Chapter-Treatises in *Ortus Medicinae*

Chapter No.	Latin Name	Modern English	English (1664 translation)	Tentative Subject Division
1	Promissa authoris	The Author Promises	The Authors Promises	Introduction
2	Confessio authoris	The Author's Confession	The Authors Confession	Autobiography
3	Studia authoris	The Author's Studies	The Authors Studies	Autobiography
4	Venatio scientiarum	The Hunting of the Sciences	The Searching Out of Sciences	Natural Philosophy
5	Causae, & initia naturalium	The Causes and Beginnings of Natural Things	The Causes and Beginnings of Natural Things	Natural Philosophy
6	Archeus faber	Archeus Faber	Archeus Faber or the Master Workman	Natural Philosophy
7	Logica inutilis	Logic is Worthless	Logick is unprofitable	Natural Philosophy
8	Physica Aristotelis & Galeni ignara	The Ignorant Natural Philosophy of Aristotle and Galen	The ignorant Natural Phylosophy of Aristotle and Galen	Natural Philosophy
9	Elementa	The Elements	The Elements	Natural Philosophy
10	Terra	The Earth	The Earth	Natural Philosophy
11	Aqua	The Water	The Water	Natural Philosophy
12	Aer	The Air	The Air	Natural Philosophy
13	Progymnasma meteori	Essay on Meteors	Essay of a Meteour	Meteorology
14	Gas aquae	The Gas of Water	The Gas of the Water	Meteorology
15	Blas meteoron	The Blas of Meteors	The Blas of Meteors	Meteorology
16	Vacuum naturae	Nature's Vacuum	A Vacuum of Nature	Meteorology
17	Meteoron anomalum	An Anomalous Meteor	An Irregular Meteor	Meteorology
18	Terrae tremor	The Earthquake	The Earth-quake	Meteorology
19	Complexionum atque mistionum elementalium figmentum	The Figment of Elemental Complexions and Mixtures	The Fiction of Elementary Complexions and Mixtures	Natural Philosophy

Chapter No.	Latin Name	Modern English	English (1664 translation)	Tentative Subject Division
20	Imago fermenti impraegnat massam semine	The Ferment's Image Impregnates the Seminal Mass	The Image of the Ferment begets the Masse with child of a seed	Natural Philosophy
21	Astra necessitant, non inclinant, nec significant de vita, corpore, vel fortunis nati	Stars Necessitate, Rather than Incline, or Signify One's Life, Body or Fortunes	The Stars do necessitate; not incline, nor signifie of the Life, Body, or Fortunes of him that is born	Natural Philosophy
22	Formarum Ortus	The Birth of Forms	The Birth or Original of Forms	Natural Philosophy
23	Magnum oportet	*Magnum Oportet*	*Magnum Oportet*, or a thing of great necessity or concernment	Natural philosophy
24	Natura contrariorum nescia	Nature Knows No Contraries	Nature is ignorant of contraries	Natural Philosophy
25	Blas humanum	The Human Blas	The Blas of Man	Natural philosophy
26	Endemica	Endemics	Endemicks	Medicine
27	Spiritus vitae	The Spirit of Life	The Spirit of Life	Medicine
28	Calor efficienter non digerit, sed tantum excitative	Heat does not digest efficiently, but only by excitation	Heat doth not digest efficiently, but excitingly onely	Medicine (Digestion)
29	Triplex Scholarum digestio	The Triple Digestion of the Schools	The threefold Digestion of the Schools	Medicine (Digestion)
30	Sextuplex digestio alimenti humani	The Six Digestions of Human Food	A sixfold Digestion of humane nourishment	Medicine (Digestion)
31	Pylorus rector	The Ruling Pylorus	*Pylorus* the Governour	Medicine (Digestion)
32	Tartari historia	A History of Tartar	A History of Tartar	Medicine (Against Tartar)
33	Tartari vini historia	A History of Wine Tartar	A History of Tartar of Wine	Medicine (Against Tartar)
34	Inventio Tartari, in morbis temeraria	The Foolish Invention of Tartar in Diseases	The rash invention of Tartar in Diseases	Medicine (Against Tartar)

Chapter No.	Latin Name	Modern English	English (1664 translation)	Tentative Subject Division
35	Alimenta Tartari insontia	Food is Free of Tartar	Nourishments are guiltlesse of Tartar	Medicine (Against Tartar)
36	Tartarus non in potu	The Tartar Cannot be Found in Drink	Tartar is not in drink	Medicine (Against Tartar)
37	Custos errans	An Erring Guardian	An erring Watchman or wandring Keeper	Medicine
38	Imago mentis	The Image of the Mind	The Image of the Mind	On the Soul
39	Demens Idea	A Mad Idea	A mad or foolish Idea	On the Soul
40	Sedes animae	The Seat of the Soul	The seat of the Soul	On the Soul
41	A sede animae ad morbos	From the Seat of the Soul to Diseases	From the seat of the Soul unto Diseases	On the Soul
42	Ius Duumviratus	The Law of the Duumvirate	The authority of the Duumvirate	On the Soul
43	Mentis complementum	The Perfecting of the Mind	The compleating or perfecting of the Mind	On the Soul
44	Scabies & ulcera Scholarum	The Scabies and Ulcers of the Schools	The Scab and Ulcers of the Schools	Diseases
45	Ignota actio regiminis	An Ill-Known Action of Regimen	An unknown action of Government	On the Soul
46	Duumviratus	Duumvirate	The Duumvirate	On the Soul
47	Tractatus de anima	Treatise on the Soul	A Treatise of the Soul	On the Soul
48	Distinctio mentis à sensitiva anima	The Distinction of Mind from the Sensitive Soul	The Distinction of the Mind from the sensitive Soul	On the Soul
49	De Anima nostra immortalitate	On the Immortality of Our Soul	Of the Immortality of the Soul	On the Soul
50	Nexus sensitiva & mentis	The Intertwining of the Sensitive Soul and the Mind	The knitting of the sensitive Soul and Mind	On the Soul
51	Asthma & Tussis	Asthma and Cough	The Asthma and Cough	Diseases
52	Latex humor neglectus	The Neglected Latex Humour	The humour Latex neglected	Diseases

Chapter No.	Latin Name	Modern English	English (1664 translation)	Tentative Subject Division
53	Cauterium	Cautery	A Cauterie	Diseases
54	Volupe viventium morbus antiquitus putatus	The Disease that was in Ancient Times Supposed to be of the Pleasurable Life	The Disease that was antiently reckoned that of delightful Livers	Diseases
55	Pleura furens	The Furious Pleurisy	A mad or raging Pleura	Diseases
56	Tria prima Chymicorum principia, neque eorundem essentias, de morborum exercitu esse	The alchemists' tria prima are neither essences nor causing diseases	That the three first Principles of the Chymists, nor the Essences of the same, are of the Army of Diseases	Diseases
57	De flatibus	On Flatus	Of Flatu's or windinesses in the Body	Diseases
58	Catarrhi deliramenta	The Error of Catarrh	The Toyes of a Catarrh or Rheum	Diseases
59	Victus ratio	On Regimen*	A Reason or Consideration of Diet	Medicine
60	Pharmacopolium ac Dispensatorium modernorum	A Modern Pharmacopolium and Dispensatory	A Modern Pharmacopolium and Dispensatory	Medicine
61	Potestas medicaminum	The Power of Medicine	The Power of Medicines	Medicine
62	Praefatio	Preface	A Preface	Diseases
63	Ignotus hospes morbus	An Unknown Guest of Disease	A Disease is an unknown Guest	Diseases
64	Ignotus hydrops	An Unknown Dropsy	The Dropsie is unknown	Diseases
65	Puerilis Humoristarum vindicta	A Childish Defence of the Humorists	A childish Vindication of the Humourists	Polemical Defence
66	Respondet author	The Answer Responds	The Author Answers	Polemical Defence
	'Tractatus de morbis'	*Treatise on Diseases*	*A Treatise of Diseases.*	
67	Introductio diagnostic	Introduction to Diagnostics	A discernable Introduction	Diseases
68	In puncto vitae subjectum inhaesionis morborum	Diseases Adhere to The Point of Life	The subject of inhering of Diseases is in the point of Life	Diseases

Chapter No.	Latin Name	Modern English	English (1664 translation)	Tentative Subject Division
69	Progreditur ad morborum cognitionem	The Author Proceeds to the Understanding of Diseases	A proceeding to the knowledge of Diseases	Diseases
70	De Ideis morbosis	On Disease-Inducing Ideas	Of the Idea's of Diseases	Diseases
71	De morbis Archealibus	On Archeal Diseases	Of Archeal Diseases	Diseases
72	Ortus imagines morbosae	The Birth of Disease-Inducing Images	The Original of a diseasie Image	Diseases
73	Aditus praeclusus ad Condum viscerum	The Blockage of Access to the Steward of the Bowels	The passage unto the Buttery of the Bowels is stopped up	Diseases
74	Confirmatur morborum sedes in anima sensitiva	The Location of Diseases in the Sensitive Soul is Confirmed	The Seat of Diseases in the sensitive Soul is confirmed	Diseases
75	Morborum phalanx, & divisio	The Cohort and Division of Diseases	The Squaldron, and Division of Diseases	Diseases
76	Recepta injecta	Things Received and Injected	Things Received that are Injected	Occult Action
77	Quaedam imperfectiora	Imperfect Writings	Some more Imperfect Works	
78	In Verbis, herbis, & lapidibus est magna virtus	In Words, Herbs and Stones is Great Power	In Words, Herbs, and Stones there is great Virtue	Occult Action
79	Butler	Butler	Butler	Medicine
80	De Injectis materialibus	On the Injection of Materials	Of Material things Injected	Occult Action
81	Injaculatorum modus intrandi	The Manner of Entry of Injected Things	The manner of entring of things Darted into the Body	Occult Action
82	De Conceptis	On Conceived Things	Of things Conceived	Occult Action
83	Vis magnetica	The Magnetic Power	A Magnetical or Attractive Power	Occult Action
84	De sympatheticis mediis	On Sympathetic Mediums	Of Sympathetical Medium's or Means	Occult Action
85	De Inspiratis	On Things Inspired	Of things Inspired	Occult Action
86	Suscepta	On Things Taken In	Things Suscepted or Undergone	Occult Action

Chapter No.	Latin Name	Modern English	English (1664 translation)	Tentative Subject Division
87	Retenta	On Things Retained	Things Retained	Occult Action
	[Liber de Vita longa]	*[Book of Long Life]*		
88	Praefatio	Preface	A Preface	De Vita Longa
89	De Tempore	On Time	Of Time	De Vita Longa
90	Vita longa, Ars brevis	Life is Long, Art is Short	Life is Long, Art is Short	De Vita Longa
91	Mortis introitus in naturam humanam decus Virginum	The Entrance of Death into Human Nature. In Praise of Virgins	The entrance of Death into humane nature, the grace of Virgins	De Vita Longa
92	Thesis	Thesis	A Position	De Vita Longa
93	Demonstratur thesis	The Thesis is Demonstrated	The Position is Demonstrated	De Vita Longa
94	Supplementum de Spadanis fontibus. Paradoxum primum.	The Supplement on the Fountains of the Spa: The First Paradox	Of the Fountains of the *Spaw:* The first Paradox	Analysis of Waters
95	Paradoxum alterum	The Second Paradox	A second Paradox	Analysis of Waters
96	Paradoxum tertium.	The Third Paradox	A third.	Analysis of Waters
97	Paradoxum quartum	The Fourth Paradox	A fourth.	Analysis of Waters
98	Paradoxum quantum	The Fifth Paradox	A fifth.	Analysis of Waters
99	Paradoxum sextum	The Sixth Paradox	A sixth.	Analysis of Waters
100	Supplementorum Paradoxum numero criticum	Paradoxical Supplements of a Critical Number	A numerocritical Paradox of Supplies.	Analysis of Waters
101	Intellectus Adamicus	The Intellect of Adam	The Understanding of *Adam*	De Vita Longa
102	Imago Dei	The Image of God	The Image of God	De Vita Longa
103	Externorum proprietas	The Property of External Things	The Property of External Things	De Vita Longa
104	Humidum radicale	The Radical Moisture	The Radical Moisture	De Vita Longa

Chapter No.	Latin Name	Modern English	English (1664 translation)	Tentative Subject Division
105	Aura vitalis	The Vital Aura	The Vital Air	De Vita Longa
106	Vita multiplex in homine	The Multiple Life of Man	A manifold Life in Man	De Vita Longa
107	Fluxus ad generationem	The Flux to Generation	A Flux unto Generation	De Vita Longa
108	Lunare tributum	A Lunar Tribute	A Lunar Tribute	De Vita longa
109	Vita	Life	Life	De Vita Longa
110	Vita brevis	Short Life	Short Life	De Vita Longa
111	Vita aeterna	Eternal Life	Eternal Life	De Vita Longa
112	Mortis occasiones	Occasions of Death	The Occasions of Death	De Vita Longa
113	De magnetica vulnerum curatione	On the Magnetic Cure of Wounds	Of the Magnetick curing of Wounds.	Weapon-Salve
114	In Sole tabernaculum	The Tabernacle in the Sun	The Tabernacle in the Sun	De Vita longa
115	Infantis nutritio ad vitam longam	The Nutrition of an Infant for Long Life	The nourishing of an Infant for Long Life	De Vita Longa
116	Arcana Paracelsi	The Arcana of Paracelsus	The Secrets of *Paracelsus*	De Vita Longa
117	Mons Domini	The Mountain of the Lord	The Mountain of the Lord	De Vita Longa
118	Arbor Vitae	The Tree of Life	The Tree of Life	De Vita Longa
	Opuscula Medica Inaudita	*Unheard-of Works of Medicine*	*Unheard of little Works of Medicine*	*(Previously published works)*
119	Tractatus de Lithiasi	A Treatise on The Disease of the Stone	Of the Disease of the Stone	
120	De Febribus	On Fevers	Of Fevers	
121	Scholarum Humoristarum passiva deceptio ac ignorantia	The Passive Deception and Ignorance of Scholastic Humorists	A passive deceiving and ignorance of the Schools the Humourists	
122	Tumulus Pestis	The Plague's Grave	The Plague-grave	

* *De victu* or *De victus ratione* was the Latin title given to Hippocrates's treatise *On Regimen*. Van Helmont is clearly drawing on Hippocrates for the title and contents.

Bibliography

Primary Sources

Agrippa, Heinrich Cornelius. *De occulta philosophia*. Lyon: Beringos Fratres, 1550.

Alsted, Johann Heinrich. *Lexicon theologicum*. Hanover: Eifridus, 1634.

Aristotle. *The Complete Works of Aristotle (Oxford Translation)*, edited by Jonathan Barnes, 2 vols. Princeton, NJ: Princeton University Press, 1984.

Artis auriferae quam chemiae vocant, 2 vols. Basel: Waldkirch, 1593.

Aquinas, Thomas. *Compendium of Theology*. Translated by Richard J. Regan. Oxford: Oxford University Press, 2009.

Aquinas, Thomas. *Summa Theologiae*. 61 vols. Cambridge: Cambridge University Press, 2006, vol. 2 (1a.2–11).

Augustine. *The Works of Saint Augustine*, edited by J.E. Rotelle. New York, NY: New City Press, 1990.

———. *The Literal Meaning of Genesis*. Translated by John Hammond Taylor. New York, NY: Paulist Press, 1982.

Basil, Saint. *Exegetic Homilies*. Translated by Sister Agnes Clare Way. Washington, DC: Catholic University of America Press, 1963.

Boehme, Jacob. *Mysterium Magnum, or an Exposition of the First Book of Moses called Genesis*. London: Lodowick Lloyd, 1656.

Bonus, Petrus of Ferrara. *Pretiosa margarita novella*, edited by Janus Lacinius. Venice: Aldus, 1526.

Bostocke, Richard. *The Difference between the Auncient Physicke . . . and the Latter Phisicke*. London: Robert Walley, 1585.

Bougerel, Joseph. *Vie de Pierre Gassendi*. Paris: J Vincent, 1737.

Boyle, Robert. *The Sceptical Chymist*. London: J. Cadwell, 1661.

———. 'An Account of the Two Sorts of the Helmontian Laudanum, Communicated to the Publisher by the Honourable Robert Boyle'. *The Philosophical Transactions* 9 (26 October 1674), 147–9.

Bracesco, Giovanni. 'Dialogo Lignum Vitae'. In *Vera Alchemiae Doctrina*, edited by Guglielmo Gratarolo, 3–47. Basel: Petrus Perna, 1561.

Broeckx, Corneille. 'Le premier ouvrage de J.B. Van Helmont, Seigneur de Merode, Royenborch, Oirschot, Pellines, etc., publié pour la première fois'. *Annales de L'Académie d'archéologie de Belgique* 10 (1853): 327–92 and 11 (1854): 119–91.

———. 'Interrogatoires du docteur J.B. van Helmont sur le magnétisme animal, publiés pour la première fois'. *Annales de l'Académie d'archéologie de Belgique* 13 (1856): 306–50.

———. 'J.B. Van Helmont: Ad Judicem neutrum causam appellat suam et suorum philadelphus'. *Annales de l'Académie d'archéologie de Belgique* 22 (1869): 65–138.

Burggravius, Johannes Ernst. *Biolychnium, seu Lucerna*. Franecker: Ulderick Dominick Balck, 1611.

Campanella, Tommaso. *De sensu rerum et magia*. Frankfurt: Egenolphus Emmelius, 1620.

Cantemir, Dimitrie. *L'immagine irrafigurabile della Scienza Sacro-Sancta*, edited by Vlad Alexandrescu. Milan: Mondadori, 2012.

Charnock, Thomas. 'The Camp of Philosophy'. In *Theatrum Chemicum Britannicum*, edited by Elias Ashmole, 306–7. London: Nathaniel Brooke, 1652.

Croll, Oswald. 'The Admonitory Preface of Oswald Crollie, Physitian: to the Most Illustrious Prince Christian Anhaltin'. In *Philosophy Reformed & Improved in Four Profound Tractates*, 1–226. Translated by Henry Pinnell. London: Lodowick Lloyd, 1657.

Cusa, Nicholas of. *The Idiot in Four Books: The First and Second Wisdome. The Third of the Minde. The Fourth of Statick Experiments, or Experiments of the Balance.* London: William Leake, 1650.

Dee, Arthur (ed.). *Fasciculus chemicus*. London: J. Flesher, 1650.

de Heer, Henri. *Observationes medicae oppido rarae in Spa et Leodii animadversae.* Liege: Van Der Aa, 1630.

d'Espagnet, Jean. *Enchyridion Physicae restitutae or The Summary of Physicks Recovered*. London: Bentley, 1651.

Dethier, Hubert, Ivan Hermans, and Sylvette Sue (eds). 'Joannis Baptistae Helmontii medici et philosophi per ignem Propositiones notatu dignae, depromptae ex ejus disputatione de magicae vulnerum curatione'. In *Johan-Baptist Van Helmont (1579–1644)*, annex I. Brussels: Vrije Universiteit Brussels, 1985.

Dorn, Gerhard. 'Speculativa philosophia'. In *Theatrum Chemicum*, 6 vols. Strasbourg: Lazarus Zetzner, 1602.

Drebbel, Cornelis. *Ein kurtzer Tractat von der Natur der Elementen*. Leiden: Henrich von Haestens, 1608.

Du Chesne, Joseph. *The Practise of Chymicall and Hermeticall Physicke*. Translated by Thomas Tymme. London: Thomas Creede, 1605.

'Ephemerides'—1650 and 1652, in *Hartlib Papers* (Ann Arbor, MI, 1995; enlarged ed., Sheffield, 2002) [on CD].

Fabre, Pierre Jean. *Alchymista Christianus*. Toulouse: Petrus Bosc, 1632.

Farmer, Sharon A., and Farmer, Stephen A. (eds). *Syncretism in the West: Pico's 900 theses (1486): the evolution of traditional, religious, and philosophical systems : with text, translation, and commentary*. Michigan: Medieval & Renaissance Texts & Studies, 1998.

Fernel, Jean. *On The Hidden Causes Of Things: Forms, Souls, And Occult Diseases In Renaissance Medicine*. Leiden: Brill, 2005.

Ficino, Marsilio, *Opera omnia*. Basel: HenricPetri, 1576.

Fludd, Robert. *Utriusque cosmi maioris, scilicet et minoris metaphysica, physica atque technical historia*, 2 vols. Oppenheim: Theodor de Bry, 1617.

Gassendi, Pierre. *Opera omnia*, 6 vols. Lyon: Annisson, 1658.

Giorgio, Francesco. *De harmonia mundi totius: cantica tria*. Venice: In Aedibus Bernardini de Vitalibus, 1525.

Goclenius, Rudolf. *Oratio qua defenditur vulnus non applicato etiam remedio citra ulium dolorem curari naturaliter posse*. Marburg: Cattorum, 1609.

———. *Synarthrosis magnetica*. Marburg: Jonas Saurium, 1617.

Goodrick-Clarke, Nicholas (ed.). *Paracelsus: Essential Readings*. Wellingborough: Aquarian Press, 1990.

Hippocrates. 'Breaths'. In *Hippocrates*, translated by W.H.S. Jones, 8 vols, II, 219–55. London: Heinemann, 1967.

Khunrath, Heinrich. *Amphitheatrum Sapientiae Aeternae*. Hanau: Wilhelm Anton, 1609.

———. *Confessio de chao physicorum*. Strasbourg: Johann Albert Dolhopff, 1609.

———. *Vom Hylealischen . . . Chaos*. Magdeburg: Johann Francken, 1616.

Leiden, Codex Vossianus Chymicus 24, fols 367r–372v.

London, British Library, MS Sloane 2879, fols 50r–74v.

London, Royal Society, MS 187, fols 35v–37r.

Martelli, Matteo (ed.). *The Four Books of Pseudo-Democritus*. Leeds: Maney Publishing, 2013.

Mechelen (Malines), Archives of the Archbishopric of Mechelen, MS 'Causa J.B. Helmontii medici', 3 vols.

Mennens, Willem. 'De aureo vellere'. In *Theatrum Chemicum*, 6 vols, V, 267–478. Strasbourg: Lazarus Zetzner, 1622.

Midrash Bereshith Rabba in Hebrew and English. Jerusalem: Shalom Publications, 1988.

Nollius, Heinrich. *Naturae sanctuarium quod est physica hermetica*. Frankfurt: Nicolas Hoffmann, 1619.

Nuysement, Clovis Hesteau de. *Traittez du vray sel, secret des philosophes, et l'esprit général du Monde*. Paris: Jérémie Perier et Abdias Buisard, 1621.

Paracelsus, Theophrastus Bombastus von Hohenheim. *Aureoli Theophrasti Paracelsi de summis naturae mysteriis libri tres*. Basel, 1570.

———. *Operum Medico-Chimicorum sive Paradoxorum*, 12 vols. Frankfurt: Collegio Musarum Palthenianarum, 1605.

———. *Sämtliche Werke: Abteilung, medizinische, naturwissenschaftliche und philosophie Schriften*, 14 vols. Edited by Karl Sudhoff. Munich and Berlin: Oldenbourg, 1922–1933.

———. *Sämtliche Werke II: Abteilung: Die theologischen und religionsphilosophischen Schriften*. Edited by Kurt Goldammer. Wiesbaden: Steiner Verlag 1955.

———. 'Seven Defensiones'. In *Four Treatises of Theophrastus von Hohenheim Called Paracelsus*. Translated by C. Lilian Temkin, edited by Henry E. Sigerist, 1–43. London: Johns Hopkins University Press, 1941.

Paracelsus, Philippus Theophrastus Bombast von Hohenheim gennant. *Dritter Theil Der Bücher und Schrifften*. Edited by Johannes Huser. Basel: Conrad Waldkirch, 1589.

———. *Sechster Theil Der Bücher und Schrifften*. Edited by Johannes Huser. Basel: Conrad Waldkirch, 1590.

———. *Achter Theil Der Bücher und Schrifften*. Edited by Johannes Huser. Basel: Conrad Waldkirch, 1590.

Parr, Bartholomew. *The London Medical Dictionary*. London: Ames Mitchell and White William Brown, 1819.

Patin, Guy. *Lettres*. Paris: J.H. Reveille-Parise, 1846.

Pico della Mirandola, Giovanni. *Commentary on a Canzone of Benivieni*, trans. by Sears Jayne. Frankfurt: Peter Lang, 1984.

———. *On the Dignity of Man, On Being and the One, Heptaplus*. Translated by Charles Glenn Wallis, Paul J.W. Miller and Douglas Carmichael. Cambridge: Hackett, 1965.

Planis Campy, David de. *Traicté de la vraye, unique, grande et universelle medecine des anciens, dite des recens Or Potable*. Paris: François Targa, 1633.

Plotinus. *Enneads*. Translated by A. H. Armstrong, 3 vols. London: Heinemann, 1966–67.

Poleman, Joachim. *Novum lumen medicum wherein the excellent and most necessary doctrine of the highly-gifted philosopher Helmont concerning the great mystery of the philosophers sulphur*. London: J.C., 1662.

Pseudo-Dionysius. *Pseudo-Dionysius—Complete Works*. Translated by Colm Luibheid. New York, NY: Paulist Press, 1987.

'Raymundi Lulli Theorica & Practica'. In *Theatrum Chemicum*, 6 vols, IV, 1–171. Strasbourg: Lazarus Zetzner, 1659.

Reuchlin, Johannes. *On the Art of the Kabbalah*. Translated by Martin and Sarah Goodman. Lincoln, NA: Abaris Books, 1983.

Roberti, Jean. 'Anatome magici libelli'. In *Theatrum sympatheticum auctum*, 226–36. Nuremberg: Sylvester Rattray, 1662.

Ruland, Martin. *Lexicon alchemiae*. Frankfurt: Johann Andrea & Wolfgang Endter, 1612.

Sendivogius, Michael (Divi Leschi Genus Amo). *De lapide philosophorum Tractatus duodecim*. Frankfurt/Prague, 1604.

Sennert, Daniel. *Hypomnemata physica*. Frankfurt: Clement Schelich, 1636.

Severinus, Petrus. *Epistola Scripta Theophrasto Paracelso*. Basel: Henricpetri, 1570/1.

———. *Idea medicinae philosophicae*. Basel: Henricpetri, 1571.

Staricius, Johannes (ed.) *Clavis, Oder, Das Zehende Buch der Archidoxen D. Philippi Theophrasti Paracelsi, Bombast von Hohenheim*. Magdeburg: Johann Francken, 1624.

Tachenius, Otto. *Epistola de famoso liquore Alcahest*. [Hamburg], 1652.

Tauler, Johann. *The Sermons and Conferences of John Tauler of the Order of Preachers*, trans. Elliott Walter. Washington, DC: Apostolic Mission House, 1910.

Trevisan, Bernard. 'Livre de venerable docteur allemant Messire Bernard Conte de la Marche Trevisane'. In *Opuscule tres-excellent de la vraye philosophie naturelle des Metaulx*, 170–280. Lyon: Benoist Rigaud, 1574.

Valentin, Michael Bernhard. 'Introductio'. In Johannes Baptista Van Helmont, *Opera Omnia*. Frankfurt: Hieronymi Christiani Paulli, 1707.

Valentine, Basil. *The Triumphant Chariot of Antimony*. London: Thomas Bruster, 1660.

Van Helmont, Jan Baptist. *Opuscula medica inaudita*. Cologne: Jodocus Kalcoven, 1644.

———. *Ortus medicinae, id est, Initia physicae inaudita: Progressus medicinae novus in morborum ultionem ad vitam longam*. Amsterdam: Ludovic Elzevir, 1648.
———. *Ortus medicinae, id est, Initia physicae inaudita: Progressus medicinae novus in morborum ultionem ad vitam longam*. Amsterdam: Ludovic Elzevir, 1652.
Van Helmont, John Baptist. *Oriatrike or Physick Refined: the Common Errors Therein Refuted and the Whole are Reformed and Rectified*. Translated by John Chandler. London: Lodowick Lloyd, 1662.
Van Helmont, Joan Baptista, *Dageraad, ofte Nieuwe Opkomst der Geneeskonst*. Rotterdam: Joannes Naeranus, 1661.
Von Helmont, Johann Baptista. *Aufgang der Artzney Kunst, das ist: Noch nie erhörte Grund-Lehren von der Natur*. Sulzbach: Johann Andreae Endters, 1683.
Van Helmont, Franciscus Mercurius. 'Amico lectori'. In Jan Baptist Van Helmont, *Ortus medicinae, id est, Initia physicae inaudita: Progressus medicinae novus in morborum ultionem ad vitam longam*. Amsterdam: Ludovic Elzevir, 1652.
Waard, Cornelis de. *Correspondance du Marin Mersenne*, 17 vols. Paris: Gabriel Beauchesne, 1936.
Weeks, Andrew (ed.). *Paracelsus: Essential Theoretical Writings*. Leiden: Brill, 2008.
Willis, Timothy. *The Search of Causes, containing a Theophysical Investigation of the Possibilitie of Transmutatorie Alchemie*. London: J. Legatt, 1616.

Secondary Sources

Abercrombie, Nigel. *The Origins of Jansenism*. Oxford: Clarendon Press, 1936.
Aertsen, Jan A. 'Truth as Transcendental in Thomas Aquinas'. *Topoi* 11 (1992): 159–71.
Albertson, David. 'That He Might Fill All Things: Creation and Christology in Two Treatises by Nicholas of Cusa'. *International Journal of Systematic Theology* 8 (2006): 184–205.
Alexandrescu, Vlad. 'Introduzione'. In Dimitrie Cantemir, *L'immagine irraffigurabile della Scienza Sacro-Santa*, 1–61. Milan: Mondadori, 2012.
Allchin, Douglas. 'Reassessing van Helmont, Reassessing History'. *BioScene: Journal of College Biology Teaching* 19 (1993): 3–5.
Allen, Michael J.B. 'The Birth Day of Venus: Pico as Platonic Exegete in *Commento* and *Heptaplus*'. In *Pico della Mirandola: New Essays*, edited by M.V. Dougherty, 81–113. Cambridge: Cambridge University Press, 2008.
Anstey, Peter R. 'John Locke and Helmontian Medicine'. In *The Body as Object and Instrument of Knowledge: Embodied Empiricism in the Early Modern Science*, edited by Charles T. Wolfe and Ofer Gal, 93–121. Dordrecht: Springer, 2010.
Arthur, Richard. 'Animal Generation and Substance in Sennert and Leibniz'. In *The Problem of Animal Generation in Early Modern Philosophy*, edited by Justin E.H. Smith, 147–74. Cambridge: Cambridge University Press, 2006.
Ashworth Jr, William B. 'Light of Reason, Light of Nature: Catholic and Protestant Metaphors of Scientific Knowledge'. *Science in Context* 3 (1989): 89–107.

Bagliani, Agostino Paravicini. 'Ruggero Bacone e l'alchimia di lunga vita: Riflessioni sui testi'. In *Il Potere Del Papa: Corporeità, autorappresentazione, simboli*, 75–96. Florence: Sismel, 2009.

Barnes, Robin Bruce. *Prophecy and Gnosis: Apocalypticism in the Wake of the Lutheran Reformation*. California, CA: Stanford University Press, 1988.

Ben-Chaim, Michael. 'Empowering Lay Belief: Robert Boyle and the Moral Economy of Experiment'. *Science in Context* 15 (2002): 51–77.

Bergamo, Mino. *L'anatomia dell'anima: De François de Sales à Fénelon*. Bologna: Il Mulino, 1991.

Bergengruen, Maximilian. *Nachfolge Christi—Nachahmung der Natur: Himmlische und Natürliche Magie*. Hamburg: Meiner Verlag, 2007.

Betz, Hans. 'The Concept of the "Inner Human Being" in the Anthropology of Paul'. *New Testament Studies* 46 (2000): 315–41.

Biographie Nationale de Belgique, 27 vols. Brussels: Thiry & Van Buggenhout, 1866–1938.

Blair, Ann. 'Mosaic Physics and the Search for a Pious Natural Philosophy in the Late Renaissance'. *Isis* 91 (2000): 32–58.

Blommestijn, Hein, Caspers, Charles, and Hofman, Rijcklof. *Spirituality Renewed: Studies on Significant Representatives of the Modern Devotion*. Leuven: Peeters, 2003.

Boantza, Victor D. 'Alkahest and Fire: Debating Matter, Chymistry and Natural History at the Early Parisian Academy of Sciences'. In *The Body as Object and Instrument of Knowledge: Embodied Empiricism in the Early Modern Science*, edited by Charles T. Wolfe and Ofer Gal, 75–92. Dordrecht: Springer, 2010.

Bono, James Joseph. *The Word of God and The Languages of Man: Interpreting Nature in Early Modern Science and Medicine*, 2 vols. Wisconsin: University of Wisconsin Press, 1995.

Browne, Alice. 'J.B. van Helmont's Attack on Aristotle'. *Annals of Science* 36 (1979): 575–91.

Bruers, Antonio. 'La "Colcodea" d'Avicenna e T. Campanella'. In Tommaso Campanella, *Del senso della cose e della magia,* 339–341. Bari:Laterza & Figli, 1925.

Bunners, Michael. *Die Abendmahlsschriften und das medizinisch-naturphilosophische Werk des Paracelsus*. Berlin: Humboldt University, 1961.

Burnell, Peter. 'Conscupiscence'. In *Augustine throughout the Ages: an Encyclopedia*, edited by Allan Fitzgerald and John C. Cavadini, 224–27. Grand Rapids, MI: Wm B. Eerdmans, 1999.

Bussanich, John. 'Goodness'. In *Augustine Throughout the Ages: an Encyclopedia*, edited by Allan Fitzgerald and John C. Cavadini, 390–91. Grand Rapids, MI: Wm B. Eerdmans, 1999.

Butterfield, Herbert. *The Origins of Modern Science, 1500–1800*. New York, NY: Macmillan, 1951.

Bynum, Caroline Walker. *The Resurrection of the Body in Western Christianity, 200–1336*. New York, NY: Columbia University Press, 1995.

Calvet, Antoine. 'Alchimie et Joachimisme dans les *Alchimica* Pseudo-Arnaldiens'. In *Alchimie et philosophie à la Renaissance*, edited by Jean-Claude Margolin and Sylvain Matton, 93–109. Paris: J. Vrin, 1993.

———. 'À la recherche de la médicine universelle: Questions sur l'élixir et la thériaque au 14e siècle'. In *Alchimia e medicina nel medioevo*, edited by Chiara Crisciani and Agostino Paravicini Bagliani, 177–217. Florence: Sismel, 2003.

Camenietzki, Carlos Ziller. 'Jesuits and Alchemy in the Early Seventeenth Century: Father Johannes Roberti and the Weapon Salve Controversy'. *Ambix* 48 (2001) 83–101.

Campbell Brown, James. *History of Chemistry*. London: J. & A. Churchill, 1913.

Casey, Michael. *A Thirst for God: Spiritual Desire in the Bernard of Clairvaux's Sermons on the Song of Songs*. Kalamazoo, MI: Cistercian Publications, 1987.

Chang, Ku-Ming (Kevin). 'Fermentation, Phlogiston and Matter Theory: Chemistry and Natural Philosophy in Georg Ernst Stahl's "Zymotechnia Fundamentalis"'. *Early Science and Medicine* 7 (2002): 31–64.

Chase, Frederic Henry. *The Lord's Prayer in the Early Church*. Cambridge: Cambridge University Press, 2004.

Clark, Mary T. *Augustine*. Washington, DC: Continuum International Publishing Group, 2005.

Clericuzio, Antonio. 'Spiritus Vitalis: Studio sulle teorie fisiologiche da Fernel a Boyle'. *Nouvelles de la Republique des lettres* (1988): 33–84.

———. 'From van Helmont to Boyle. A Study of the Transmission of Helmontian Chemical and Medical Theories in Seventeenth-Century England'. *The British Journal for the History of Science* 26 (1993): 303–34.

———. *Elements, Principles and Corpuscles*. London: Kluwer, 2000.

Collingwood, Robin George. *The Idea of Nature*. Oxford: Oxford University Press, 1960.

Cooper, John. 'Aristotle on Natural Teleology'. In *Language and Logos*, edited by Malcolm Schofield and Martha Nussbaum, 197–222. Cambridge: Cambridge University Press, 1981.

Cooper, John W. *Panentheism: The Other God of the Philosophers: from Plato to Present*. Grand Rapids, MI: Baker Academic, 2006.

Coudert, Allison. *The Impact of the Kabbalah in the Seventeenth Century: The Life and Thought of Francis Mercury Van Helmont (1614–1698)*. Leiden: Brill, 1999.

Crisciani, Chiara. 'The Conception of Alchemy as Expressed in the "Pretiosa Margarita Novella" of Petrus Bonus of Ferrara'. *Ambix* 20 (1973): 165–81.

———. 'Note Sull'Alchemia "Francescana" nel sec XIII'. In *Atti del XXV Congresso Nazionale di Filosofia* 214–20 (Roma: Societa filosofica italiana, 1980).

———. *Il Papa e l'Alchimia*. Rome: Viella, 2002.

———. 'Il farmaco d'oro, alcuni testi tra i secoli XIV e XV'. In *Alchimia e medicina nel medioevo*, edited by Chiara Crisciani e Agostino Paravicini Bagliani, 217–47. Florence: Sismel, 2003.

Crisciani, Chiara and Michela Pereira. *L'Arte del Sole e della Luna: Alchimia e filosofia nel Medioevo*. Spoleto: Centro Italiano di Studi sull'Alto Medioevo, 1996.

Crisciani, Chiara and Michela Pereira. 'Black Death and Golden Remedies: Some Remarks on Alchemy and the Plague'. In *The Regulation of Evil: Social and*

Cultural Attitudes to Epidemics in the Later Middle Ages, edited by Agostino Paravicini Bagliani and Francisco Santi, 7–41. Florence: Sismel, 1998.

Daniel, Dane T. 'Paracelsus on Baptism and the Acquiring of the Eternal Body'. In *Paracelsian Moments*, edited by Gerhild Scholz Williams & Charles D. Gunnoe, Jr, 117–35. Kirksville, MI: Sixteenth Century Journal Publishers, 2002.

———. 'Paracelsus' Astronomia Magna (1537/38): Bible-Based Science and the Religious Roots of the Scientific Revolution'. PhD diss., Indiana University, 2003.

———. 'Invisible Wombs: Rethinking Paracelsus's Concept of Body and Matter', *Ambix* 53 (2006): 129–42.

———. 'Medieval Alchemy and Paracelsus' Theology: Pseudo-Lull's *Testamentum* and Paracelsus' *Astronomia Magna*'. *Nova Acta Paracelsica* 22/23 (2008/2009): 121–35.

Dear, Peter. *Mersenne and the Learning of the Schools*. London: Cornell University Press, 1988.

Debus, Allen. 'Fire Analysis and the Elements in the 16th and 17th Centuries'. *Annals of Science* 23 (1967): 127–47.

———. *The French Paracelsians*. Cambridge: Cambridge University Press, 1991.

———. *The Chemical Philosophy: Paracelsian Science and Medicine in the 16th and 17th Centuries*, 2nd edition. Mineola, NY: Dover Publications, 2002.

Deghaye, Pierre. 'La Lumière de Nature chez Paracelse'. In Lucien Braun et al., *Paracelse*, 53–89. Paris: Albin Michel, 1980.

Delio, Ilia. *Simply Bonaventure: An Introduction to his Life, Thought and Writings*. New York, NY: New City Press, 2001.

Des Chene, Dennis. *Physiologia: Natural Philosophy in Late Aristotelian and Cartesian Thought*. Ithaca, NY: Cornell University Press, 1996.

———. *Life's Form: Late Aristotelian Conceptions of the Soul*. London: Cornell University Press, 2000.

Desmarez, G. 'L'État civil de J.-B. Van Helmont'. *Annales de la société d'archéologie de Bruxelles* 21 (1907): 107–23.

DeVun, Leah. *Prophecy, Alchemy, and the End of Time: John of Rupescissa in the Late Middle Ages*. New York, NY: Columbia University Press, 2009.

De Waele, Henri. *J-B Van Helmont*. Bruxelles: Office de Publicité, 1947.

Drevillon, Herve. *Lire et écrire l'avenir: l'astrologie dans la France du Grand Siècle, 1610–1715*. Paris: Champ Vallon, 1996.

Driesch, Hans. *History and Theory of Vitalism*. London: Macmillan, 1914.

Ducheyne, Steffen. 'A Preliminary Study of the Appropriation of Van Helmont's Oeuvre in Britain in Chymistry, Medicine and Natural Philosophy'. *Ambix* 55 (2008): 122–35.

Duclow, Donald. *Masters of Learned Ignorance: Eriugena, Eckhart, Cusanus*. Aldershot: Ashgate, 2006.

Dunn, James D.G. *The Theology of Paul the Apostle*. Edinburgh: T&T Clark, 1998.

Eberl, Jason, T. 'Aquinas on the Nature of Human Beings'. *The Review of Metaphysics* 58:2 (2004): 333–65.

Egan, Harvey D. *Soundings of the Christian Mystical Tradition*. Collegeville, MI: Liturgical Press, 2010.
Epstein, S.R. 'Craft Guilds, Apprenticeship, and Technological Change in Preindustrial Europe'. *The Journal of Economic History* 58 (1998): 684–713.
Faivre, Antoine. 'Elie Artiste, ou le Messie des Philosophes de la Nature'. *Aries* 2 (2002): 119–52.
Fattori, Marta and Massimo Bianchi (eds). *Spiritus: IV colloquio internazionale Roma, 7–9 gennaio 1983*. Roma: Edizioni dell'Ateneo, 1984.
Feuillet, A. *Le Christ Sagesse de Dieu d'après les Épitres pauliniennes*. Paris: Études Bibliques, 1966.
Figala, Karin and Ulrich Neumann. '"Author Cui Nomen Hermes Malavici": New Light on the Bio-bibliography of Michael Maier (1569–1622)'. In *Alchemy and Chemistry in the 16th and 17th Centuries*, edited by Piyo Rattansi and Antonio Clericuzio, 121–47. Dordrecht: Kluwer, 1994.
Fitzmeyer, Joseph A. *The Gospel According to Luke, X-XXIV*. Garden City, NY: Doubleday, 1985.
Forshaw, Peter J. 'Oratorium—Auditorium—Laboratorium: Early Modern Improvisations on Cabala, Music, and Alchemy'. *Aries* 10:2 (2010): 169–95.
———. 'Vitriolic Reactions: Orthodox Responses to the Alchemical Exegesis of Genesis'. In *The Word and the World: Biblical Exegesis and Early Modern Science*, edited by Kevin Killeen and Peter J. Forshaw, 111–36. Basingstoke: Palgrave Macmillan, 2007.
Foucault, Michel. *The Order of Things*. London: Routledge, 2002.
Fransen, Sietske. 'Jan Baptista van Helmont and his Editors and Translators in the Seventeenth Century'. PhD diss., Warburg Institute, 2014.
Freudenthal, Gad. *Aristotle's Theory of Material Substance: Heat and Pneuma, Form and Soul*. Oxford: Clarendon Press, 1995.
———. 'The Medieval Astrologization of Aristotle's Biology: Averroes on the Role of the Celestial Bodies in the Generation of Animate Beings'. *Arabic Sciences and Philosophy* 12 (2002): 111–37.
Friedman, Lionel. 'Occulta Cordis'. *Romance Philology* 11(1957): 103–11.
Funkenstein, Amos. *Theology and the Scientific Imagination: From the Middle Ages to the Seventeenth Century*. Princeton, NJ: Princeton University Press, 1986.
Garcia, Jorge J.E. 'Does Philosophy Tolerate Christening? Thomas Aquinas and the Notion of Christian Philosophy'. In *Philosophy of Religion for a New Century: Essays in Honor of Eugene Thomas Long*, edited by Jeremiah Hackett and Jerald Vallulis, 37–63. Dordrecht: Kluwer, 2004.
Gause, Ute. 'On Paracelsus's Epistemology in his Early Theological Writings and in his *Astronomia Magna*'. In *Paracelsus: The Man and his Reputation, His Ideas and their Transformation*, edited by Ole Peter Grell, 207–21. Leiden: Brill, 1998.
Getz, Faye. 'Roger Bacon and Medicine: The Paradox of the Forbidden Fruit and the Secrets of Long Life'. In *Roger Bacon and the Sciences: Commemorative Essays*, edited by Jeremiah Hackett, 337–64. Leiden: Brill, 1997.

Giglioni, Guido. 'Per una storia del termine Gas da Van Helmont a Lavoisier: Constanza e variazione del significato'. *Annali della Facolta di Lettere e Filosofia dell'Università di Macerata* 25–26 (1992–1993): 431–67.

———. *Immaginazione e malattia: Saggio su Jan Baptista Van Helmont*. Milan: Francoangeli, 2000.

———. 'Life and its Animal Boundaries: Ethical Implications in Early Modern Theories of Universal Animation'. In *Ethical Perspectives on Animals in the Renaissance and Early Modern Period*, edited by C. Muratori and B. Dohm, 111–37. Florence: SISMEL, 2013.

Gijsen, Annelies van. 'Isaac Hollandus Revisited'. In *Chymia: Science and Nature in Medieval and Early Modern Europe*, edited by Miguel López Pérez, Didier Kahn and Mar Rey Bueno, 310–30. Newcastle upon Tyne: Cambridge Publishers, 2010.

Gilly, Carlos. *Adam Haslmayr. Der erste Verkünder der Manifeste der Rosenkreuzer.* Amsterdam: Bibliotheca Philosophica Hermetica, 1994.

———. '*Theophrastia sancta*: Paracelsianism as a Religion in Conflict with the Established Churches'. In *Paracelsus: the Man and His Reputation, His Ideas and their Transformation*, edited by Ole Peter Grell, 151–87. Leiden: Brill, 1998.

Gilson, Etienne. *Introduction a l'étude de Saint Augustin*. Paris: Vrin, 1929.

———. 'Notes bibliographiques pour servir à l'histoire de la notion de philosophie chrétienne'. In *L'esprit de la philosophie medievale*, 2nd ed., 413–440. Paris: Vrin, 1944.

———. *The Christian Philosophy in the Middle Ages*. London: Sheed and Ward, 1955.

———. *The Christian Philosophy of St Augustine*. New York, NY: Random House, 1960.

Giulea, Dragos. *Noetic Paschal Anthropos: Genesis 1:27 and the Theology of the Divine Image in Early Paschal Literature* (unpublished doctoral thesis, Marquette University, 2010).

Goethals, Felix Victor. *Lectures relatives à l'historie de sciences, des arts, des letters, des moeurs et de la politique en Belgique*, 2 vols. Brussels: Vandooren, 1837.

Goldammer, Kurt. 'III. Einleitungen zu den Schriften dieses Bandes, 2. De genealogia Christi'. In Theophrastus Paracelsus, *Sämtliche Werke II: Abteilung: Die theologischen und religionsphilosophischen Schriften*. Wiesbaden: Steiner Verlag 1955.

———. 'Lichtsymbolik in philosophischer Weltanschauung, Mystik und Theosophie vom 15. bis zum 17. Jahrhundert'. *Studium Generale* 13 (1960): 670–82.

———. *Paracelsus in neuen Horizonten: Gesammelte Aufsätze*. Vienna: Verband der wissenschaftlichen Gesellschaften Osterreichs, 1986.

———. *Der Göttliche Magier und die Magierin Natur.* Stuttgart: Franz Steiner Verlag, 1991.

Grant, Edward. *Much Ado about Nothing: Theories of Space and Vacuum from the Middle Ages to the Scientific Revolution.* Cambridge: Cambridge University Press, 1981.

———. *Science and Religion: 400 BC—AD 1500, From Aristotle to Copernicus*. Baltimore, MA: Johns Hopkins University Press, 2006.

Grell, Ole Peter. 'Plague, Prayer and Physic: Helmontian Medicine in Restoration England'. In *Religio Medici: Medicine and Religion in Seventeenth-Century England*, edited by Ole Peter Grell and Andrew Cunningham, 204–27. Aldershot: Ashgate, 1996.

———. 'The Acceptable Face of Paracelsianism: The Legacy of *Idea medicinae* and the Introduction of Paracelsianism in Early Modern Denmark'. In *Paracelsus: The Man and his Reputation, His Ideas and their Transformation*, edited by Ole Peter Grell, 245–69. Leiden: Brill, 1998.

Guinsburg, Arlene Miller. 'Paracelsian Magic and Theology: A Case Study of the Matthew Commentaries'. *Medizinisches Journal* 16 (1981): 125–39.

———. 'The Counterthrust to Sixteenth Century Misogyny: The Work of Agrippa and Paracelsus', *Historical Reflections* 8 (1981): 3–28.

Hall, Alexander W. *Thomas Aquinas and John Duns Scotus: Natural Theology in the High Middle Ages*. King's Lynn: Continuum, 2007.

Halleux, Robert. 'Helmontiana I'. In *Academiae analecta*. 33–63. Brussels: Koninklijke Academie voor Wetenschappen, Letteren en Schone Kunsten van België, 1983.

———. 'Helmontiana II'. In *Academiae analecta*, 17–36. Brussels: Koninklijke Academie voor Wetenschappen, Letteren en Schone Kunsten van België, 1987.

———. 'Theory and Experiment in the Early Writings of Johan Baptist Van Helmont'. In *Theory and Experiment*, edited by D. Batens and J. P. van Bendegem, 93–101. Dordrecht: Kluwer, 1988.

———. 'Le procès d'inquisition du chimiste Jean-Baptiste Van Helmont (1578–1644): les enjeux et les arguments'. *Académie des inscriptions et belles-lettres* 148 (2004) : 1059–86.

———. 'Nouveaux matériaux sur la cour savante d'Ernest de Bavière', *Cinquième Congrès de l'Association des Cercles Francophones d'Histoire et d'Archéologie de Belgique*, 394–404.

Halleux, Robert and Bernès, Anne-Catherine. 'La cour savante d'Ernest de Bavière'. *Archives Internationales d'Histoire des Sciences* 45 (1995): 4–29.

Hannaway, Owen. *The Chemists and the Word*. Baltimore, MA: Johns Hopkins University Press, 1975.

Harline, Craig. *Miracles at the Jesus Oak*. New Haven, CT: Yale University Press, 2011.

Harrison, Peter. *The Fall of Man and the Foundations of Science*. Cambridge: Cambridge University Press, 2007.

Harrison, Peter. 'Voluntarism and Early Modern Science'. *History of Science* 40 (2002): 63–89.

Hasse, Dag Nikolaus. 'Avicenna's "Giver of Forms" in Latin Philosophy, especially in the Works of Albertus Magnus'. In *The Arabic, Hebrew and Latin Reception of Avicenna's Metaphysics*, edited by Dag Nikolaus Hasse and Amos Bertolacci, 225–51. Berlin: Walter de Gruyter, 2012.

Hasse, Dag Nikolaus. 'Plato Arabico-Latinus'. In *The Platonic Tradition in the Middle Ages: A Doxographic Approach*, edited by Stephen Gersh and J.F.M. Hoenen, 31–66. Berlin: Walter de Gruyter, 2002.

Hedesan, Georgiana D. 'Reproducing the Tree of Life: Radical Prolongation of Life and Interpretations of the Bible in Seventeenth-Century Alchemy'. *Ambix* 60 (2013): 341–60.

———. 'The Mystery of *Mysterium Magnum*: Paracelsus's Interpretation of Creation in *Philosophia ad Athenienses* and its Early Modern Commentators'. In *Hidden Truths from Eden: Esoteric Interpretations of Genesis 1–3*, edited by Caroline Vander Stichele and Susanne Scholz, 141–62. Atlanta: Society of Biblical Literature, 2014.

———. 'Paracelsian Medicine and Theory of Generation in "Exterior Homo", a Manuscript Probably Authored by Jan Baptist Van Helmont (1579–1644)'. *Medical History* 58 (2014): 375–96.

———. 'Alchemy'. In *The Occult World*, edited by Christopher Partridge, 552–63. Oxford: Routledge, 2015.

———, 'From Interpretation to Integration: Van Helmont and His Medical-Alchemical Hippocratism', in *Alchemy and Medicine from Antiquity to the Enlightenment*, edited by Jennifer M. Rampling and Peter M. Jones (Aldershot: Ashgate, forthcoming).

Heinecke, Berthold. 'The Mysticism and Science of Johann Baptista Van Helmont (1579–1644)'. *Ambix* 42 (1995): 65–78.

———. *Mystik und Wissenschaft bei Johan Baptista Van Helmont (1579–1655)*. Bern: Peter Lang, 1995.

Henne, Alexandre and Wauters, Alphonse Guillaume Ghislain. *Historie de la ville de Bruxelles*. Brussels: Librairie Encyclopedique de Perichon, 1845.

Henry, John. 'Animism and Empiricism: Copernican Physics and the Origins of William Gilbert's Experimental Method'. *Journal of the History of Ideas* 62 (2001): 99–119.

———. *The Scientific Revolution and the Origins of Modern Science*, 2nd edition. Basingstoke: Palgrave Macmillan, 2001.

———. 'Voluntarist Theology at the Origins of Modern Science: A Response to Peter Harrison'. *History of Science* 47 (2009): 79–113.

Hershey, David R. 'Digging Deeper into Helmont's Famous Willow Tree Experiment'. *American Biology Teacher* 53 (1991): 458–60.

———. 'Pseudohistory and Pseudoscience: Corrections to Allchin's Historical, Conceptual and Educational Claims'. *Science & Education* 15 (2006): 121–25.

Hirai, Hiro. *Le concept de semence dans les théories de la matière à la Renaissance, de Marsile Ficin à Pierre Gassendi*. Turnhout: Brepols, 2005.

———. '"Prisca Theologia" and Neoplatonic Reading of Hippocrates in Fernel, Cardano and Gemma'. In *Cornelius Gemma: Cosmology, Medicine and Natural Philosophy in Renaissance Louvain*, 91–104. Rome: Serra, 2008.

———. 'The World-Spirit and Quintessence in the Chymical Philosophy of Joseph Du Chesne'. In *Chymia: Science and Nature in Medieval and Early Modern Europe*, ed. Miguel López Pérez, Didier Kahn and Mar Rey Bueno, 247–61. Newcastle upon Tyne: Cambridge Scholars, 2010.

———. *Medical Humanism and Natural Philosophy: Renaissance Debates on Matter, Life and Soul.* Leiden: Brill, 2011.

Hoff, Hebbel E. 'Nicolaus of Cusa, van Helmont and Boyle: The First Experiment of the Renaissance in Quantitative Biology and Medicine'. *Journal of the History of Medicine and Allied Sciences* 19 (1964): 99–117.

Holmyard, E.J. *Alchemy.* Harmondsworth: Penguin, 1957.

Hooykaas, Reijer. 'Die Elementenlehre der Iatrochemiker'. *Janus* 41 (1937): 1–28.

Howe, H.M. 'A Root of van Helmont's Tree'. *Isis* 56 (1965): 409–19.

Hunter, William B. Jr. 'The Seventeenth Century Doctrine of Plastic Nature'. *Harvard Theological Review* 43 (1950): 197–213.

Idel, Moshe. *Kabbalah in Italy (1280–1510): A Survey.* New Haven, CT: Yale University Press, 2011.

Janacek, Bruce. *Alchemical Belief: Occultism in the Religious Culture of Early Modern England.* University Park: University of Pennsylvania Press, 2011.

Jansen, F.X. *Baius et le Baianisme: Essai théologique.* Brussels: Dewit, 1927.

Javelet, Robert. *Image et resemblance au douzieme siècle*, 2 vols. Strasbourg: Université de Strasbourg, 1967.

Joly, Bernard. 'La réception de la pensée de Van Helmont dans l'oeuvre de Pierre Jean Fabre'. In *Alchemy Revisited: Proceedings of the International Conference on the History of Alchemy at the University of Groningen 17–19 April 1989*, edited by Z.R.W.M. von Martels, 206–14. Leiden: Brill, 1990.

———. *La rationalité d'alchimie au XVIIe siècle.* Paris: Vrin, 1992.

———. 'L'alkahest, dissolvent universel, ou quand la théorie rend pensable une pratique impossible'. *Revue d'histoire des sciences* 49 (1996): 305–44.

Kahn, Didier. 'L'intérpretation alchimique de la Genèse chez Joseph Du Chesne dans le context de ses doctrines alchimiques et cosmologiques'. In *Scientiae et artes: Die Vermittlung alten und neuen Wissens in Literatur, Kunst und Musik*, edited by Barbara Mahlmann-Bauer, 641–92. Wiesbaden: Harrassowitz, 2004.

———. *Alchimie et Paracelsisme in France.* Paris: Droz, 2007.

Karpenko, Vladimir Karpenko. 'Alchemy as Donum Dei'. *Hyle* 4 (1998): 63–80.

Kennedy, Thomas. 'Teleology'. In *The Encyclopedia of Christianity*, 5 vols, V, 326–8. Grand Rapids, MI: Eerdmans & Brill, 2008.

Kieckhefer, Richard. 'The Notion of Passivity in the Sermons of Tauler'. *Recherches de theologie ancienne et medievale* 48 (1981): 198–211.

Kristeller, Paul Oskar. *The Philosophy of Marsilio Ficino.* New York, NY: Columbia University Press, 1943.

———. *Renaissance Concepts of Man and Other Essays.* New York: Harper Row, 1972.

———. 'Thomism and the Italian Thought of the Renaissance'. In *Medieval Aspects of Renaissance Learning: Three Essays by Paul Oskar Kristeller*, edited and translated by Edward P. Mahoney, 29–91. New York, NY: Columbia University Press, 1992.

Lang, Helen S. *The Order of Nature in Aristotle's Physics*. Cambridge: Cambridge University Press, 1998.

Leclerq, Jean. 'Influence and Noninfluence of Dionysius in the Western Middle Ages'. In Pseudo-Dionysius, *The Complete Works*, translated by Colm Luibheid, 25–32. New York: Paulist Press, 1987.

Leijenhorst, Cees. *The Mechanisation of Aristotelianism: The Late Aristotelian Setting of Thomas Hobbes' Natural Philosophy*. Leiden: Brill, 2002.

Lindberg, Carter. *European Reformations*. Oxford: WileyBlackwell, 2009.

Lindberg, David C. 'The Genesis of Kepler's Theory of Light: Light Metaphysics from Plotinus to Kepler'. *Osiris* 2 (1986): 4–42.

Long, Pamela. *Openness, Secrecy, Authorship: Technical Arts and the Culture of Knolwedge from Antiquity to the Renaissance*. London: Johns Hopkins University, 2001.

Lossky, Vladimir. *The Mystical Theology of the Eastern Church*. Bucharest: Bonifaciu, 1998.

Louth, Andrew. *Denys the Areopagite*. London: Continuum Books, 1989.

Lubac, Henri de. *Augustinianism and Modern Theology*, trans. Lancelot Sheppard. London: Geoffrey Chapman, London, 1969.

Machielsen, Jan. *Martin Delrio: Demonology and Scholarship in the Counter-Reformation*. Oxford: Oxford University Press, 2015.

Mandosio, Jean Marc. 'L'alchimie dans les classifications des sciences et des arts à la Renaissance'. In *Alchimie et Philosophie à la Renaissance*, edited by Jean-Claude Margolin and Sylvain Matton, 11–41. Paris: Vrin, 1993.

Maspero, Giulio. *Trinity and Man: Gregory of Nyssa's Ad Ablabium*. Leiden: Brill, 2007.

Maxwell-Stuart, P.G. 'Introduction'. In *Martin Del Rio: Investigations into Magic*, edited and translated by P.G. Maxwell-Stuart, 1–15. Manchester: Manchester University Press, 2000.

McDonough, Jeffrey K. 'The Heyday of Teleology and Early Modern Philosophy'. *Midwest Studies in Philosophy* 35 (2011): 179–204.

McEvoy, James. 'The Metaphysics of Light in the Middle Ages'. *Philosophical Studies,* 26 (1979): 126–45.

———. *The Philosophy of Robert Grosseteste*. Oxford: Clarendon, 1982.

McGinn, Bernard. 'The Human Person as Image of God: II, Western Christianity'. In *Christian Spirituality I: Origins to the Fifth Century*, edited by Jean Leclercq, Bernard McGinn, John Meyendorff, 312–30. New York: Crossroad, 1985.

———. *Growth of Mysticism: Gregory the Great through the Twelfth Century*. New York, NY: Crossroads Publishing, 1994.

———. *The Harvest of Mysticism in Medieval Germany*. New York, NY: Crossroads Publishing, 2005.

McGinnis, Jon. *Avicenna*. Oxford: Oxford University Press, 2010.

McKenna, Stephen. 'Introduction'. In Saint Augustine, *The Trinity*, vii–xvii. Washington, DC: Catholic University of America, 1981.

Mendelsohn, Everett. 'Philosophical Biology vs Experimental Biology: Spontaneous Generation in the Seventeenth Century'. In *Topics in the Philosophy of Biology*, edited by Marjorie Grene, 37–65. Dordrecht: Kluwer, 1976.

Mendelsohn, J. A. 'Alchemy and Politics in England (1649–1665)'. *Past and Present* 135 (1992): 30–78.

Meyer, Susan Sauve. 'Aristotle, Teleology and Reduction'. *The Philosophical Review* 101 (1992): 791–825.

Michael, Emily. 'Renaissance Theories of Body, Soul and Mind'. In *Psyche and Soma: Physicians and Metaphysicians on the Mind-Body Problem from Antiquity to Enlightenment*, edited by John P. Wright and Paul Potter, 147–73. Oxford: Clarendon Press, 2000.

Milton, J. R. 'Locke, Medicine and the Mechanical Philosophy'. *British Journal for the History of Philosophy* 9:2 (2001): 221–43.

Moran, Bruce. *The Alchemical World of the German Court: Occult Philosophy and Chemical Medicine in the Circle of Moritz of Hessen (1572–1632)*. Stuttgart: Franz Steiner, 1991.

Moran, Bruce T. *Andreas Libavius and the Transformation of Alchemy: Separating Chemical Cultures with Polemical Fire*. Sagamore Beach: Science History Publications, 2007.

Morris, Paul and Deborah Sawyer. *A Walk in the Garden: Biblical, Iconographical and Literary Images of Eden*. Sheffield: Sheffield Academic Press, 1992.

Nadler, Steve. *Occasionalism: Causation amongst the Cartesians*. Oxford: Oxford University Press, 2011.

Nautin, P. 'Genèse 1, 1–2, de Justin à Origène'. In *In principio. Interpretations des premiers versets de la Genèse*, 61–94. Paris: Centre d'Etudes des religions du livre, 1973.

Nauwelaers, J. 'La Maison de Van Helmont à Vilvorde'. In *Le Folklore brabançon: Numéro spécialement consacré à J. B. van Helmont, 1579–1644*, 167–78. Brussels, 1933.

Newman, William R. 'L'influence de la *Summa Perfectionis* du Pseudo-Geber'. In *Alchimie et Philosophie à la Renaissance: Actes du Colloque International de Tours (4–7 Décembre 1991)*, edited by Jean-Claude Margolin and Sylvain Matton, 65–77. Paris: Vrin, 1993.

———. 'The Corpuscular Theory of J. B. Van Helmont and its Medieval Sources'. *Vivarium* 21(1993): 161–91.

———. *Promethean Ambitions: Alchemy and the Quest to Perfect Nature*. Chicago, IL: University of Chicago Press, 2004.

Newman, William and Lawrence M. Principe. 'Alchemy vs Chemistry: The Etymological Origins of a Historiographic Mistake'. *Early Science and Medicine* 3 (1998): 32–65.

Newman, William and Lawrence M. Principe. *Alchemy Tried in the Fire: The Fate of Helmontian Chymistry*. Chicago, IL: Chicago University Press, 2002.

Niebyl, Peter. 'Old Age, Fever and the Lamp Metaphor'. *Journal of the History of Medicine* 26 (1971): 351–68.

Norris, John A. 'Mineral Exhalation Theory of Metallogenesis in Pre-Modern Mineral Science'. *Ambix* 53 (2006): 43–65.

Norris, John A. 'Early Theories of Aqueous Mineral Genesis in the Sixteenth Century'. *Ambix* 54 (2007): 69–86.

Novak, B.C. 'Giovanni Pico della Mirandola and Jochanan Alemanno'. *Journal of the Warburg and Courtauld Institutes* 45 (1982): 125–47.

Nummedal, Tara. 'Introduction', Special Issue on Alchemy and Religion in Christian Europe. *Ambix* 60 (2013), 311–22.

Obrist, Barbara. *Les débuts de l'imagerie alchimique*. Paris: Le Sycomore, 1982.

O'Collins, Gerald. *Christology: A Biblical Historical and Systematic Study of Jesus*. Oxford: Oxford University Press, 2009.

Osler, Margaret. 'The Intellectual Sources of Robert Boyle's Philosophy of Nature: Gassendi's Voluntarism and Boyle's Physico-Theological Project'. In *Philosophy, Science and Religion in England*, 1640–1700, edited by Richard Kroll, Richard Ashcraft and Perez Zagorin, 178–98. Cambridge: Cambridge University Press, 1992.

———. *Divine Will and Mechanical Philosophy: Gassendi and Descartes on Contingency and Necessity in the Created World*. Cambridge: Cambridge University Press, 1994.

Ozment, Steven E. *Mysticism and Dissent: Religious Ideology and Social Protest in the Sixteenth Century*. London: Yale University Press, 1973.

Pagel, Walter. 'Some Religious and Philosophical Aspects of Van Helmont's Science and Medicine'. *Supplements to the Bulletin of the History of Medicine* 2. Baltimore: Johns Hopkins Press, 1944.

———. 'J. B. Van Helmont's Reformation of the Galenic Doctrine of Digestion—and Paracelsus'. *Bulletin of the History of Medicine* 29 (1955): 563–8.

———. 'Medieval and Renaissance Contribution to the Knowledge and Philosophy of the Brain'. In *The History of the Brain and its Functions*, edited by E.N.L. Poynter. Oxford: Oxford University Press, 1958.

———. *Paracelsus: an Introduction to Philosophical Medicine in the Era of the Renaissance*. Basel: S. Karger, 1958.

———. 'Prima Materia of Paracelsus'. *Ambix* 9 (1961): 117–35.

———. *Joan Baptista Van Helmont: Reformer of Science and Medicine*. Cambridge: Cambridge University Press, 1982.

———. *Smiling Spleen: Paracelsianism in Storm and Stress*. Basel: Karger, 1984.

Papy, Jan. 'The Attitude towards Aristotelian Biological Thought in the Louvain Medical Treatises during the Sixteenth and Early Seventeenth Century: The Case of Embryology'. In *Aristotle's Animals in the Middle Ages and the Renaissance*, 318–37. Leuven: Leuven University Press, 1999.

Pardoe, Julia. *The Life of Marie de Medicis, Queen of France*, 3 vols. London: Colburn, 1852.

Park, Katharine and Lorraine Daston (eds). *Cambridge History of Science: Vol. 3 Early Modern Science*. Cambridge: Cambridge University Press, 2006.

Partington, James R. *A History of Chemistry*, 4 vols. London: Macmillan, 1961.
Patterson, T.S. 'Van Helmont's Fire and Water Experiments'. *Annals of Science* 1 (1936): 463–4.
Pelikan, Jaroslav. *The Growth of Medieval Theology (600–1300)*. London: University of Chicago Press, 1978.
———. *Jesus Through the Centuries*. New Haven, CT: Yale University Press, 1985.
Pereira, Michela. *Oro dei Filosofi: Saggio sulle idee di un alchimista del Trecento*. Spoleto: Centro Italiano di Studi sull'Alto Medioevo, 1992.
———. 'Un tesoro inestimabile: elixir e "prolongatio vitae" nell'alchimia del '300', *Micrologus* 1 (1993): 161–87.
———. 'Teorie dell'Elixir nell'alchimia latina medievale'. *Micrologus* 3 (1995): 103–48.
———. 'L'alchimista come medico perfetto nel "Testamentum" pseudo-Lulliano'. In *Alchimia e medicina nel medioevo*, edited by Chiara Crisciani e Agostino Paravicini Bagliani, 77–109. Florence: Sismel, 2003.
Pichot, André. *Histoire de la notion de vie*. Paris: Gallimard, 1993.
Pinckaers, Servais. *The Sources of Christian Ethics*, 3rd edition. Washington, DC: Catholic University of America, 1995.
Pirenne, Henri. *Histoire de Belgique*, 7 vols. Brussels: Henri Lamertin, 1920.
Poma, Roberto. *Magie et guérison: La rationalité de la médecine magique (XVIe-XVIIe)*. Paris: Orizons, 2009.
Poole, William. 'Francis Lodwick's Creation: Theology and Natural Philosophy in the Early Royal Society'. *Journal of the History of Ideas* 66 (2005): 245–63.
Popkin, Richard Henry. *The History of Skepticism from Erasmus to Spinoza*. Berkeley, CA: University of California Press, 1979.
Porto, Paulo. *Van Helmont e o conceito de gás: química e medicina no século XVII*. São Paolo: EDUC-EDUSP, 1995.
———. '"*Summus atque felicissimus salium*": The Medical Relevance of the Liquor Alkahest'. *Bulletin of the History of Medicine* 76 (2002): 1–29.
Principe, Lawrence M. *The Aspiring Adept: Robert Boyle and his Alchemical Quest*. Princeton, NJ: Princeton University Press, 1998.
Quispel, Gilles. 'Hermes Trismegistus and the Origins of Gnosticism'. In *From Poimandres to Jacob Boehme: Gnosis, Hermetism and the Christian Tradition*, edited by Roelof van der Broeck and Cis van Heertum, 145–67. Amsterdam: Bibliotheca Philosophica Hermetica, 2000.
Rattansi, Piyo. 'Newton's Alchemical Studies'. In *Science, Medicine and Society in the Renaissance: Essays to Honor Walter Pagel*, 2 vols, edited by Allen Debus. New York: Science History Publications, 1982, II, 175–80.
Reeds, K. Meier. 'Renaissance Humanism and Botany'. *Annals of Science* 33 (1976): 519–42.
Renan, Ernest. *Averroes et averroisme: essai historique*. Paris: Michel Levy, 1866.
Reti, Ladislao. 'Van Helmont, Boyle and the Alkahest'. In Ladislao Reti and William C. Gibson, *Some Aspects of Seventeenth-Century Medicine and Science*, 3–19. Los Angeles: William Andrews Clark Memorial Library, 1969.

Rice, Eugene F. 'Nicholas of Cusa's Idea of Wisdom'. *Traditio* 13 (1957): 345–68.

Richardson, Alan and John Bowden (eds). *The Westminster Dictionary of Christian Theology*. Philadelphia, PA: Westminster Press, 1983.

Rist, John M. *Augustine: Ancient Thought Baptized*. Cambridge: Cambridge University Press, 1995.

Roger, Jacques. *Les sciences de la vie dans la pensée française du XVIIIe siècle*, 2nd edition. Paris: Albin Michel, 1993.

Roodnat, Albert. 'J.B. Van Helmont'. In *Dictionary of Gnosis and Western Esotericism*, edited by Wouter J. Hanegraaff, 469–74. Leiden: Brill, 2006.

Roos, Anna-Marie. *The Salt of the Earth: Natural Philosophy, Medicine, and Chymistry in England, 1650–1750*. Leiden: Brill, 2007.

Rorem, Paul. *Pseudo-Dionysius: A Commentary on the Texts*. Oxford: Oxford University Press, 1993.

Rossi, Paolo. *The Birth of Modern Science*. Oxford: Blackwell, 2000.

Rudolph, Hartmut. 'Kosmosspekulation und Trinitätslehre: Ein Beitrag zur Beziehung zwischen Weltbild und Theologie bei Paracelsus'. *Salzburger Beiträge zur Paracelsusforschung* 21 (1980), 32–47.

———. 'Hohenheim's Anthropology in the Light of his Writings on the Eucharist'. In *Paracelsus: The Man and his Reputation, His Ideas and their Transformation*, edited by Ole Peter Grell, 187–207. Leiden: Brill, 1998.

Ruska, Julius. 'Die Vision des Arislei'. In *Historische Studien und Skizzen zu Natur- und Heilwissenschaft*, edited by G. Sticker, 22–6. Berlin: Julius Springer, 1930.

Rutherford, Donald R. (ed.). *The Cambridge Companion to Early Modern Philosophy*. Cambridge: Cambridge University Press, 2006.

Sambursky, Samuel. *Physics of the Stoics*. Westport, CN: Greenwood Press, 1959.

Schipperges, Heinrich. 'Paracelsus and His Followers'. In *Modern Esoteric Spirituality*, edited by Antoine Faivre and Jacob Needleman, 154–85. New York: Crossroad, 1992.

Schmidt-Biggemann, Wilhelm. *Philosophia Perennis: Historical Outlines of Western Spirituality in Ancient, Medieval and Early Modern Thought*. Dordrecht: Springer, 2004.

Schmitt, Charles. '*Prisca Theologia* e *Philosophia Perennis*: due temi del Rinascimento italiano e la loro fortuna'. In *Il pensiero italiano del Rinascimento e il tempo nostro: Atti del V Convegno Internazionale di Centro di studi umanistici, Montelpuciano, 1968*, 211–36. Florence: Leo S. Olschki, 1970.

———. *Aristotle in the Renaissance*. Boston, MA: Harvard University Press, 1983.

Shackelford, Jole. 'The Chemical Hippocrates: Paracelsian and Hippocratic Theory in Petrus Severinus' Medical Philosophy'. In *Reinventing Hippocrates*, edited by David Cantor, 59–88. Aldershot: Ashgate, 2002.

———. *A Philosophical Path for Paracelsian Medicine: The Ideas, Intellectual Context, and Influence of Petrus Severinus, 1540–1602*. Copenhagen: Museum Tusculanum Press, 2004.

Shapin, Steven. *The Scientific Revolution*. Chicago, IL: University of Chicago Press, 1996.

Shapin, Steven and Simon Schaffer. *Leviathan and the Air-Pump*. Princeton: Princeton University Press, 1985.

Smith, Justin E.H. 'Leibniz and the Cambridge Platonists in the Debate over Plastic Natures'. In *Leibniz and the English-Speaking World*, edited by Pauline Phemister and Stuart Brown, 97–110. Dordrecht: Kluwer, 2007.

Spink, John S. *French Free Thought from Gassendi to Voltaire*. London: Athlone Press, 1960.

Staden, Heinrich von. 'Physis and Techne in Greek Medicine'. In *The Artificial and the Natural: an Evolving Polarity*, edited by Bernadette Bensaude-Vincent and William Newman, 21–51. Cambridge, MA: MIT, 2007.

Steneck, Nicholas H. *Science and Creation in the Middle Ages: Henry of Langenstein (d. 1397) on Genesis*. London: University of Notre Dame Press, 1976.

Stone, M.W.F. 'Michael Baius (1513–1589) and the Debate on "Pure Nature": Grace and Moral Agency in Sixteenth-Century Scholasticism'. In *Moral Philosophy on the Threshold of Modernity*, edited by Jill Kraye and Risto Saarinen, 51–90. Dordrecht: Springer, 2005.

Stroobant, Louis. 'Les origins du Docteur van Helmont'. In *Le Folklore brabançon: Numéro spécialement consacré à J. B. van Helmont, 1579–1644*, 140–66. Brussels, 1933.

Stump, Eleonore and Kretzmann, Norman. 'Being and Goodness'. In *Being and Goodness: The Concept of the Good in Metaphysics and Philosophical Theology*, edited by Scott MacDonald, 98–128. Ithaca, NY: Cornell University Press, 1991.

Sudhoff, Karl. 'Vorwort zum 14. Band'. In Theophrastus von Hohenheim gen. Paracelsus, *Sämtliche Werke*, I, 14 vols, 5–35. Munchen and Berlin: Oldenbourg, 1933.

Szulakowska, Urszula. *The Alchemy of Light: Geometry and Optics in Late Renaissance Alchemical Illustration*. Leiden: Brill, 2000.

———. *The Sacrificial Body and the Day of Doom*. Leiden: Brill, 2006.

Taylor, Denys. *The Darkness of God: Negativity in Christian Mysticism*. Cambridge: Cambridge University Press, 1998.

Taylor, F. Sherwood. 'The Idea of the Quintessence'. In *Science, Medicine and History*, edited by E. Ashworth Underwood, 2 vols, I, 247–65. London: Oxford University Press, 1953.

Teske, Roland. 'Augustine's Theory of the Soul'. In *The Cambridge Companion to Augustine*, edited by Eleonore Stump and Norman Kretzmann, 116–24. Cambridge: Cambridge University Press, 2006.

Thorndike, Lynn. *History of Magic and Experimental Science*, 8 vols. New York: Columbia University Press, 1923–1956.

Trinkaus, Charles. *In Our Image and Likeness: Humanity and Divinity in Italian Humanist Thought*, 2 vols. London: Constable, 1970.

Tsourkas, Cléobule. *Les debuts de l'enseignement philosophique et de la libre pensée dans les Balkans: La vie et l'oeuvre de Theophyle Corydalée (1570–1646)*. Thessaloniki: Institute for Balkan Studies, 1967.

Van der Meer, Matthieu. 'Nicholas of Cusa's View of Time and Eternity'. In *Christian Humanism: Essays in Honours of Arjo Vanderjagt*, 317–39. Leiden: Brill, 2009.

van der Poel, Marc. *Cornelius Agrippa: The Humanist Theologian and His Declamations*. Leiden: Brill, 1997.

Van Eijl, E.J.M. 'La controverse louvaniste autour de la Grace et du libre arbitre à la fin du XVIe siècle'. In *L'Augustinisme à l'ancienne faculté de théologie de Louvain*, edited by M. Lamberigts and L. Kenis, 207–83. Leuven: University Press, 1994.

Van Kooten, Geurt. *Paul's Anthropology in Context: the Image of God, Assimilation to God, and Tripartite Anthropology in Ancient Judaism, Ancient Philosophy and Early Christianity*. Tübingen: Mohr Siebeck, 2008.

Van Roon, A. 'The Relationship between Christ and the Wisdom of God according to Paul'. *Novum Testamentum* 16 (1974): 207–39.

Vanpaemel, G.H.W. 'Jesuit Science in the Spanish Netherlands'. In *Jesuit Science and the Republic of Letters*, edited by Mordechai Feingold, 389–433. London: MIT Press, 2003.

Vasoli, Cesare. 'L'amorosa filosofia: dall'amore platonico all' universale *philautia*'. In *Francesco Patrizi da Cherso*. Rome: Bulzoni, 1989.

Waddell, Mark A. 'The Perversion of Nature: Johannes Baptista Van Helmont, the Society of Jesus, and the Magnetic Cure of Wounds'. *Canadian Journal of History* 38 (2003): 179–97.

Walker, D.P. *The Ancient Theology*. London: Duckworth, 1972.

Walton, Michael T. 'Genesis and Chemistry in the Sixteenth Century'. In *Reading the Book of Nature*, edited by Allen G. Debus and Michael T. Walton, 1–14. Kirksville, IL: Truman State, 1998.

———. *Genesis and the Chemical Philosophy: True Christian Science in the Sixteenth and Seventeenth Centuries*. New York: AMS Press, 2011.

Wear, Sarah K. and Dillon, John M. *Dionysius the Areopagite and the Neoplatonist Tradition*. Aldershot: Ashgate Publishing, 2007.

Webster, Charles. 'Water as the Ultimate Principle of Nature: the Background to Boyle's *Sceptical Chymist*'. *Ambix* 13 (1966): 96–107.

———. *The Great Instauration*. London: Duckworth, 1975.

———. *From Paracelsus to Newton: Magic and the Making of the Modern Science*. Cambridge: Cambridge University Press, 1984.

———. *Paracelsus: Medicine, Magic & Mission at the End of Time*. New Haven, CT: Yale University Press, 2008.

Weeks, Andrew. *Paracelsus: Speculative Theory and the Crisis of the Early Reformation*. Albany, NY: SUNY Press, 1997.

Weijers, Olga, 'Contribution à l'histoire des termes "natura naturans" et "natura naturata" jusqu'à Spinoza'. *Vivarium* 16 (1978): 70–80.

Westman, Robert. 'Nature, Art and Psyche: Jung, Pauli and the Kepler-Fludd Polemic'. In *Occult and Scientific Mentalities in the Renaissance*, edited by Brian Vickers, 177–229. Cambridge: Cambridge University Press, 1984.

Williams, Rowan. 'De Trinitate'. In *Augustine Throughout the Ages: an Encyclopedia*, edited by Allan Fitzgerald and John C. Cavadini, 845–51. Grand Rapids, MI: Wm B. Eerdmans, 1999.

Windisch, H. 'Die göttliche Weisheit der Juden und die paulinische Christologie'. In *Neutestamentliche Studien für Georg Heinrici*. Leipzig: J.C. Hinrichs,1914.

Wojcik, Jan W. *Robert Boyle and the Limits of Reason*. Cambridge: Cambridge Univ. Press, 1997.

Woodward, Walter W. *Prospero's America: John Winthrop, Jr, Alchemy, and the Creation of New England Culture, 1606–1676*. Chapel Hill: University of North Carolina Press, 2010.

Index

FSC
www.fsc.org
MIX
Papier | Fördert gute Waldnutzung
FSC® C083411

Kolibri 360 GmbH
Ferdinand-Jühlke-Straße 7
99095 Erfurt, Deutschland
produktsicherheit@kolibri360.de